P9-DVU-361

Species Identity and Attachment

Garland Series in Ethology

Series Editor: Gordon M. Burghardt

Primate Ecology and Human Origins: Ecological Influences on Social Organization
Edited by: Irwin S. Bernstein and Euclid O. Smith

The Development of Behavior: Comparative and Evolutionary Aspects
Edited by: Gordon M. Burghardt and Marc Bekoff

The Behavioral Significance of Color
Edited by: Edward H. Burtt, Jr.

The Behavior and Ecology of Wolves
Edited by: Erich Klinghammer

Handbook of Ethological Methods
By: Philip N. Lehner

Cat Behavior: The Predatory and Social Behavior of Domestic and Wild Cats
By: Paul Leyhausen

Dominance Relations: An Ethological View of Human Conflict and Social Interaction
Edited by: Donald R. Omark, F. F. Strayer, and Daniel G. Freedman

Species Identity and Attachment: A Phylogenetic Evaluation
Edited by: M. Aaron Roy

Other Garland Books in Ethology

Primate Bio-Social Development: Biological, Social, and Ecological Determinants
Edited by: Suzanne Chevalier-Skolnikoff and Frank E. Poirier

The Face of Man: Expressions of Universal Emotion in a New Guinea Village
By: Paul Ekman

The Dog: Its Domestication and Behavior
By: Michael W. Fox

Ethological Dictionary: In English, French, and German
By: Armin Heymer

Species Identity and Attachment:

A Phylogenetic Evaluation

edited by

M. Aaron Roy

Northern Michigan University
Marquette, Michigan

Garland STPM Press
New York & London

15 14 13 12 11 10 9 8 7 6 5 4 3 2 1

Library of Congress Cataloging in Publication Data

Main entry under title:

Species identity and attachment.

(Garland series in ethology)
Bibliography
Includes index.
1. Social behavior in animals. 2. Critical periods (Biology) 3. Imprinting (Psychology) 4. Species. I. Roy, M. Aaron. II. Series.

QL775.S68 596′.05 79-14540
ISBN 0-8240-7052-6

Published by Garland STPM Press
136 Madison Avenue, New York, New York 10016

Printed in the United States of America

**To Robin and Amy,
the girls in my life**

Contents

Preface

This volume is the result of work in three phases. The first was my familiarization with various mammals who were not reared with their own species and with some early infantile autistic children. They gave the impression that their atypical interactions with conspecifics signified that they did not fully "know" to which species they belonged. This appeared to be the one important element that differentiated them from their normal species mates and hindered the development of normal social interactions with their conspecifics.

With the encouragement of Paul Scott, I refined my thinking and solicited comments from those who I felt could critique my ideas. The culmination of this second phase was the presentation of a symposium entitled "Early Experience and Its Effect on the Formation of a 'Species Identity'" at the 1977 American Psychological Association meeting. Its participants (Gordon Gallup, Gil Meier, Paul Scott, Jim Shapiro, Larry Slobodkin, and I) formed the nucleus for contributions to this volume.

The final phase involved contacting interested individuals who could discuss the concept of a species identity in species with which they were familiar. Writers of these core chapters used a draft of this book's introductory chapter and copies of the APA presentations for direction. Given the subjectiveness of the area, I decided that it would be valuable to include a number of chapters that would critique the core chapters. Authors of these commentaries received nearly final drafts of the core chapters, and they were free to choose which chapters they would comment on as well as whether they would include new information.

I am pleased with the outcome of this volume. The use of core and commentary chapters has provided depth. The contributors represent various disciplines and have had quite different interactions with the species they discuss. The viewpoints about species identity are not the same, given the relative newness of the species identity concept. A range of species is discussed. Most importantly, the contributors have addressed an area that needs attention and understanding.

All of us have enjoyed working on this volume and, as editor, I have refined and extended my thinking about the issue of species identity. We hope that you, the reader, find our work both enjoyable and enlightening.

List of Contributors

Mary D. S. Ainsworth

University of Virginia
Charlottesville, Virginia

David P. Barash

University of Washington
Seattle, Washington

Thomas L. Bennett

Colorado State University
Fort Collins, Colorado

Inge Bretherton

University of Colorado
Boulder, Colorado

Nancy G. Caine

California Primate Research Center
University of California
Davis, California

Fred Cooke

Queens University
Kingston, Ontario
Canada

Henry A. Cross Jr.

Colorado State University
Fort Collins, Colorado

Michael W. Fox

Institute for the Study of Animal Problems
Division of the Humane Society of the United States
Washington, D.C.

Arnold D. Froese

Sterling College
Sterling, Kansas

Gordon G. Gallup, Jr.

State University of New York
Albany, New York

Gary W. Guyot
West Texas State University
Canyon, Texas

Eckhard H. Hess
University of Chicago
Chicago, Illinois

Peter H. Klopfer
Duke University
Durham, North Carolina

Lee I. McCann
University of Wisconsin–Oshkosh
Oshkosh, Wisconsin

Gilbert W. Meier
Caribbean Primate Research Center
The University of Puerto Rico
San Juan, Puerto Rico

Victoria Dudley-Meier
Caribbean Primate Research Center
The University of Puerto Rico
San Juan, Puerto Rico

G. Mitchell
University of California
Davis, California

Slobodan B. Petrovich
University of Maryland
Baltimore, Maryland

M. Aaron Roy
Northern Michigan University
Marquette, Michigan

J. P. Scott
Bowling Green State University
Bowling Green, Ohio

L. James Shapiro
The University of Manitoba
Winnipeg, Manitoba
Canada

Lawrence B. Slobodkin

State University of New York
Stony Brook, New York

Valerie L. Smith

St. Louis Zoological Park
St. Louis, Missouri

Jane W. Temerlin

Clinical Psychology Consultants, Inc.
Oklahoma City, Oklahoma

Part I

The Nonprimate Vertebrates

Chapter 1

An Introduction to the Concept of Species Identity

M. Aaron Roy

When most people are affectionate, lonely, sick, or troubled, they usually seek other people for help, attention, or love. Attracted to these people in turn are those who can offer this aid, understanding, or companionship. This type of affiliation between people is so common and "natural" that most of us fail to pose an important and basic question: Why does an organism affiliate with members of its own species rather than with members from other species? This question is more than academic. Without normal attractions between conspecifics, a species' ability to procreate would be threatened, and most importantly, those species that did manage to exist would surely show impairments in any type of social interaction.

To answer that a particular mechanism (e.g., imprinting) is the basis for intraspecies attractions and interactions does little to explain the phenomenon. It remains unclear what mediates these preferences in various species: a constellation of innate behaviors elicited by particular stimuli, cognitive processes, or both. Furthermore, the use of a descriptive mechanism (e.g., species recognition) does not clarify to what degree, if at all, an organism's species-typical social preferences are influenced by previous interactions with conspecifics.

Species Identity as a Concept

An interest in social preferences in an organism or within a species is not new, yet the description of these preferences has varied considerably.

Hediger (1955) has described an "assimilation tendency" in mammals which is directed toward species or nonspecies members following rearing with them. Scott (1968, 1972) states that rearing without conspecific contacts leads not only to the lack of "social attachments" toward the biological species, but that it also prevents the development of allelomimetic behavior. Terminology such as "assimilation tendency" and "social attachment" is connotatively similar to the concepts of "filial bonds" and "filial attachments" used by primatologists (e.g., Mason, Hill, and Thomsen, 1971, 1974; Mason and Kenny, 1974). Mitchell (1973) defines an affiliation as an "attachment to (or affection for) fellow members of one's own species." Cooke (1977), like Lorenz (1935), mentions the ability of individuals to respond to "supra-individual species-specific cues" of conspecifics. Lehrman and Rosenblatt (1971) favor "species recognition." Brown (1975) uses "species recognition" and "species identifications" (from Gottlieb, 1971) when referring to the general ability of organisms to appropriately respond to cues which are part of, or emitted by, conspecifics. Gottlieb (1978) in turn discusses a "species identification" and "species-specific perceptions" which develop through the interaction of innate and experiential actions. Fox (1971) uses the concept "species identity."

One may ask if there is a single appropriate description or expression to utilize when discussing social preferences in any species. A number of alternatives are available: species orientation, species identification, species attachment, species affiliation, species attraction, species recognition, species awareness, and/or any arrangement of more specific phrases (e.g., supra-individual species-specific cues). One alternative would be to use any term that is most suited to the particular species being discussed. This would be advantageous until more is learned about the similarities and differences in various organisms. Unfortunately, the use of many terms may hamper communication and unintentionally divert attention from an issue which has not received the attention it deserves. Conversely, a single term may convey both an impression of understanding the phenomenon or of some agreed upon commonality between species. Neither of these is the case, as will be seen in the following chapters which discuss species ranging from fish to humans. Nevertheless, some term or phrase should be utilized, and as consistently as possible. The term "species identity" is suggested in this chapter, though this selection is not without some personal bias and not without some reservations.

Species identity is a theoretical construct without individual meaning but rich in relations with other kinds of behavior. As with other constructs such as love, intelligence, or emotion, to measure a species identity is to infer it through the presence of other behaviors. *A species identity is considered to exist in a target organism if this organism shows a preference for a group of other organisms which have similar physical and behavioral characteristics among themselves, but who may or may not be similar in characteristics to the target organism. Both the presence of the species identity and the class of objects to which the preference is directed can be inferred very generally from the target organism's behavior in social situations. More specifically, a species identity can be measured by the following types of behavior with the species with which an identification has occurred: an ability to maintain proximity (in gregarious species) or distance (in solitary or territorial species); courting and/or attempted breeding; imitation or copying; the ability to communicate; a degree of mutual grooming or other reciprocal interactions; the ability to engage in group defense or group predation; and/or the presence of allelomimetism.*

A species identity is believed to exist in individuals of many species. It remains to be clarified (1) which species do or do not possess such an identity, (2) what mechanism(s) mediate its effect on behavior, and in those species that do have a species identity, (3) how it develops if it is not innate.

It should be simple to determine whether the environment influences the development of an organism's species identity. One would have to prevent the subject from having contact with its own kind, and one would then observe its interactions with conspecifics when introduced to them and with the species (if any) with which it has been reared. Many such cases have been presented in the literature and popular press, especially those reports of home-reared chimpanzees (Hayes, 1952; L. Hess, 1954; Kellogg and Kellogg, 1933; Temerlin, 1975). Worthy of discussion are three cases with organisms from different positions on the phylogenetic scale—a gorilla, a sheep, and a bird.

The Primate

Hoyt raised an infant female gorilla named "Toto" in and around her home for nine years (Hoyt, 1941). In many respects, especially socially, Toto behaved like a human toward other humans: "She wanted to share our human life fully" (p. 139). Toto imitated humans and demonstrated both sympathy and great affection for them. For example, Hoyt tells of an injury Toto inflicted on her.

> We were playing pleasantly and happily in her enclosure when Toto, in an excess of gaiety, jumped into her swing, gave it a long push backwards and rushing at me pushed me. It happened so quickly that I wasn't able to get out of the way. I fell over backwards, my hands coming down hard between two flagstones, and, in an excruciating flash of pain, knew that I had broken both of my wrists. Instantly, Toto, seeing that I was hurt and knowing that it was she who had hurt me, was all contrition. She came toward me wanting to kiss me, but knowing that there was nothing that I could do to keep her from handling my wrists, I ordered her off (p 163).

When the wrist cast had been removed,

> . . . I saw clearly how conscious she was of the fact that she had hurt me. She was quieter and more restrained, looked at me sadly and frequently hung her head as if in shame. Time and time again . . . she would take my hands gently in hers, turning them over with the utmost care, looking at my wrists carefully, blowing on them and kissing them as she had done so many times when I had had little scratches (p 165).

Mrs Hoyt also comments on Toto's affections for the opposite sex.

> Our difficulties were always tremendously increased during two or three days of every month when Toto seemed to fall in love with one of the men. Sometimes it was one of the gardeners . . . sometimes the second chauffeur . . . sometimes the butler. During these periods, she would follow the object of her affections about the grounds, sitting and staring at him with lovesick eyes whenever he was working, wanting to touch him—an expression [we] prevented, never feeling sure what the consequences might be. And although we were always successful in keeping her at a distance from the man she had chosen for special adoration, it was only at the cost of constant and varied ingenuity and Toto's continuous displeasure (pp 181–182).

The Ovine

The example of Scott and his wife hand-raising a lamb is fairly typical.

> We took a female lamb from its mother at birth and raised it on a bottle for the first ten days of life. After that we put it out in the field where other sheep were grazing. We continued the bottle feeding and made no attempt to force the lamb into contact with the flock. When it approached the other sheep curiously, the mothers drove it away. The orphan followed an entirely different rhythm in its grazing pattern and had almost no contact with the flock, although they all stayed in the same small field. When it came into estrus, it submitted to being mounted by the rams but also stood still when caught by the human observer, presumably giving a sexual reaction. Even after a period of several years this sheep displayed a great deal of independence from the rest of the flock, not running when they were frightened and staying away from them on most occasions (1972, pp 200–201).

When this ewe did conceive and give birth she "would allow her own offspring to nurse but did not call them when they strayed away nor did she respond to their calls" (1968, p 107).

The Avian

In addition to the great ape and the lamb, the behavior of a jackdaw named "Jock" is relevent (Lorenz, 1935, 1952). Jock's behavior is even more interesting since she had access to both her own species and the host species, whereas the lamb and gorilla did not. Though the cognitive and behavioral repertoire of a jackdaw is obviously inferior to that of an ape, one finds many similarities between these species when reared by humans.

Jock was raised by hand, kept at home, and permitted to fly freely both inside and outside. Lorenz (1952) states:

> It must have been owing to my gift of imitating its call that it soon preferred me to any other person. I could take long walks and even bicycle rides with it and it flew after me, faithful as a dog. Although there was no doubt that it knew me personally and preferred me to anybody else, yet it would desert me and fly after some other person if he was walking much faster than me, particularly if he overtook me. As soon as he had left me, Jock would notice his error and correct it, coming back to me hurriedly. As he grew older, he

> learned to repress the impulse to pursue a stranger, even one walking very fast indeed. Yet even then I would often notice his giving slight start or a movement indicative of flying after the faster traveller (p 131). . . .
>
> When Jock reached maturity, he fell in love with our housemaid, who just then married and left our service. A few days later, Jock discovered her in the next village two miles away, and immediately moved into her cottage, returning only at night to his customary sleeping quarters. In the middle of June, when the mating season of jackdaws was over, he suddenly returned home to us . . . the kind of advances which Jock made to our housemaid, slowly but surely divulged the fact that "he" was a female! She reacted to this young lady exactly as a normal female jackdaw would to her mate (p 135).

When discussing a second hand-raised jackdaw, Lorenz continues:

> Another tame adult male jackdaw fell in love with me and treated me exactly as a female of his own kind. By the hour, this bird tried to make me creep into the nesting cavity of his choice, a few inches in width. . . . The male jackdaw became most importunate in that he continually wanted to feed me with what he considered the choicest delicacies. Remarkably enough, he recognized the human mouth in an anatomically correct way as the orifice of ingestion and he was overjoyed if I opened my lips to him, uttering at the same time an adequate begging note. You will understand that I found it difficult to cooperate with the bird in this manner every few minutes! But if I did not, I had to guard my ears against him, otherwise, before I knew what was happening, the passage of one of these organs would be filled right up to the drum with warm worm pulp, for jackdaws, when feeding their female or their young, push the food mass, with the aid of their tongue, deep down into the partner's pharynx (pp 135–136).

Although considerable species differences exist between jackdaws, sheep, and gorillas, there are definite similarities in these three cases. First and foremost is the direction of the social preferences as determined by the object of the species-typical social behaviors. The object in all cases was man, the species present when the subject was an infant. Second, there was a deviation of species-typical behaviors that were directed toward the biological species in the jackdaw, the sheep, and possibly the gorilla.[1]

The behavior of these atypically reared animals should not be surprising. Many of us know of instances in which infants of wild and domesticated species have been reared as pets, in pens with other species, or in nurseries (e.g., zoos or research institutes) alone or with infants of another species. Under such conditions the infant not only becomes tame, but in adulthood a full range of social (Fox, 1968; Newton and Levine, 1968) and sexual (Maple, 1977) behaviors may be directed toward another species.

Issues Relating to Species Identity

The case histories of Jock, Toto, and Scott's lamb are more than interesting reading. They suggest that an infant reared apart from its species' members will be significantly affected in its orientation toward its conspecifics. Under such circumstances, it is the social and reproductive, not other behaviors (e.g., nest building, food procurement and eating, self-grooming), which are often distorted. Depending upon its postnatal experiences, the organism at maturity may identify (as shown by affiliative, social, and sexual behaviors) in one of the following ways: (1) solely toward the host species which reared it, (2) solely toward the species with which it was confined, (3) with some combination of preference for both host and confinement species, or (4) at best, to both its biological species and the species which served as the host or confinement mate. For many species, early experiences apparently have considerable effect on adult social preferences.

A number of issues relating to species identity arise from reading about atypically reared organisms. These concerns will be addressed within the chapters of this text and need only to be highlighted at this point.

Does every living organism possess a species identity? Surely there are many species of invertebrates and perhaps some vertebrates which do not possess a species identity. Where one draws the line for certain families and species living in particular ecological systems must be determined. If recognition of conspecifics signals the presence of a species identity, then one would conclude that such an identity is found in many different organisms. Conspecific recognition has been shown in damselfish (Thresher, 1976), crickets (Otte and Cade, 1976), and ants (Cammaerts-Tricot, 1975). Individual recognition has been reported, for example, in wood lice (Linsenmaier and Linsenmaier, 1971) and banded shrimp (Johnson, 1977). Intrinsic to this issue is how one views an organism's species identity: it may be considered to influence the recognition of conspecifics by means of a cognitive process or by means of some basic stimulus–response relationship.

What are the genetic influences on the development of a species identity? Innate preferences or dispositions for particular stimuli are found in a range of species, and the sign stimuli or social releasers for these are usually present in conspecifics. Avians show these preferences both before and after hatching (see Hess and Hess, 1969; Lorenz, 1952; Gottlieb, 1971; Marler and Mundinger, 1971). Sackett's (1966) report that pictures of infant rhesus monkeys generate more viewing time in infant rhesus isolates than do equally complex pictures which are not of monkeys could be interpreted as suggesting innate preferences in primates. Also relevant is the selective attention shown by infants when viewing schematic face patterns (see Fantz, 1965; Lewis, 1969). The study by Meltzoff and Moore (1977) showing imitation of gestures by neonates may be viewed as supportive. Innate predispositions to react to stimuli which are characteristic of the biological species, however, imply neither that the development of a species identity will occur with conspecifics, nor that there exists an innate species identity. The work showing that there is a neural substrate for affiliative behaviors (Kling and Steklis, 1976) and for innate social displays (MacLean, 1972) in nonhuman primates does little to clarify how the target or recipient of these behaviors evolves in each individual.

What role do early social experiences have in the development of a species identity? Some conclusions have been based on insufficient data, and systematic research needed to clarify this question has not been conducted. Lorenz (1935), for example, concluded that curlews (*Numenius*) and great godwits (*Limosa*) have an innate "schema" of the parent because any incubator-reared fledgling flees at first contact with a human. This may be true, though to equate fleeing from a nonbiological species member with an innate schema for the biological species is risky. One may assume that sea turtles do not require early conspecific social interactions since they go to sea after hatching and apparently have no contact with peers or parents until maturity. Recent observations question this assumption. Swimming far out to sea, the hatchlings approach floating sargassum weeds for both food and protection (Fletemeyer, 1978; Frick, 1976), thus enhancing the probability that they will be exposed to peers during their first year of life.

Research pertaining to this issue is both equivocal and contradictory. Male guinea pigs, though reared with chicks, show a preference for female conspecifics (Beauchamp and Hess, 1971), while those reared with rats apparently do not (Beauchamp and Hess, 1973). Russock and Schein (1978) report that cichlids raised with visual isolation until maturity still recognize and prefer conspecifics. Isolate-reared giant African snails will follow mucus trails laid down by conspecifics (Chase, Pryer, Baker, and Madison, 1978). Work with locusts has found that gregarious behaviors in adults are influenced by early contacts with conspecifics (Ellis, 1963; Gillett, 1973). Gottlieb (1978) reports distortions in species-specific perceptions and preferences in aurally deprived ducklings. Voles (McDonald and Forslund, 1978) and mice (McCarty and Southwick, 1977) will prefer nonconspecifics and their odors if they are reared with that species. Studies with atypically reared mammals (e.g., canines, felines, and many nonhuman primates), which are discussed in subsequent chapters, suggest that early experiences can modify adult social preferences. The example of adult, surrogate-reared rhesus monkeys who emit copulatory behaviors toward cloth-covered models but not sexually receptive and experienced conspecifics (Deutsch and Larsson, 1974) further supports a position that postnatal experiences determine social preferences.

Is a species identity acquired gradually or during a particular period? Various reports suggest that a group of phases or stages may be necessary in some species in order for a complete species identity to evolve. Jock, the jackdaw reared by Lorenz, associated with humans except during two instances—when it was instinctively caring for other young jackdaws and when flying with a flock. Group flying in Jock:

> had been conditioned to hooded crows when its group instinct matured. They were the first flying *Corvidae* the jackdaw ever saw. Later, when it shared its attic with a whole flock of other jackdaws, it still kept flying with the free hooded crows. It did not regard the other jackdaws as flight companions (Lorenz, 1957, p 108).

Klinghammer (1967), working with altricial birds (i.e., Columbidae), concludes that pair formation at sexual maturity is influenced by two particular mechanisms: one a function of preweaning experiences and the other of postweaning, but pre-adult, experiences. The work of Schutz (1965) on periods of sexual imprinting in precocial fowl is also pertinent here. Thus species-typical attractions, as measured by social and sexual preferences, may be acquired during different periods in some species.

Are there differences in the degree to which each sex identifies with its conspecifics? Preliminary work suggests differences in some species, notably avians (see Schutz, 1971). Leyhausen (1973) reports that knowledge of the *male* mallard is innate for both the female and male of the species. If, however, a male mallard is reared without conspecific contacts, it will try to mate with any *female* of many duck species. Apparently, knowledge of what the opposite sex of the biological species looks like is not innate. Though Kerfoot (1965) has questioned a sex difference in pigeons, work by Warriner and colleagues (1960, 1963) suggests that mate preferences in male ring pigeons are influenced by early rearing experiences whereas those in females are not. Sex differences in response to cross-species fostering has been reported for zebra finches (Immelmann, 1965, 1972) and for guinea pigs (Beauchamp and Hess, 1971, 1973). It is not clear if these sex differences imply differential innate control over preferences in the sexes, different periods of susceptibility which were not controlled for, or another explanation.

Does an organism identify with only one species? Identifications with one species appear to be the norm in nature, but not under manipulated rearing conditions. The story of Jock suggests that an identity with more than one species is possible. Scott (1973) reports that "puppies which have contact with both dogs and people during the period of socialization become attached to both species" (p 209), but that contact with only one of these species during this period generates an exclusive attachment to that particular species. A common assumption is that parasitic egg layers randomly lay their eggs in various nests for rearing by various hosts. This is *not* the case in those parasitic species which have been systematically studied; many lay their eggs in nests of certain host species. Host specificity exists for many cuckoos, some cowbirds, widow birds, and the honey guides (Chance, 1922; Friedman, 1929, 1963, 1966; Nicolai, 1964, 1974). An explanation for the parasitic egg layer's dual attraction to both the biological species and the host species may be found in multiple identifications, an identity formed with the host species during rearing and one with the biological species formed during a postfledgling period.

Will an organism which identifies with one species interact with biologically related species? It is likely that through stimulus generalization or commonality of stimuli (e.g., coloration size, marking, movements), cross-species interactions will occur (Richard, 1970; Rose, 1977; van Lawick-Goodall, 1968). Should this happen when other isolating mechanisms are insignificant, hybridization or attempted crossbreeding could even occur in nature (Bernstein, 1966) or in captivity when confinement brings species together (see the *International Zoo Yearbook* for examples). Hybridization, nevertheless, is uncommon in the wild (Mayr, 1963) or when species are confined together. Hediger (1955) for example, states:

> In this connection, W. H. Hodge's account (1946, p 656) of a cross between an alpaca (*Lama pacos*) and a vicuna (*Lama vicũgna*) is most interesting. This crossbreeding has never been observed in a zoo. It can only occur when a newborn male vicũna has been caught. The Peruvians kill at the same time a newborn alpaca and cover the young vicũna with its skin. The latter is then accepted by the mother alpaca and reared. Only vicũnas of this kind later pair with alpacas and produce paca-vicũnas, i.e., hybrids. In other words, the vicũnas must be "imprinted" on alpacas, and then become quite assimilated and pair (p 85).

Do different species acquire their species identity in different ways? It is highly probable that species or subspecies differences do exist in how a species identity develops. Differences are expected as to (a) the type of interaction that is needed with species' members (e.g., parents and siblings, only parents or siblings, a particular parent); (b) the duration of contact needed (e.g., hours, days, or months during a sensitive period); (c) the period and its length during which the species identity is acquired (e.g., soon after hatching as in precocial fowl, longer and/or later in altricial types, and even longer and/or later in mammals as compared to altricial fowls); and (d) the mode(s) of interaction most desirable (i.e., tactile–kinesthetic, auditory, visual, olfactory, or even gustatory).

What is the relationship between a species identity and a self-identity? The great apes and humans have the ability for self-recognition, as implied by their appropriate use of mirrors, whereas monkeys and nonprimates apparently do not (Bertenthal and Fischer, 1978; Gallup, 1977). This would imply that in some species, an awareness of the "self" is not a prerequisite for the presence of a species identity. Yet it remains unclear how a species identity is related to a self-identity in higher primates, and if one precedes, follows, or parallels the development of the other.

Species Identity and Its Generality

The attempt by this text's authors to discuss species identity within a phylogenetic framework has inherent difficulties. It is fortunate, however, that the following limitations focus around our relative naiveté about early developmental processes in many species; thus, these limitations are temporary if this volume's evaluation of species identity meets its aims of focusing attention, stimulating interest, and generating much needed research.

1. Systematic research which has attempted to investigate species identifications is scarce. As a result, a discussant of this concept is forced to rely on experimental data which have been gathered primarily for other purposes, subjective observations both in and outside of the laboratory, and speculations. This approach may not provide the reader with many answers, but it will more than satisfy the aims of this text.
2. An organization of presentations along taxonomic lines implies, at least to some, commonality of behaviors within classes or that a degree of generality across species is permissible and correct. This may be the case, but it is far from proven. Ecological systems interact with morphological characteristics to create similar behaviors among divergent species and vice versa. Much of our knowledge about a particular genus has come from work with only a portion of its available species during a small time sample of the evolutionary process. There is, however, always a need to synthesize what data we have, make cross-species comparisons, and, until classification systems are modified (see Klopfer, 1977), continue to work within the current phylogenetic framework.
3. The attempt to utilize a single term to characterize social preferences in divergent species will generate different connotations and usages of the species identity concept. This apparent limitation is actually beneficial for it will readily point out which species cannot be so classified and help to determine the generality of the concept.

A Research Strategy

The task of studying species identifications systematically will not be easy. There is a considerable "gray area" which is open to interpretation. A report on cross-species affinities in macaques by Chamove and Harlow (1975) illustrates this all too well. Forty-six macaques representing three subspecies (rhesus, stumptail, and pigtail) were social isolate-reared in a colony room, having visual and auditory contact with rhesus monkeys from birth to 3 months of age. They then received different rearing experiences from 3 to 9 months (period I); from 9 to 15 months (period II); and from 15 to 21 months (period III). The following six groups were established: (1) subjects without any cagemates but having playroom contact with species' members for 30 minutes a day; (2) subjects with no cagemates but having visual and auditory contact with another species for periods I and II; (3) subjects paired with a mate of the same species for period I, but with a mate of another species for period II; (4) subjects reared with another species for both periods, but with different mates in each period; (5) subjects housed with nonconspecific mates in periods I and II, but with a different species in each period; and (6) subjects raised with a nonconspecific mate for the first period but with a conspecific mate for the second period. Social preferences were observed while all subjects responded in a Sackett Self-Selection Circus and when 12 of the subjects were housed in a mixed-species social group during period III. Three statements summarize the results:

1. Subjects housed with a mate of a different species during period I significantly preferred that species over their own species.
2. Subjects housed with mates from two different species during periods I and II showed no preference for any species, even if one of the mates was a conspecific.
3. During period II, monkeys slept huddled with species they were housed with during period II, but they played with their own species (even if they had never been paired with them previously).

Chamove and Harlow conclude that

> It is clear that animals and certainly primates develop an attachment for an alien species when given exclusive experience with

> that species early in life. The specificity, malleability, or permanence of that preference is not clear (p 136).

This study is very suggestive. It implies that early experiences influence the acquisition of a species identity in macaques. The results are, however, difficult to interpret given the available data, and certain questions arise. Initially, were the sleeping partners in period III prior cagemates (which could mean that a sleeping partner "habit" was present) or were the sleeping partners new (which could represent a preference for that particular species)? Second, did playing with species members reflect some degree of innate preference for the biological species, did it occur because the biological species provided more reciprocal play activities due to their similar motor and communication patterns, and/or did it occur because play was discouraged by nonspecies members of the group? Finally, did subjects prefer particular species during social-preference testing, or were they actually avoiding certain choice subjects or species?

We hope that this volume's evaluation of species identity will help to clarify issues of generality, specificity, malleability, and permanence by stimulating research with many species.

Note

1. Toto at age nine was given to a circus and presented to Gargantua, estimated to be 11 or 12 years old. Their first meeting went as follows:

> In a special enclosure, Toto's air-conditioned wagon was wheeled up close to but not touching Gargantua's. Then the steel doors of the two wagons slid back, leaving the steel bars and sufficient distance between them so that they could not reach each other, even by extending their arms.
>
> For a moment they stared at each other in surprise, then Toto barked in rage. I have never been able to understand why, but I'm sure that no swain ever received a more discouraging welcome than Gargantua did that day. Perhaps in this, the first creature like herself which she had seen since the days that preceded her earliest memory, she sensed a rival for the human affection and the attention which she had always received. Perhaps in the strange regions of her subconscious mind she associated that gorilla face and body with the reflection in her mirror in Cuba which she had come to resent because she had been taken away from it so often when she was smaller. Perhaps it was simply that she didn't like Gargantua's

face, the mouth of which is set in what seems to be a permanent sneer as the result of acid burns which he received as a baby on the boat that brought him from Africa to New York.

Whatever was the reason, she would have none of him. When he thrust his hands through the bars, attempting to reach her in apparent friendliness, she stamped and barked furiously. When he threw a stalk of celery, left over from his breakfast, into her car as a peace offering she hurled it back into his face. Tomas in her cage told her to throw a kiss at Gargy, but she only stamped the more furiously (Hoyt, 1941, pp 214–216).

References

Beauchamp, G. K. and E. H. Hess. 1971. The effects of cross-species rearing on the social and sexual preferences of guinea pigs (*Cavia porcellus*). *Z. Tierpsychol. 28*, 69–76.

Beauchamp, G. K. and E. H. Hess. 1973. Abnormal early rearing and sexual responsiveness in male guinea pigs. *J. of Comp. Physiol. Psychol. 85*, 383–396.

Bernstein, I. S. 1966. Naturally occurring primate hybrid. *Science 154*(3756), 1559–1560.

Bertenthal, B. I. and K. W. Fischer. 1978. Development of self-recognition in the infant. *Dev. Psychol. 14*(1), 44–50.

Brown, J. L. 1975. *The evolution of behavior.* New York: Norton.

Cammaerts-Tricot, M. C. 1975. Ontogenesis of the defense reactions in the workers of *Myrmica rubra*. L. (Hymenoptera: Formicidae). *Anim. Behav. 23*, 124–130.

Chamove, A. S. and H. F. Harlow. 1975. Cross-species affinity in three macaques. *J. Behav. Sci. 2*(3), 131–136.

Chance, E. 1922. *The cuckoo's secret.* London: Sidgwick & Jackson.

Chase, R., K. Pryer, R. Baker, and D. Madison. 1978. Responses to conspecific chemical stimuli in the terrestrial snail. *Achantina fulica* (Pulmonata: Sigmurethra). *Behav. Biol. 22*, 302–315.

Cooke, F. 1977. Evolutionary consequences of early learning in the dimorphic bird *Anser caerulesceus.* Paper presented at the International Ethology Congress, Bielefeld, Germany.

Deutsch, J. and K. Larsson. 1974. Model-oriented sexual behavior in surrogate-reared rhesus monkeys. In W. Riss (Ed.), *Brain, behavior, and evolution.* Basel: Karger.

Ellis, P. 1963. Changes in social aggregation of locust hoppers with changes in rearing conditions. *Anim. Behav. 11*, 152–160.

Fantz, R. L. 1965. Visual perception from birth as shown by pattern selectivity. *Ann. N. Y. Acad. of Sci. 118*, 793–814.

Fletemeyer, J. R. 1978. The lost year. *Sea Frontiers 24*(1), 23–26.

Fox, M. W. 1968. *Abnormal behavior in animals*. Philadelphia: Saunders.

Fox, M. W. 1971. *Integrative development of brain and behavior in the dog*. Chicago: Univ. Chicago Pr.

Frick, F. 1976. Orientation and behavior of hatchling green turtles (*Chelonia mydas*) in the sea. *Anim. Behav. 24*, 849–857.

Friedmann, H. 1929. *The cowbirds: A study in the biology of social parasitism*. Springfield, Illinois: Thomas.

Friedmann, H. 1963. Host relations of the parasitic cowbirds. *Bull. U.S. Nat. Mus. 233*, 1–276.

Friedmann, H. 1966. Additional data on the host relations of the parasitic cowbirds. *Smithson. Misc. Collect. 149*(11), 1–12.

Gallup, Jr., G. G. 1977. Self recognition in primates: A comparative approach to the bidirectional properties of consciousness. *Am. Psychol. 32*(5), 329–339.

Gillett, S. D. 1973. Social determinants of aggregation behavior in adults of the desert locust. *Anim. Behav. 21*, 599–606.

Gottlieb, G. 1971. *Development of species identification in birds*. Chicago: Univ. Chicago Pr.

Gottlieb, G. 1978. Development of species identification in ducklings: IV. Change in species-specific perception caused by auditory deprivation. *J. Comp. Physiol. Psychol. 92*(3), 375–387.

Hayes, C. 1952. *The ape in our house*. New York: Harper.

Hediger, H. 1955. *Studies of the psychology and behavior of captive animals in zoos and circuses*. London: Butterworths Sci. Publ.

Hess, E. H. and D. B. Hess. 1969. Innate factors in imprinting. *Psychonomic Science 14*, 129–130.

Hess, L. 1954. *Christine, the baby chimp*. London: Bell.

Hodge, W. H. 1946. Camels of the clouds. *Nat. Geogr. Mag. 89*, 641–656.

Hoyt, A. M. 1941. *Toto and I: A gorilla in the family*. Philadelphia: Lippincott.

Immelmann, K. 1965. Objektfixierung geschlechtlicher Triebhandlung bei Prachtfinken. *Die Naturwissenschaften 52*, 169.

Immelmann, K. 1972. Sexual and other long term aspects of imprinting in birds and other species. In D. S. Lehrman, R. A. Hinde, and E. Shaw (Eds.), *Advances in the study of behavior*. Vol. 4. New York: Academic Pr.

Johnson, V. R., Jr. 1977. Individual recognition in the banded shrimp *Stenopus hispidus* (Oliver). *Anim. Behav. 25*, 418–428.

Kellogg, W. N. and L. A. Kellogg. 1933. *The ape and the child*. New York: McGraw-Hill.

Kerfoot, E. M. 1965. Some aspects of mate selection in domestic pigeons. *Diss. Abstr. 25*, 4250.

Kling, A. and H. D. Steklis. 1976. A neural substrate for affiliative behavior in nonhuman primates. *Brain Behav. Evol. 13*, 216–238.

Klinghammer, E. 1967. Factors influencing choice of mate in altricial birds. In H. S. Stevenson, E. H. Hess, and H. L. Rheingold (Eds.), *Early behavior: Comparative and developmental approaches.* New York: Wiley.

Klopfer, P. H. 1977. Social Darwinism lives! (Should it?) *Yale J. Biol. Med. 50,* 77–84.

Lehrman, D. S. and J. S. Rosenblatt. 1971. The study of behavioral development. In H. Moltz (Ed.), *The ontogeny of vertebrate behavior.* New York: Academic Pr.

Lewis, M. 1969. Infants' responses to facial stimuli during the first year of life. *Dev. Psychol. 1,* 75–86.

Leyhausen, P. 1973. On the choice of a sexual partner by animals. In K. Lorenz and P. Leyhausen (Eds.), *Motivation of human and animal behavior.* New York: Van Nostrand.

Linsenmaier, K. E. and C. Linsenmaier. 1971. Paarbildung und Paarzusammenhalt bei der Monogamen Wustenassel *Hemilepistus reaumuri* (Crustacea Isopoda Oniscoidga). *Z. Tierpsychol. 29*(2), 134–155.

Lorenz, K. 1935. Der Kumpan in der Umwelt des Vogels. *J. Ornithol. 2,* 80. (Reprinted in K. Lorenz (Ed.). 1970. *Studies in animal and human behaviour.* Vol. 1. Cambridge, Mass.: Harvard Univ. Pr.)

Lorenz, K. 1939. Comparative study of behavior. Paper presented at the Zoological Convention, 1939. (Printed in C. H. Schiller (Ed.). 1957. *Instinctive behavior.* New York: International Univ. Pr.)

Lorenz, K. 1952. *King Solomon's ring.* New York: Crowell.

MacLean, P. D. 1972. Cerebral evolution and emotional processes: New findings on the striatal complex. *Ann. N. Y. Acad. Sci. 193,* 137–149.

Maple, T. 1977. Unusual sexual behavior of nonhuman primates. In J. Money and H. Musaph (Eds.), *Handbook of sexology.* Amsterdam: North Holland Pr.

Marler, P. and P. Mundinger. 1971. Vocal learning in birds. In H. Moltz (Ed.), *The ontogeny of vertebrate behavior.* New York: Academic Pr.

Mason, W. A., S. D. Hill, and C. E. Thomsen. 1971. Perceptual factors in the development of filial attachments. In *Proceedings of the 3rd International Congress of Primatology, Zurich, 1970.* Basel: Karger.

Mason, W. A., S. D. Hill, and C. E. Thomsen. 1974. Perceptual aspects of filial attachment in monkeys. In N. F. White (Ed.), *Ethology and psychiatry.* Toronto: Univ. Toronto Pr.

Mason, W. A. and M. D. Kenney. 1974. Redirection of filial attachments in rhesus monkeys: Dogs as mother surrogates. *Science 22*(March), 1209–1211.

Mayr, E. 1963. *Animal species and evolution.* Cambridge, Mass.: Belknap Pr. (Harvard Univ. Pr.).

McCarty, R. and C. H. Southwick. 1977. Cross-species fostering: Effects on the olfactory preference of *Onychomys torridus* and *Percomyscus oeucopus. Behav. Biol. 19,* 255–260.

McDonald, D. L. and L. G. Forslund. 1978. The development of social preferences in the voles *Microtus montanus* and *Microtus canicaudus*: Effects of cross-fostering. *Behav. Biol. 22*, 497–508.

Meltzoff, A. N. and M. K. Moore. 1977. Imitation of facial and manual gestures by human neonates. *Science 198*(4313), 75–78.

Meyer-Holzapfel, M. 1968. Abnormal behavior in zoo animals. In M. W. Fox (Ed.), *Abnormal behavior in animals.* Philadelphia: Saunders.

Mitchell, G. 1973. Comparative development of social and emotional behavior. In G. Bermant (Ed.), *Perspectives on animal behavior.* Glenview, Ill.: Scott, Foresman.

Newton, G. and S. Levine. 1968. *Early experience and behavior.* Springfield, Ill.: Thomas.

Nicolai, J. 1964. Der Brutparasitismus als ethologisches Problem. *Z. Tierpsychol. 21*, 129–204.

Nicolai, J. 1974. Mimicry in parasitic birds. *Sci. Am. 12*, 93–98.

Otte, D. and W. Cade. 1976. On the role of olfaction in sexual and interspecies recognition in crickets (*Acheta* and *Gryllus*). *Anim. Behav. 24*, 1–6.

Richard, A. 1970. A comparative study of the activity pattern and behavior of *Alouatta villosa* and *Ateles geoffroyi. Folia Primatol. 12*, 241–263.

Rose, M. D. 1977. Interspecific play between free ranging guerezas (*Colobus guereza*) and vervet monkeys (*Ceropithecus aethiops*). *Primates 18*(4), 957–964.

Russock, H. I. and M. W. Schein. 1978. Effect of socialization on adult social preferences in *Tilapia mossambica (Sarotherodon Mossambicus*); Pisces: Cichlidae. *Anim. Behav. 26*, 148–159.

Sackett, G. P. 1966. Monkeys reared in isolation with pictures as visual input: Evidence for an innate releasing mechanism. *Science 154*, 1468–1473.

Sackett, G. P. 1973. Innate mechanisms in primate social behavior. In C. R. Carpenter (Ed.), *Behavioral regulators of behavior in primates.* Lewisburg, Pa.: Bucknell Univ. Pr.

Schutz, F. 1965. Sexuelle Prägung bei Anatiden. *Z. Tierpsychol. 22*, 50–103.

Schutz, F. 1971. Prägung des Sexualverhaltens von Enten und Gansen durch Socialeindrucke wahrend der Jugendphase. *J. Neuro-Visc. Relat. Suppl. 10*, 339–357.

Scott, J. P. 1968. *Early experience and the organization of behavior.* Belmont, Cal.: Wadsworth Publ.

Scott, J. P. 1972. *Animal behavior*, 2nd edition. Chicago: Univ. Chicago Pr.

Temerlin, M. K. 1975. *Lucy: Growing up human.* Palo Alto, Cal.: Science and Behavior Books.

Thresher, R. E. 1976. Field experiments on species recognition by the threespot damselfish, *Eupomacentrus planierons* (Pisces: Pomacentridae). *Anim. Behav. 24*, 562–569.

van Lawick-Goodall, J. 1968. The behavior of free-living chimpanzees in the Gombe Stream Reserve. *Anim. Behav. Monogr. 1,* 161–311.

Warriner, C. C. 1960. *Early experience as a variable in mate selection among pigeons.* Ann Arbor, Mich.: University Microfilms.

Warriner, C. C., W. B. Lemmon, and T. S. Ray. 1963. Early experience as a variable in mate selection. *Anim. Behav. 11,* 221–224.

Chapter 2

Species Identification in Fish

Lee I. McCann

The presence of mechanisms by which fish identify their own species is apparent in the very existence of such social interactions as schooling, sexual behavior, and aggression. Recognition of the existence of such a capability does not imply, unfortunately, any basic understanding of the nature of the underlying processes. The exact sensory mechanisms and stimuli involved, the degree to which heredity and experience contribute and interact, and the behaviors which most accurately reflect the existence and nature of species identification are all topics which require further study and understanding.

The concept of species identity has only recently appeared in the literature dealing with fish behavior (McCann and Matthews, 1974; Hopkins, 1977). There is, however, an extensive behavioral literature dealing with fish (McCann, 1978) from which considerable relevant information may be inferred. Many of these studies deal specifically with the effects of early rearing conditions upon social behavior, allowing a preliminary review and evaluation of how hereditary and experiential factors may contribute to species identification in fish.

Fish as Behavioral Subjects

Most modern fish descend from a group with cerebral development in the visual rather than the more evolutionarily significant olfactory region. The modern teleost fish, which serve as the subjects for a majority of current studies, developed from an earlier group of visually oriented fish. The first land-living vertebrates are believed to have evolved from nose-oriented predators of the crossopterygian group (Romer, 1958), a distinct fish line. Therefore, fish species used in most current work are not representative of the ancestral strains which led to the more complex vertebrates upon which most current behavioral reserach is carried out. Generalization of the current data must therefore be based upon the common nature of solutions to problems imposed by the environment to separate lines, rather than upon any presumption of a common linear ancestry which might otherwise be assumed to provide the basis of behavioral similarity between fish and terrestrial vertebrates.

A further limitation is imposed by the realization that relatively few of the thousands of different species of fish have been studied with any intensity, and that many of those species which have been studied were chosen because of commercial importance or because of their convenient availability as "pets." Such species as the common goldfish, guppy, and Siamese fighting fish are the result of intense selective breeding favoring morphological characteristics. The reader should thus be alerted to the potential roles of selection and domestication as factors in the behavior of the readily available fish species frequently chosen as research subjects.

One characteristic of most fish species which must influence our assumptions regarding the contributions of heredity and environment to their social behavior is the lack of any parent–young relationship. When combined with the knowledge that many young fish, particularly the fry of egg layers, bear minimal resemblances to either juveniles or adults, this fact has lent credence to the long-standing assumption that fish behavior is basically innate (Breder and Halpern, 1946). These characteristics do not eliminate experience as a factor in fish behavior by any means, but they do restrict the ways in which experience might contribute.

In all studies of the behavior of fish, the influence of methodology upon results and conclusions is a factor which must be considered. Since the effect of atypical rearing conditions upon subsequent social behavior is a common basis for many of the studies to be considered, the general effects of rearing environment, the experimental stimuli, and potential confounding factors affecting the behaviors studied are all crucial methodological considerations which must be explored.

Specific behaviors such as sexual interaction, aggression, schooling, or general aggregation are commonly studied in fish. With respect to the concept of species identification being considered here, it is necessary to be aware that the general affiliation involved in these behaviors does not imply specific species identification nor, for that matter, personal or individual identity (unlikely in fish in any case). Fear responses, an attraction to motion, or an undifferentiated reaction to another organism may be involved, particularly in an animal which has been completely isolated from fish of any species prior to the moment of testing (Pinckney and Andersen, 1967).

The concept of a species identity is not directly tested in most studies of these various behaviors, and we must infer its influence. The chain of inference thus created becomes progressively and unpleasantly lengthy. We assume that species identity must underlie and influence most social behaviors in fish, and thus if atypical rearing conditions or other early experiences influence the development of species identity they should, in turn, influence those social behaviors based upon species identification. It is, however, also possible that the various early experiences might act primarily upon the social behaviors themselves and/or their associated motivational complexes and have little to do with species identity.

Species Identification

The single study which has specifically investigated a process labeled "species identification" in fish was done by McCann and Matthews (1974). It provided a choice between similar stimuli at testing, as compared to the no-choice test situation common in schooling studies, and used the term "species identification" to describe the phenomenon being investigated. Zebra fish (*Brachydanio rerio*, striped) were raised from the egg in isolated conditions, and species identification was tested by providing subjects with a choice between groups of zebra fish and pearl danios (*Brachydanio albolineatus*, nonstriped) of equal physical size and number. The investigators found that isolates spent less time in nonpolarized schooling (aggregations) with their own kind than did group-reared subjects. The results indicated that isolated animals displayed decreased species identification when compared to controls, but retained a preference for their own kind over nonconspecifics. The authors interpret this as an indication that both *innate and experiential factors* contribute to the species identity of *B. rerio*, but make no attempt to specify the exact mechanisms involved.

Other studies have also directly approached the subject of species identification in fish through measurement of what may be best described as species preference or recognition. Three species of platies (genus *Xiphophorus*) were isolated for periods ranging from one to three days, and the time spent adjacent to stimulus animals of each species which were presented simultaneously was measured (Cross et al., 1969). Subjects overwhelmingly chose their own kind, with the short-term isolates showing no difference from animals maintained in a group of nonconspecifics. The authors related their results to a "biological selectivity," and did not discuss species identity or recognition as such. A comparison of these results with McCann and Matthews is difficult because of both the short-term isolation and the lack of true control animals maintained in a group of conspecifics.

McCann, Koehn, and Kline (1971) presented adult zebra fish with a choice between photographs of groups of normal zebra fish, zebra fish with stripes reversed, and those with no stripes. Each of these three groups was reproduced at three different sizes of stimulus fish. The time subjects spent adjacent to each photo was recorded and compared, the results indicating that zebra fish prefer striped fish to nonstriped but do not distinguish between normal and reverse striping. These results can be generally interpreted as species recognition, if one remembers that the choice allowed was between pictures quite similar to normal zebras (normal and reverse striping) and nonstriped fish with the zebra body configuration. McCann et al. (1971) also found that their subjects definitely preferred photographs with stimulus fish equal to their own size, reinforcing the importance of size in aggregation and schooling found in many of the schooling studies soon to be discussed.

These studies strongly indicate that fish can identify their own kind as indicated by objective behavioral measurement. They do not indicate the relative contributions and combinations of heredity and experience with any certainty, but the results of

McCann and Matthews (1974) plus those of McCann et al. (1971) do seem to show that vision alone provides sufficient sensory information to elicit the identification in *B. rerio*.

A different variety of species identification is reported by a number of authors who have studied the recognition of fry by parents of various species of cichlid fish. Noble and Curtis (1939) first reported that inexperienced cichlid parents could not distinguish fry of their own species from the fry of other cichlids, but parents with previous brooding experience definitely made the distinction when the fry were free-swimming. Confirmation of these results by Baerends and Baerends-Van Roon (1950) led to the conclusion that these studies were an example of adult imprinting (Tinbergen, 1951; Thorpe, 1956). Further work by Myrberg (1961) supported the imprinting hypothesis, but found that fry were distinguished during the wiggler stage which occurs prior to free swimming. He also noted that adults imprinted only on young of their own species and not on foster young which they had previously reared.

Contradictory results indicating that experienced parents will accept foreign fry have also been reported (Greenberg, 1961, 1963a, 1963b; Collins and Braddock, 1962), with Greenberg (1963b) suggesting that Myrberg's results were due to incubation-period differences in the fry used for testing. Myrberg (1966) reviews these various findings and disagrees with Greenberg on several methodological points. He suggests that chemoreception is the dominant modality which cichlids use in tending wiggling fry, and that vision is the dominant sense with free-swimming young. The possibility of differences between species is also apparent in his review.

The conclusions which may be drawn from this work are clouded by possible species differences and methodological problems. It does seem clear, however, that a process similar to imprinting is occurring with some consistency, although the change from chemical to visual stimulation as the young develop and various other factors have so far obscured the details of the process. This "imprinting" does suggest the probability of an experiential contribution to the identification of fry. While the recognition of fry is potentially quite different from the identification of adults, an identification process is occurring within a species with experience involved—results which are generally similar to those of McCann and Matthews (1974). The identification of chemoreception as a sensory factor is also noteworthy, since vision has been the predominant factor in the studies discussed previously.

The possibility that stimuli other than vision may influence species identification is also suggested by the work of Hopkins (1977), who demonstrated species differences in the electrical signals generated by various species of electric fish. It is probable that further research will suggest the involvement of a variety of other sensory mechanisms. Davis and Pilotte (1975) have shown that paradise fish (*Macropodus opercularis*) are attracted to chemicals in the water in which both conspecifics and nonconspecifics have been kept, and fish are well known to be sensitive to various types of vibratory stimulation as well.

Further, though indirect, evidence regarding species identification in fish can also be found in studies of specific social behaviors which are presumably based to some degree upon the ability to identify one's own species.

Schooling and Aggregation

Schooling is one of the most frequently studied behaviors in fish. It is of interest here in that the joining of a school of conspecifics implies an identification of one's own kind, even though there is typically no choice among species allowed. There have been a number of studies in which isolation and early experience were involved, with the initial conclusions favoring innate mechanisms as the basis of schooling and the more recent work providing evidence of the role of experience. Breder and Halpern (1946) reared two zebra fish in isolation and found that they immediately joined an aggregation (nonpolarized school) of that species as adults. No alternate species was available to provide a choice, so these results indicate only a preference to join a group rather than the actual identification of conspecifics. Breder and Halpern conclude that aggregation is based on an "innate mechanism." These authors and Shaw (1970), in separate reviews, also conclude that vision is essential to the formation and maintenance of a school.

The most detailed investigation of the role of early experiences in schooling behavior is reported by Williams and Shaw (1971). They reared three groups of a saltwater minnow (*Menidia menidia*), one in groups of conspecifics, one in isolation, and another group of isolates given a single brief exposure to conspecifics prior to the thirty-fifth day of life. They observed the development of schools among isolates, but reported that such schools lacked the smooth organization and coordination found in normal schools. A typical withdrawal behavior seen in young animals schooling for the first time appeared to persist in the young isolates in the face of continued schooling experience. The isolates with brief exposure to conspecifics also tended to delay the onset of schooling behavior, a result confirming the earlier work of Shaw (1961). Williams and Shaw interpret their results as an indication that schooling behavior is modified by atypical environmental conditions, and conclude that the behavior cannot be meaningfully described in terms of an innate–acquired dichotomy. Their feeling that a total developmental system must be hypothesized supports the heredity–environment interaction model suggested by McCann and Matthews (1974) in relation to species identification.

The fact that isolates will school has been repeatedly demonstrated. Jorné-Safriel and Shaw (1966) observed such behavior in isolates of atherinids (*Atherina mochon*) while Dambach (1963) reported similar results with various species of cichlid (*Tilapia*). Shaw also reported schooling in isolates of saltwater minnows in separate studies in 1960 and 1961. She noted the gradual nature of the development of schooling behavior in contrast to the earlier assumption that schooling occurred immediately after hatching (Morrow, 1948), thus laying the groundwork for her later work demonstrating the modifiability of schooling behavior. These several reports of schooling in the isolates of various species draw attention to a basic contribution of innate factors to schooling and, by probable extension, to other social behaviors in fish in which heredity might also be involved to greater or lesser degrees.

The demonstration of an experiential component in schooling by Williams and Shaw, which supports Shaw's (1970) suggestion of combined maturational–experiential factors following her general review of the schooling literature, lends fairly direct support to McCann and Matthews' (1974) conclusions regarding species identification. This consistency of moderate experiential effects across behaviors and species supports the assumption of significant experiential involvement in what initially had been considered essentially innate behaviors of fish.

Social activities such as aggression, display, and reproductive behavior also involve an interaction between the subject and another member, or members, of the same species. Once again the involvement of some degree of species identification is implicit in such behaviors and relevant studies, particularly those involving early environment, should provide at least indirect information about the nature of species identity.

Mate Selection

The data on courtship behavior suggest the existence of a critical period for a process similar to imprinting which begins at sexual maturity. Thus we again find experiential factors involved in social behavior, with the imprinting model best describing their involvement in this case. Weber and Weber (1976) found that the color of conspecifics with which male zebra cichlids (*Cichlosoma nigrofasciatum*) had been reared had no effect on mate selection measured at sexual maturity. Instead, female dominance was the most important factor. Ferno and Sjolander (1973) reared various color variants of platies (*Xiphophorous maculatus*) with single-color groups. They found, as did Weber and Weber (1976), that the color of the animals with which the subject was raised to sexual maturity had no effect on mate selection; and isolation-reared subjects also showed no color preference. However, male subjects reared with a multicolor group and then placed for two months following sexual maturity with a group of a color different from their own showed a distinct preference for mates of the other-color group. Attempts to reverse the apparent "imprinting" were successful in some cases but not in others. In no case did females show a color preference.

Haskins and Haskins (1950) reared male guppies (*Lebistes reticulatus*) in isolation until they reached sexual maturity. The subjects were then placed with a female of a different color for periods exceeding one month and tested with females of various colors for sexual preference. They preferred the color of the female with which they had been placed at sexual maturity. Attempts to recondition these males to other colors succeeded only one-third of the time, a result similar to those reported by Ferno and Sjolander. The results of these studies all suggest no genetic determination of color preference (not surprising within a species, particularly when most of the color variants involved arise from selective breeding), but do suggest the presence of a critical period for a process similar to imprinting (or a stable variety of learned behavior) which begins at sexual maturity. Regardless of the specific interpretation, experience is obviously a contributing factor to the selection of a mate in those male animals with experience subsequent to sexual maturity.

In a related study Kassel and Davis (1975) reared paradise fish in isolation, with conspecifics, or with nonconspecifics. As "young adults" the cross-reared subjects performed social displays to mirror images and to live conspecific stimuli less frequently than did either controls or isolates, which did not differ from each other. Why rearing experience with nonconspecifics should reduce display frequency more than complete isolation is not clear, but the results are not entirely inconsistent with those previously reported. The reduced display performance of cross-reared subjects could be a consequence of some imprinting or experiential effect initiated at sexual maturity and detracting from, or interfering with, reactivity to conspecifics. Other interpretations are obviously possible as well.

The fact that social behavior was developed by these isolates is, however, consistent with the results of the studies cited previously and with the schooling studies.

Other Social Behaviors

The fact that social behavior does develop in isolates has also been reported in jewel fish (*Hemichromis bimaculatus*) by Noble and Curtis (1935/36) and in various species of paradise fish by Kuo (1960). Aggressive behavior appeared in isolated Siamese fighting fish (*Betta splendens*; Braddock and Braddock, 1958) and sexual behavior in isolated platies (*Xiphophorus maculatus*; Shaw, 1962). In addition Cullen (1960) reports that rearing three-spined sticklebacks (*Gasterosteus aculeatus*) in isolation does not interfere with courtship, fighting, or nesting in most subjects, specifically ruling out experience with a red-colored father as a factor in the development of these behaviors.

Russock and Schein (1974) also describe an innate response in fry to a maternal mouth model in mouthbreeders (*Tilapia mossambica*). Animals reared normally with their mother, in groups separated from the mother, and isolates responded equally to the model. The lack of an experiential effect in this study is not really surprising, since the behavior studied is apparent only during the initial three weeks of life.

The conclusion that innate processes exert a major effect on these various sexual and social behaviors is obvious. However, the fact that the behaviors in question may be altered through differential experience is also apparent. The modifications described here are most similar to imprinting, a result not found with adult species identity or with schooling. This pattern of a basically innate response being moderately altered by experience remains consistent across the studies of species identification, schooling, and sexual or reproductive behavior, and it must be considered in any description of the social behavior of fish, including the role and characteristics of species identification.

Other Isolation Effects

It is very probable that a multitude of effects are produced by the isolated rearing condition. We must consider the nature of these effects in arriving at a fair evaluation of the results of all isolation studies, and we must be particularly concerned with the possibility of any contamination or confounding of the behavioral measures used. Since most of the social behaviors discussed are measured by the approach to a stimulus object or by a choice between such objects, any possible effects of isolation on the general "emotionality" or arousal of the subject, or on the general patterns of activity displayed, could produce either facilitory or inhibitory effects upon the response in question. Potential motivational effects resulting from social isolation must also be considered.

Pinckney and Anderson (1967) reared guppies in isolation and reported that isolates initially avoided other animals in open-field testing but spent increasing time with them as the experiment progressed. Their controls' behavior was exactly the opposite, leading the authors to hypothesize "anxiety" as a basis for the difference. They explained that strange animals would provoke an anxiety reaction in isolates which would dissipate with familiarity, but produce avoidance while present. Control subjects would presumably find the open-field situation itself anxiety-producing and seek other animals to relieve this reaction, which would also dissipate with time. If this hypothesis were correct, and if the results were replicated in other studies and species, this anxiety reaction would have to be dealt with by every isolation study. Kassel and Davis (1975), however, found no effect of isolation on frequency of air gulp and observed no freezing or hyperactivity in their subjects, suggesting that isolation does not influence arousal and leaving the issue unresolved.

Activity is frequently used as an index of emotion in psychological research, and Kassel and Davis did note lowered activity in their isolated subjects. Other activity measures of isolated subjects have dealt with the short-term isolation of adults, and thus may demonstrate the motivational effects of social deprivation more than any of the long-term isolation effects of interest here. Any such drive effects are, however, important in that they could produce general effects on any behavior being studied.

Escobar (1936) found that the activity of goldfish was greatest in isolation; followed by heterotypic and then homotypic groups. In contrast, McCann and Matthews (1974) cited unpublished data on adult zebra fish isolated for one week which showed no difference in activity or attraction to movement as compared to controls. In a more detailed study now being prepared for publication, McCann, Burkhardt, and Stedman isolated zebra fish for periods of one to five days and observed a monotonic increase in position preference in an open field. Isolated subjects showed greater preference for positions in the unsheltered center of the tank with increasing isolation time, while both individual and grouped controls preferred the perimeter. They could demonstrate no increase or decrease in activity, however.

Unfortunately these emotionality and activity studies show no clear pattern. The question of the general effects of long-term isolation are thus unresolved, as are the potential influences of these effects on social behavior in general, including species identification. The short-term isolation data do show an effect of isolation on activity which seems similar to the motivational effects of deprivation, and a variety of studies testify to the existence of deprivation effects in fish. Such deprivation effects lend support to the possibility of a drivelike result of isolation and are thus important here.

The most relevant of these studies (Cross et al., 1969) has been previously mentioned. Several species of adult platies were isolated and tested for species preference with a choice allowed between three strains. The results indicated a strong preference for conspecifics and no curiosity about or attraction to the other animals present. Other deprivation–drive relationships which have been demonstrated include effects on male displays in paradise fish (Davis, 1975), sexual responsiveness in swordtails (*Xiphophorus helleri;* Franck and Geissler, 1973), aggression in Siamese fighting fish (Hinkel and Maier, 1974), and sexual behavior in guppies (Breder and Coates, 1935). In light of these various indications we are forced to assume that any isolation situation will have arousal and drive enhancement as one of its consequences, with unpredictable effects on the emotional reactivity and general activity of the subjects. Results of the various studies of isolation and its effect on social behavior which have been considered must, therefore, be interpreted conservatively. It is likely that the various approach and choice behaviors which have been used as dependent variables are influenced by these relatively uncontrolled effects of isolation as a treatment.

Conclusions

The results of the few studies of species identification in fish and the indications from work involving the effects of isolation on various social behaviors produce a picture that is fairly consistent both across studies and with general trends across phylogeny. Species identification in fish is a fact, supported by every instance of social interaction with a conspecific. This species identity is strongly controlled by innate factors and thus heredity, which makes it relatively primitive in comparison to what is seen in creatures of greater behavioral plasticity. However, it is there.

Nearly all indications support the assumption that species identification in fish is primarily an innate mechanism, a conclusion which most probably applies to all their social behaviors. The work of McCann and Matthews (1974) on species identification, the various studies of schooling by Shaw, and the results of many studies of sexual behavior all indicate that the behaviors in question appear spontaneously in animals isolated from the egg. These several studies also indicate that the basically innate social behaviors of fish are modifiable by environmental and experiential factors, although the degree and detail of this modifiability vary considerably. Further investigation is warranted.

Since studies of more complex organisms have generally produced interactionist models of the relationship between heredity and environment, and since the contribution of heredity is generally found to be proportionally greater in the simpler and more "primitive" organisms (Dethier and Stellar, 1970; Denny and Ratner, 1970), this pattern of results in fish is not surprising. The most surprising aspect, in fact, is the degree and consistency with which modification has been demonstrated. Although one might anticipate the involvement of experience in combination with heredity, it is unusual to find the effects of experience so readily and consistently apparent in such a comparatively simple creature.

Any conclusion about the nature of species identification in fish must stress its probable involvement in all social behaviors, and the degree to which it appears genetically based but modifiable. Even the most extraordinary deviation of the environment from that which is typical will not severely distort a fish's species identity. Details of the behavior in question may be distorted by changes in arousal and stimulus responsiveness, and social experience *within* a species may produce imprinting to particular classes of sex stimulus, but the basic ability to respond to conspecifics in preference to nonconspecifics will remain essentially stable.

Thus species identification is basically a genetic predisposition to respond to particular patterns of stimuli, and the species identity of a fish is primarily a consequence of hereditary factors. All of this fits with our knowledge of the general level of behavioral complexity and plasticity in fish. The modifiability of social behavior which may be attributed to environment and experience suggests a modest interaction between heredity and environment and cautions that the innate appearance of many behaviors may be in part a consequence of consistent rearing conditions in an environment typical and usual for the species. While imprinting (Scholz et al., 1976) and the ability to learn (McCann, 1978, reports numerous studies) have been consistently demonstrated in fish, they do not appear to have a primary role in species identity or in other social behaviors.

There is no recent record of any data regarding or speculation concerning a concept of personal or individual identity in fish. While such a concept is of great interest and concern to those studying primate behavior, its inclusion in any discussion of fish would be quite inappropriate. The species identity of fish is best described as the capacity for identifying and responding appropriately to others of the same species. Heredity provides this capability, and experience may slightly alter or modify its expression. Even in the face of the most extraordinary and atypical rearing conditions, this basic ability to recognize others of its own species will appear, and the resultant behavior of the fish will approximate normality.

References

Baerends, G. P. and J. M. Baerends-Van Roon. 1950. An introduction to the study of the ethology of cichlid fishes. *Behaviour Suppl. 1,* 1–242.

Braddock, J. C. and Z. I. Braddock. 1958. Effects of isolation and social contact upon the development of aggressive behavior in the Siamese fighting fish, *Betta Splendens. Anim. Behav. 6,* 249.

Breder, C. M. and C. W. Coates. 1935. Sex recognition in the guppy, *Lebistes reticulatus* Peters. *Zoologica 14*(5), 187–207.

Breder, C. M. and F. Halpern. 1946. Innate and acquired behavior affecting the aggregation of fishes. *Physiol. Zool. 19,* 154–190.

Collins, H. L. and J. C. Braddock. 1962. Notes on fostering experiments with cichlid fishes, *Tilapia sparrmani* and *Acquidens portalegrensis. Am. Zool. 2,* 400.

Cross, H. A., L. J. Laux, J. C. Wright, V. J. Penzoldt, J. J. Lowenstein, T. D. Vincent, and N. W. King. 1969. Behavioral selectivity in tropical fish. *Psychonomic Sci. 17*(5), 297–298.

Cullen, E. 1960. Experiment on the effect of social isolation on reproductive behaviour in the three-spined stickleback. *Anim. Behav. 8,* 235.

Dambach, M. 1963. Vergleichende Untersuchungen über das Schwarmverhalten von *Tilapia*—Jungfischen (Cichlidae, Teleosti). *Z. Tierpsychol. 20*(3), 267–296.

Davis, R. E. 1975. Readiness to display in the paradise fish *Macropodus opercularis,* L., Belontiidae: The problem of general and specific effects of social isolation. *Behav. Biol. 15*(4), 419–433.

Davis, R. E. and N. J. Pilotte. 1975. Attraction to conspecific and nonconspecific chemical stimuli in male and female *Macropodus opercularis* (Teleostei Anabantoidei). *Behav. Biol. 13*(2), 191–196.

Denny, M. R. and S. C. Ratner. 1970. *Comparative psychology: Research in animal behavior,* 2nd ed. Homewood, Ill.: Dorsey.

Dethier, V. G. and E. Stellar. 1970. *Animal behavior,* 3rd ed. Englewood Cliffs, N.J.: Prentice-Hall.

Escobar, R. A., R. P. Minahan, and R. J. Shaw. 1936. Motility factors in mass physiology: Locomotion activity of fishes under conditions of isolation, homotypic grouping, and heterotypic grouping. *Physiol. Zool. 9,* 66–78.

Ferno, A. and S. Sjolander. 1973. Some imprinting experiments on sexual preferences for colour variants in the Platy fish (*Xiphophorous maculatus*). *Z. Tierpsychol. 33*(34), 417–423.

Franck, D. and U. Geissler. 1973. Short-term social isolation and changes of sexual responsiveness in the male *Xiphophorus helleri. Z. Tierpsychol. 33*(34), 408–416.

Greenberg, B. 1961. Parental behavior and imprinting in cichlid fishes. *Am. Zool. 1,* 450.

Greenberg, B. 1963. Parental behavior and imprinting in cichlid fishes. *Behavior 21*, 127–144. (a)

Greenberg, B. 1963. Parental behavior and recognition of young in *Cichlosoma biocellatum. Anim. Behav. 11*, 578–582. (b)

Haskins, C. D. and E. F. Haskins. 1950. Factors governing sexual selection as an isolating mechanism in three species of peociliid fishes *Lebistes reticulatus. Proc. Nat. Acad. Sci. 36*, 464–476.

Hinkel, T. J., and R. Maier. 1974. Isolation and aggression in Siamese fighting fish (*Betta Splendens*). *Psychol. Rep. 34*(3, Pt 2), 1323–1326.

Hopkins, C. D. 1977. Electroreception and species recognition among electric fish. *Society for Neuroscience Meeting*, Anaheim, Calif.

Jorné-Safriel, O. and E. Shaw. 1966. The development of schooling in the atherinid fish, *Atherina mochon* (Cuvier). *Pubbl. Stn. Zool. Napoli 35*, 76–88.

Kassel, J. and R. E. Davis. 1975. Early behavioral experience and adult social behavior in the paradise fish, *Macropodus opercularis* L. Behav. Biol. 15(3), 343–351.

Kuo, Z. Y. 1960. Studies on the basic factors in animal fighting: V. Interspecies coexistence in fish. *J. Genet. Psychol. 97*, 181–194.

McCann, L. I. 1978. Behavioral studies of fish: Indexed bibliography (1967–1977). *JSAS Catalog of Selected Documents in Psychology 8*, 59, Ms. 1714.

McCann, L. I., P. Burkhardt, and M. E. Stedman. The effect of short-term isolation on open-field position preference in the zebra fish (*Brachydanio rerio*). In preparation.

McCann, L. I., D. J. Koehn, and N. J. Kline. 1971. The effect of body size and body markings on nonpolarized schooling behavior of zebra fish (*Brachydanio rerio*). *J. Psychol. 79*, 71–75.

McCann, L. I. and J. J. Matthews. 1974. The effects of lifelong isolation in zebra fish (*Brachydanio rerio*). *Dev. Psychobiol. 7*(2), 159–163.

Morrow, J. E., Jr. 1948. Schooling behavior in fishes. *Q. Rev. Biol. 23*(1), 27–38.

Myrberg, A. A., Jr. 1961. *An analysis of the preferential care of eggs and young by cichlid fishes.* Doctoral dissertation, Zoology Department, Univ. California, Los Angeles.

Myrberg, A. A., Jr. 1966. Parental recognition of young in cichlid fishes. *Anim. Behav. 14*, 565–571.

Noble, G. K. and B. Curtis. 1935/36. Sexual selection in fishes. *Anat. Rec. 64*, 84–85.

Noble, G. K. and B. Curtis. 1939. The social behavior of the jewel fish, *Hemichromis bimaculatus* Gill. *Bull. Am. Mus. Nat. Hist. 76*, 1–46.

Pinckney, G. A. and L. E. Anderson. 1967. Rearing conditions and sociability in *Lebistes reticulatus. Psychonomic Sci. 9*, 591–592.

Romer, A. S. 1958. Phylogeny and behavior with special reference to vertebrate evolution. In A. Roe and G. G. Simpson (Eds.), *Behavior and evolution.* New Haven: Yale Univ. Pr., 48–75.

Russock, H. I. and M. W. Schein. 1974. Effect of early experience and age on initial responsiveness of *Tilapia mossambica* fry to a maternal model. *Anim. Learn. Behav.* 2(2), 111–114.

Scholz, A. T., R. M. Horrall, J. C. Cooper, and A. D. Hasler. 1976. Imprinting to chemical areas: The basis for home stream selection in salmon. *Science* 192(4245), 1247–1249.

Shaw, E. 1960. The development of schooling behavior in fishes. *Physiol. Zool.* 33, 79–86.

Shaw, E. 1961. The development of schooling in fishes II. *Physiol. Zool.* 34, 263–279.

Shaw, E. 1962. Environmental conditions and the appearnace of sexual behavior in the platyfish. In E. Bliss (Ed.), *Roots of behavior.* New York: Harper.

Shaw, E. 1970. Schooling in fishes: Critique and review. In L. R. Aronson, E. Tobach, D. S. Lehrman, and J. S. Rosenblatt (Eds.), *Development and evolution of behavior: Essays in memory of T. C. Schneirla.* San Francisco: Freeman, 452–480.

Stockard, C. R. 1931. *The physical basis of personality.* New York: Norton, 109.

Thorpe, W. H. 1956. *Learning and instinct in animals.* Cambridge, Mass.: Harvard Univ. Pr.

Tinbergen, N. 1951. *The study of instinct.* Oxford: Clarendon Pr.

Weber, P. G. and S. P. Weber. 1976. The effect of female color size dominance and early experience upon mate selection in male convict cichlids *Cichlosoma Nigrofasciatum* Pisces Cichlidae. *Behaviour* 56(1–2), 116–135.

Williams, M. M. and E. Shaw. 1971. Modifiability of schooling behavior in fishes: The role of early experience. *Am. Mus. Novit.* 2448, 1–19.

Chapter 3

Reptiles

Arnold D. Froese

A phylogenetic review on any given topic should cover as broad a range of species as possible. As species having more remote relationships to humans are considered, however, we have difficulty describing the phenomena we observe in ways which parallel our descriptions of human behavior. Species awareness or species identity exemplifies this problem. While humans can verbalize their affinity for other humans, they must infer such conspecific affinity from behavior of a different kind in other species. We should, therefore, specify the problems we study in ways which can be investigated across a broad range of species. If species awareness is our concern, we must specify general behavioral criteria which indicate such awareness in a variety of species. In this way, comparative statements may not be limited to those few species whose behaviors we think we intuitively recognize. The problem can also be illustrated by the difficulty in understanding human neonate awareness. We cannot ask neonates about their awareness, yet they behave in ways which suggest that they are aware (Meltzoff and Moore, 1977).

The common data base for a phylogenetic investigation of species identity can be found in behavioral interactions among animals. Since we cannot directly determine what most animals "know," we are restricted to investigating what they do. The terminology we use ought to reflect this level of analysis. For this reason the term "species-identification processes" is preferred over "species identity" and "species awareness." The former connotes an analysis of organismic interactions, while the latter two connote a hypothetical property which an organism possesses.

Several different terms have been used in reptile studies to refer to species-identification processes. The term "species discrimination" has generally been used in cases where a reptile was tested for its ability to respond differentially to conspecifics versus nonconspecifics (e.g., Pyburn, 1955; Auffenberg, 1965; Hunsaker, 1962). A somewhat more specific term, "interracial discrimination" (Ferguson, 1969), indicated the phylogenetic relationship among the reptiles being tested. "Species identification" and its synonym "species recognition" have been applied to a broad range of observations and speculations regarding reptilian interactions (e.g., Auffenberg, 1965; Hunsaker, 1962; Jenssen and Hover, 1976; Noble and Clausen, 1936; Williams and Rand, 1977). These terms include discrimination processes as described above, but their use is not restricted to specific discrimination tests. These terms imply the operation of receiver characteristics in the identification process. At a broader level, the term "species identification processes" includes receiver and signal characteristics. The reciprocal interaction of animals contains many components which may be important for the identification process. This term is preferred, then, for referring to the major topic under consideration because it suggests a broader empirical investigation of that entire process.

Species-identification processes are a component of biological isolating mechanisms (e.g., Hunsaker, 1962; Rand and Williams, 1970). Factors which keep species from interacting at all (e.g., geographical or ecological separation) are unrelated to species identification processes. Other factors which allow for interaction between species, but insure differential responding to conspecifics versus nonconspecifics, are the essence of species-identification processes.

The terms described above suggest different levels of specificity in the processes being considered. They can be arranged in a hierarchical order from terms which are broad to terms which refer to narrower aspects of the investigation of species identification processes. Such an order is illustrated in Figure 1. Other terms which suggest different kinds of restrictions could be inserted as our data base broadens. While some terms appear to be synonymous in the reptile literature (e.g., species identification and species recognition), the various levels of specificity possible in the research suggest that we continue to use a variety of terms

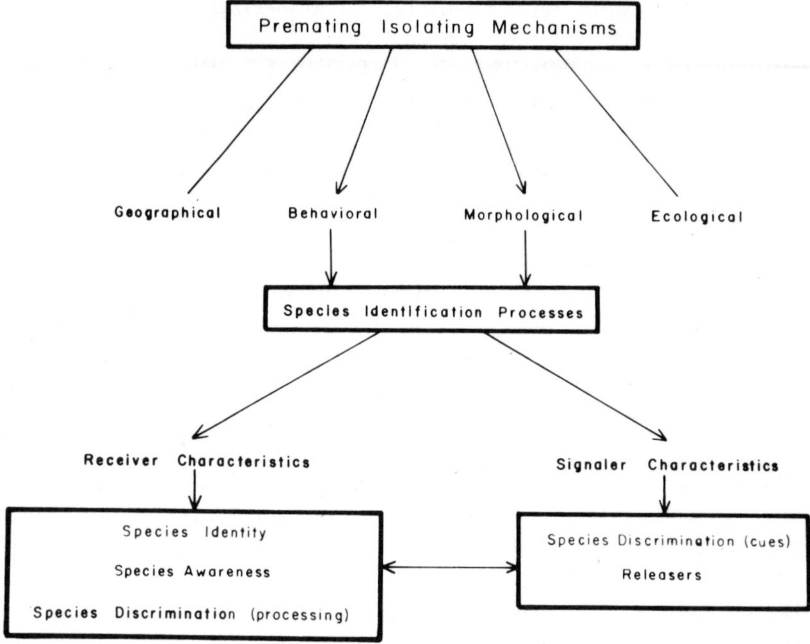

Figure 1. A hierarchical representation of terminology appropriate in studies of species identification processes. Specific terms appear in boxes, and the list could be expanded. The horizontal arrow at the bottom of the figure represents reciprocal interactions between signaler and receiver.

to describe research on species-identification processes. These different terms may help to portray the appropriate level of specificity for a given research project. Beer (1977) warned that "rigour of definition risks *rigor mortis* of concept" (p 155). The terminological flexibility suggested previously leaves the concept of species identity open for revision.

Reptiles

Wickler (1976) recently pointed out that ethology and comparative psychology have concentrated on only 2 of the 26 phyla of modern animals—the arthropods and the vertebrates. While reptiles are vertebrates, they have received minimal attention in behavioral research. Reptiles are generally considered to be slow and stupid.[1] Some mature slowly, and breeding in captivity presents additional problems which are not encountered in the typical psychology rat lab. The simple fact that reptiles are ectothermic has produced problems for unwary researchers studying their learning ability or social behavior (Brattstrom, 1974). Burghardt, Greene, and Rand (1977) nevertheless suggested that reptiles might provide a key to our understanding of the ontogeny and phylogeny of social behavior in birds and mammals. Their suggestion is based on the observation of coordinated social behaviors in animals which had no opportunity to perfect their behavior during a probationary childhood and adolescence.

The class Reptilia includes a diverse group of animals. About 5000 species of lizards, snakes, turtles, and crocodilians, along with the tuatara, exist today. Some of these reptiles are nocturnal, others are diurnal. Some are largely terrestrial while others are aquatic or fossorial. There are carnivores, herbivores, and scavengers among the reptiles. Within all of this diversity, however, one aspect of their life history is relatively constant. In most species, no parental care is known after the female deposits and conceals the eggs. In those species which have been reported to show parental care, neither the extent of that care nor its importance for normal development of the offspring is clearly known. Parental care is certainly a different matter in reptiles than in birds or mammals. This difference is especially striking in view of the extent to which socialization is presumed to modify behavior. One of the important questions in this volume concerns the effects of early rearing conditions on the development of species identification processes. Rearing under parental care is not the norm for reptiles. Yet, as in birds and mammals, they exhibit degrees of individual recognition, dominance, territoriality, and other social behaviors. Certainly, reptiles deserve special attention in a phylogenetic review of species-identification processes which includes birds and mammals.

EVIDENCE FOR SPECIES IDENTIFICATION. The mere fact that a sexually reproducing species exists suggests that some identification processes have occurred during courtship. Anatomical specializations serve to reproductively isolate one species from another. A snake and a turtle simply do not match. Species identification among closely related species, such as members of the same genus, may be more confusing. Often such closely related species are isolated by geographical barriers or by ecological differences. If they are not thus isolated, discrimination among conspecifics and nonconspecifics is required. The following sections will review laboratory and field evidence for these discrimination abilities in lizards, snakes, turtles, and crocodilians.

Lizards

Evidence for Identification Processes

Species discrimination abilities in lizards have only recently been tested in an experimental manner. Pyburn (1955) conducted an early study on the Texas spiny lizard (*Sceloporus olivaceus*) and the sympatric crevice spiny lizard (*Sceloporus poinsetti*). Pyburn placed wild male and female adults of both species in a four-compartment test cage. Each compartment had a door into adjacent compartments. The frequency of each possible combination of species and sex of the animals in each compartment was recorded. Conspecific pairs were observed together more than twice as often as nonconspecific pairs. In several instances, a mixed-species trio was observed together. The odd-species female was found to be gravid, and Pyburn speculated that she may have been avoiding conspecific males. This study confirmed the ability of some spiny lizards to distinguish between conspecifics and sympatric congeners.

Hunsaker (1962) recorded the frequency of occurrence of conspecific and nonconspecific pairs in an open arena and found distinct conspecific aggregations in several spiny lizard species (*Sceloporus torquatus*, *S. poinsetti*, and *S. cyanogenes*). Spiny lizards did not interact much with two whiptail lizards (genus *Cnemidophorus*). Isolated spiny lizards also responded to their mirror images in a typical conspecific fashion. Hunsaker was aware of behavioral display differences among lizards, and he constructed a choice apparatus to test the ability of spiny lizards to discriminate between conspecific and nonconspecific head-bob displays. Tested lizards could remain in a neutral area, they could approach a model on one side which simulated a conspecific head-bob display, or they could approach a model on the opposite side which bobbed in a random fashion. Two tested species (*S.*

torquatus and *S. mucronatus*) were attracted to conspecific head-bob displays and reacted indifferently to the randomly bobbing control displays. This differential behavior in response to isolated visual cues suggests that vision at least partially mediates species-identification processes. It is interesting to note that spiny lizards responded to the movement cues of the model in spite of other visual differences between conspecifics and the model. Since only specific visual cues seem necessary for species identification, head-bobbing patterns may function as releasers of further social interaction.

Ferguson (1966) found that male side-blotched lizards (*Uta stansburiana*) responded with species-typical displays to two other species, the northern earless lizard (*Holbrookia maculata*) and the tree lizard (*Urosaurus ornatus*), which were similar in body form to the side-blotched lizards. In field tests, however, side-blotched males courted conspecific females significantly more often than female sagebrush lizards (*Sceloporus graciosus*), juvenile fence lizards (*S. undulatus*) or tree lizards (Ferguson, 1972). In another study, Ferguson (1969) found that male side-blotched lizards even differentiated among conspecifics introduced from different localities. Free-living males from Colorado responded differently to tethered females from Colorado and Texas. Captive Texas males responded differently to Colorado and Texas females.

Jenssen (1971) has reported differences in behavioral displays in geographically separated populations of anoles (*Anolis nebulosus*). Whether these differences lead to differential responses to conspecifics from separate populations was not determined; but the behavioral differences are implicated as significant in species-identification processes.

Ferguson (1971) also studied sagebrush lizards (*S. graciosus*) and fence lizards (*S. undulatus*) which occur sympatrically in Utah. About twice as many encounters among conspecifics were observed in the field as among nonconspecifics. The fact that interactions among nonconspecifics occurred suggests that interspecific interactions may progress to some extent with apparent confusion of species identity. In another study, however, Ferguson (1973) found that in areas of sympatry, the species-typical displays of the sagebrush lizard and the fence lizard were more divergent than the displays of the same species from regions where they were not sympatric. This is a classic case of character displacement, and indicates that species-isolating mechanisms can develop to reduce species-identity confusion.

At a symposium on lizard ecology in 1965 (see Milstead, 1967) several researchers addressed the question of species-identification and species-isolating mechanisms. Carpenter (1967) commented that of all the lizards studied, no two species exhibited the same species-specific head-bob display, and that interspecific interactions were infrequently observed when several species were placed together. Ruibal (1967) filmed the display patterns of anoles (*Anolis allisoni*) to two sympatric congeners (*Anolis sagrei* and *Anolis porcatus*). Differences in the displays of anoles to conspecifics and nonconspecifics indicated that species discrimination readily occurred.

Jenssen has conducted a series of species discrimination studies in anoles. He showed (1970) that in a two-choice situation, female anoles (*A. nebulosus*) chose a normal conspecific filmed display significantly more often than a reversed or edited display. However, gravid females often made no choice. This lack of choice seems to be a function of the reproductive condition of the female rather than of any inability to discriminate conspecifics from nonconspecifics. This finding supports Pyburn's (1955) speculation that gravid females avoided conspecific males. It also emphasizes that negative results in choice tests do not indicate that an animal cannot discriminate between members of its own species and members of other species.

It appears that in the lizard species studied, species identity can be readily observed in that an animal discriminates conspecifics from nonconspecifics. In some instances, however, an animal may respond to a nonconspecific. Such responses have been briefly described for whiptail lizards by Maslin and Walker (1973) and for iguanid lizards (genus *Tropidurus*) by Carpenter (1977a). These interspecific interactions do not necessarily indicate any failure in species identification. They may simply express the interspecific competition for resources required by both species or preliminary behavioral interactions which permit further identification processes.

Cases of hybridization in natural populations suggest that identification processes have been inadequate. Several hybrid anoles have been reported in the recent literature (Gorman and Atkins, 1968; Jenssen, 1977a). Jenssen's hybrid anole (*Anolis grahami* × *Anolis lineatopus neckeri*) failed to show prolonged interaction with either parent species. While these cases of hybridization are interesting, conclusions about failure of identification mechanisms need to be tempered when one considers the rarity of hybridization. In addition, at least some cases of hybridization have apparently occurred following artificial introduction of new species (Gorman et al., 1971).

Some indirect evidence for species identification in lizards comes from observations of aggregations of conspecifics. While aggregations may form as lizards individually respond to environmental variables (e.g., lizards will aggregate on the only rock available for basking), there are reports of coordinated movements and aggregations which appear to be related to conspecific "attractiveness." Henderson (1974) reported concentrated groups of juvenile green iguanas (*Iguana iguana*) on bushes and trees. Burghardt, Greene, and Rand (1977) reported group movements in this species as lizards emerged from their nests following hatching.

The experimental evidence supports the conclusion that lizards possess mechanisms for determining whether other animals they meet are conspecifics.

Mechanisms of Species Identification

Most of the research on the mechanisms employed by lizards in species identification suggests that visual cues are of prime importance. Body form (Ferguson, 1966; Ruibal, 1967), color (Clark and Hall, 1970; Ferguson, 1966; Ruibal, 1967; Williams and Rand, 1977) and head-bob displays have been suggested as visual mechanisms involved in species-identification processes in lizards. When visual cues are isolated from all other cues, species identification still occurs in many cases (Hunsaker, 1962; Jenssen, 1970).

Head-bobbing displays have been studied in great detail because of their potential role as behavioral isolating mechanisms. While species-specific displays have been described for many lizard species (see Carpenter, 1967, for a review), individual variations exist in the display patterns. Jenssen (1971) reported that different populations of anoles (*A. nebulosus*) showed differences in the sequence of the display pattern. This corresponds with Ferguson's (1973) description of population differences in fence lizards (*S. undulatus*). Crews (1975) reported that green anoles (*A. carolinensis*) showed individual differences in displays which could convey information useful in individual recognition. Jenssen and Hover (1976) reported similar variability for another anole species (*A. limifrons*). Stamps and Barlow (1973) noted variation in several display patterns of another anole (*A. aeneus*).

The type of display given depended, in part, on the distance from an approaching conspecific. Some displays were variable between lizards, while others were invariant. The "signature display" seemed relatively constant. These head-bobbing displays seem to carry enough information to permit species identification, but given that other visual factors such as body form or color may also be sufficient for proper identification, the display may not be necessary for such identification. On the other hand, behavioral displays may be critical in the identification process when morphological differences between species are minimal.

Rand and Williams (1970) emphasized the species identification function of dewlap color in sympatric anoles. They found redundancy in species-specific signal characteristics and suggested that amount of redundancy should be correlated with faunal complexity in a given area. Williams and Rand (1977) confirmed this suggestion with information on dewlap characteristics for various fauna from different islands. They also recognized that species identity could be coded in other ways besides dewlap characteristics. Each additional code simply increased the redundancy of the species identity signal, thereby reducing the chances for species confusion in complex faunas.

Vision is such a predominant sense in many lizards that other potential cues to species identity have received little attention. DeFazio et al. (1977) suggest that chemical cues may be important in lizard behavior. While providing few details, Hunsaker (1962) reported that the crevice spiny lizard (*S. poinsetti*) and the blue spiny lizard (*S. cyanogenys*) approached nonconspecifics if secretions from conspecifics were smeared on the nonconspecifics. This suggests that some olfactory cues might play a role in species identification. In contrast, Ferguson (1966) failed to find that odor cues were important in conspecific interactions of the desert side-blotched lizard. The difference between these two reports may reflect species differences or methodological differences. Certainly, different species have different life histories and utilize different cues in identifying conspecifics (see Williams and Rand, 1977). Whether this is the case above has not been determined.

Auditory cues may also be used by lizards in species identification. Some geckos are nocturnal lizards, and while their vision is excellent, visual cues might not be expected to predominate in their social interactions. Geckos are known to vocalize, and this vocalization has been observed when males sight distant geckos (Marcellini, 1974). Frankenberg (1974) analyzed the calls of male fan-toed geckos (*Ptyodactylus hasselquistii*) and reported that the calls in four subspecies were taxon-specific. In addition, calls in the presence of a female were different than calls in the presence of a male. Marcellini (1977a) reported that male, but not female, geckos (*Hemidactylus frenatus*) avoided an area from which a conspecific call was emanating. The females gave no reliable approach or avoidance responses to the calls. Marcellini suggested that calls served to establish and maintain territories. However, his data do not exclude a species identification function for vocalizations. Calls may be a mechanism of species and sex determination, but additional studies are needed to confirm this auditory function (Marcellini, 1977b).

Origin of the Identification Mechanisms

Several authors have attempted to evaluate the source of these behavioral isolating mechanisms. This inquiry has focused on the importance of genetic factors in producing species-typical behavioral displays. The evidence suggests that Carpenter (1967) was correct when he said, "the display–action–pattern is an instinctive behavior pattern and is therefore an inheritable trait" (p 102). Ruibal (1967) in the same volume pointed out that "we have no direct evidence for the genetic basis of behavior in *Anolis*. We assume that these behavioral patterns are genetically determined because of the complexity of the response and the uniformity of the pattern within each species" (p 134). Stamps and Barlow (1973) agree that at least the signature bob, a portion of the entire display repertoire, is invariant, and may fit the criteria for a fixed-action pattern.

There is evidence that the display patterns occur without prior experience. Noble and Bradley (1933) and Jenssen and Hover (1976) stated that the adult patterns appeared shortly after hatching, and Hunsaker (1962) noted that spiny lizards (*S. torquatus*) which were less than one day old and had been isolated from birth bobbed in a species-typical fashion. Done and Heatwole (1977) reported that skinks (*Sphenomorphus kosciuskoi*) began displaying to each other a few minutes after birth. Burghardt, Greene, and Rand (1977) observed social display behavior in newly hatched green iguanas (*I. iguana*), but no comparison of these displays with adult displays were reported.

Another line of evidence for the genetic basis of display behavior comes from cases of hybridization between species which have different display patterns. Gorman (1969) concluded that genetic control of displays was indicated by the intermediate display of a hybrid anole (*A. trinitatis* × *A. aeneus*). Jenssen (1977a), however, did not find intermediate display characteristics in another hybrid anole (*A. grahami* × *A. lineatopus neckeri*). Both of the hybrids tested above were adults. They were caught wild and their previous experience with parental species was unknown. It is possible that previous experiences influenced display type. However, the evidence basically suggests that the form of species-typical displays is genetically determined.

Greenberg (1977) recently reported a possible neural center for species recognition in the green anole. Anoles with lesions of the paleostriatum acted as if they did not recognize conspecifics.

While the form of some lizard display behaviors may be genetically based, little information is available on the source of the orientation of these displays. Upon emerging, the green iguanas observed by Burghardt, Greene, and Rand (1977) did not interact with hatchling crocodiles or turtles found in the same area. This suggests that lizards direct their display behaviors to conspecifics without prior experience with the conspecifics. Ferguson (1969), however, suggested that prior experience might play a role in identification mechanisms in the side-blotched lizard. Male lizards who court females from similar species might be rejected by all but conspecific females. They could thereby learn to discriminate between conspecific and nonconspecific females as a result of their differential reinforcement by conspecifics. This theory places the burden of species identification on the female, who may have a greater investment in correct identification (Jenssen, 1977b). The question then becomes how she acquires the ability to identify conspecific males. Reinforcement history seems at best an incomplete explanation of the development of the orientation of species-typical displays. The genetic basis of orientation to conspecifics, however, needs further experimental verification.

The above research suggests that lizards use a variety of mechanisms in species identification. In one lizard different mechanisms may even be employed, depending on environmental circumstances and the similarity of the approaching animal to conspecifics. Morphological characteristics may, in themselves, be enough to allow a lizard to identify another lizard. Behavioral displays may also permit appropriate identification. Determining whether another lizard is a conspecific or not certainly seems to involve more than one process and one decision point. Species differences in mechanisms of identification appear to be related to the number of closely related sympatric species (Williams and Rand, 1977). A species-typical display pattern may not even allow for correct identification if such a display is given to a species which has been artificially introduced in the field. Species-identification processes in lizards, therefore, seem to be uniquely adapted to the life-history requirements of a given species in its natural habitat.

Snakes

Evidence for Identification Processes

There are fewer direct accounts in the literature of species discrimination for snakes than for lizards. Nevertheless, the basic discrimination abilities appear similar in both groups, and some of the classic work on species discrimination has been performed with snakes.

Noble and Clausen (1936) conducted a series of experiments on aggregative behavior of several snake species which provided evidence of species-discrimination abilities. Young and mature brown snakes (*Storeria dekayi*) moved together to form large masses when frightened. In the field brown snakes were often found in contact with conspecifics when active as well as when hibernating. In the laboratory, isolated brown snakes came to rest in front of mirrors. They failed to aggregate with ribbon snakes (*Thamnophis sauritus*) and even preferred stained conspecifics to stained ribbon snakes. When mixed groups of brown snakes and ribbon snakes were placed together, they moved into species-segregated aggregations, indicating that they could distinguish conspecifics from nonconspecifics. Interestingly, gestating females consistently remained isolated. This finding corresponds with Pyburn's (1955) speculation about spiny lizards (p 43) and with Jenssen's (1970) findings with anoles (p 45). Possibly gravid reptiles remove themselves from the courting population, thereby conserving their own energy as well as that of males who might court them or compete for them. It would be interesting to see how widespread this phenomenon is in reptiles. Accounts of aggregations of gravid females (Gregory, 1975) may represent a balance between advantages of aggregation and advantages of separation from the courting population.

Other species are also known to aggregate (see Gregory, 1975 for a literature review). Burghardt (1977) described significantly greater conspecific aggregations than expected by chance in newly hatched garter snakes (*Thamnophis sirtalis*) and brown snakes when these species were placed in a common enclosure. Brown, Parker, and Elder (1974), however, found racers (*Coluber constrictor*) and garter snakes (*Thamnophis elegans*) hibernating in nonsegregated clumps. While conspecific aggregations suggest species-identification abilities, mixed species aggregations do not mean that the animals are incapable of species discrimination.

Porter and Czaplicki (1974) have directly recorded species differences in response to conspecific chemical cues. While garter snakes (*Thamnophis radix*) spent significantly more time in the half of a cage that had been occupied previously by a conspecific than in the clean half (suggesting an aggregative response), water snakes (*Natrix rhombifera*) preferred the clean side over the side previously inhabited by a conspecific (suggesting an avoidance response). Neither species responded differentially when one side of the cage had been previously inhabited by a sympatric non-conspecific. These results clearly indicate species discrimination, but they also suggest differences in species-typical responses to conspecifics.

Noble (1937) observed that even though two species of garter snakes (*T. sirtalis* and *T. butleri*) were caged together, no cross-species courtship occurred. In fact, if a receptive female non-conspecific was substituted for a female which was being courted, males quickly ceased courtship. In comparison with lizards, Noble wrote, "This rarity of courtship between two different species of snakes stands in striking contrast to the behavior of lizards. In these . . . there are many records of one species attempting to mate with another." It is suspected that the contrast between successful courtship in lizards and snakes is far less striking. The apparent difference may simply be due to different species-identification mechanisms employed by the two groups.

Bogert (1941) evaluated a different aspect of species identification by rattlesnakes (genus *Crotalus*). King snakes (genus *Lampropeltis*) eat other snakes, including rattlesnakes. Rattlesnakes assume a typical defensive posture when approached by king snakes, and Bogert demonstrated that this posture was performed whether sympatric or allopatric king snakes approached. In one instance this defensive response was given to a coachwhip (*Masticophis flagellum*), a snake which occasionally eats other snakes.

Rattlesnakes gave no defensive responses to a series of snakes from about 15 other genera. Carpenter and Gillingham (1975) confirmed Bogert's (1941) findings, and extended to other crotaline species this ability to identify king snakes. The defensive response to king snakes is quite different from the combat response among conspecific rattlesnakes (see Carpenter et al., 1976 for a recent description).

Mechanisms of Species Identification

Noble and Clausen (1936) found that brown snakes with blocked olfactory senses aggregated in the light but not in darkness, suggesting that visual cues mediated aggregation. The brown snakes did not aggregate with several kinds of motionless models, further suggesting that movement was a critical component of the visual cue. The importance of movement was demonstrated by placing an aggregated mass of brown snakes in a glass enclosure. Several conspecifics placed outside of the small glass enclosure moved to the aggregated mass. In another test, color of conspecifics did not seem to be important, as stained animals were still approached by conspecifics. Movement cues (Noble, 1937) and animal size (Hawley and Aleksuik, 1976) have also been suggested as important for garter-snake social interactions.

While visual cues play a role in species identification in at least some snakes, they are not solely responsible for that identification. Noble and Clausen's (1936) brown snakes, when blindfolded, still aggregated with conspecifics. Blindfolded male garter snakes still courted conspecific females if they happened to meet them (Noble, 1937). Blindfolded rattlesnakes responded defensively when king snakes were introduced into their cages (Bogert, 1941).

Chemical cues have been found to play a significant role in species identification by snakes. Eliminating chemical reception inhibited or eliminated aggregation in brown snakes and garter snakes (*T. sauritas*) (Noble and Clausen, 1936). These snakes followed odor trails made by moving the body of a conspecific over a glass plate. Noble (1937) found further that snake body odors were sex-specific, as well as species-specific, at least during the mating season.

This role of chemical cues in mediating behavioral reactions to conspecific snakes was confirmed by Porter and Czaplicki (1974) for garter snakes and water snakes. Chemical cues were most important in the interspecific identification of king snakes by rattlesnakes (Bogert, 1941). Rattlesnakes performed the typical king-snake defensive posture when placed in glass jars which had previously been occupied by king snakes. Sticks which had been rubbed on the back of king snakes elicited the defensive reactions of the rattlesnakes. Many snake species thus seem to be highly sensitive to species-typical body chemicals.

While no systematic discrimination studies are available on other sense modalities, it has been suggested that tactile cues might play a role in snake communication (Carpenter, 1977b). Noble (1937) recognized that tactile cues had to be involved in courtship to produce proper alignment of the animals for copulation. Gillingham (1974) described rhythmic jerking movements during courtship in the western fox snake (*Elaphe v. vulpina*) and suggested that these might be important for species recognition. These jerking movements, or at least body undulations, seem to be widespread among copulating snakes (Carpenter, 1977b). Tactile cues occur after the animals have made contact. In the cases cited by Noble (1937) species and sex recognition occurred prior to contact, so the tactile cues themselves were superfluous to species identification. It seems unlikely that, given the sensitivity of olfactory cues, courtship would progress to the body undulation stage. The possibility that particular movement patterns are used in species identification should not be ruled out, however, because of the limited number of species for which we have good data.

The snakes studied identify conspecifics and behave differentially toward conspecifics and nonconspecifics. Different species also show different responses to conspecific cues. While snakes, like lizards, interact with nonconspecifics, there are typical responses which occur only with conspecifics. Many of these are seen most clearly during courtship, and some seem masked by variables like the reproductive state of the female.

As with lizards, several mechanisms may be employed for species identification in snakes. The most common sense modalities used by snakes are olfaction and vision, and perceptual shifting between these modalities seems to occur as interactions among snakes continue. Visual and olfactory cues alone are insufficient to produce species-typical aggregation and courtship if one of the animals is anesthetized or dead. A view of species identification as resulting from interaction among conspecifics is thus suggested.

The fact that some young (Noble and Clausen, 1936) and newly hatched snakes (Burghardt, 1977) aggregate more with conspecifics than with nonconspecifics suggests that species identification may occur without benefit of experience. Innate components of species identification in snakes and the role of experience in modifying those innate components might be profitably investigated by manipulating early experience with conspecifics, rearing animals in isolation, and systematically recording responses of additional newly hatched snakes to conspecific versus nonconspecific cues.

Turtles

Evidence for Species Identification

Several reports describe turtle responses to conspecifics and nonconspecifics. Evans (1956) noticed that male box turtles (*Terrapene c. carolina*) would challenge their mirror images the way they challenged conspecifics. Auffenberg (1965) reported that individuals of two sympatric South American tortoise species (genus *Geochelone*) challenged all turtlelike objects. Responses to the challenge determined the nature of continued interaction. Responses were species-specific, so proper species identification occurred. While male tortoises of another species (*Geochelone travoncorica*) mounted tortoises of almost any genus, morphological differences prohibited cross-species copulation with the only other naturally sympatric tortoise (Auffenberg, 1964). Gopher tortoise males (*Gopherus polyphemus*) approached turtles of other species, but failed to continue interactions with nonconspecifics (Auffenberg, 1966). Weaver (1970) also found that male gopher tortoises would not court a nonconspecific female. However, these males challenged and rammed South American tortoises (genus *Geochelone*) in typical conspecific combat fashion. Weaver also observed that the male desert tortoise (*Gopherus agassizi*) would court female Texas tortoises (*Gopherus berlandieri*). These two species are considered to be more closely related to each other than either species is to the gopher tortoise, and there are accounts of hybridization between them (Woodbury, 1952).

Mahmoud (1967) observed no interspecific courtship among the stinkpot (*Sternotherus odoratus*), the razor-backed musk turtle (*Sternotherus c. carinatus*), or the Mississippi mud turtle (*Kinosternon subrubrum hippocrepis*). However, a male yellow mud turtle (*Kinosternon f. flavescens*) courted male and female Mississippi mud turtles. While failing to demonstrate discrimination at the species level, the yellow mud turtle did not court any of the available turtles of another genus (*Sternotherus*). Harless and Lambiotte (1971) reported that ornate box turtles (*Terrapene ornata*) ignored wood turtles (*Clemmys insulpta*) kept in the same enclosure. Jackson (1977) kept several turtle species in a large aquarium and observed that a pair of Florida red-bellied turtles located each other for courtship.

Davis and Jackson (1973) staged nonconspecific encounters between a male slider (*Chrysemys scripta taylori*) and male and/or female turtles from six other species. The tested male courted any of the turtles, as long as they did not approach him rapidly. The tested male was from a geographically isolated species, and its courtship behavior appeared to be composed of primitive precursors of courtship in some of the other species. While the male was relentless in courtship, red-eared turtle females (*Chrysemys scripta elegans*) would not permit copulation unless they were lightly anesthetized. The responses of the other species to the courtship were not reported.

Legler (1955) observed that male snapping turtles (*Chelydra serpentina*) mounted conspecifics and nonconspecifics which touched their legs or plastron. Legler also reported courtship and apparent copulation between a male Florida softshell turtle (*Trionyx ferox hartwegi*) and a female smooth softshell turtle (*Trionyx muticus*). This mating occurred in the laboratory, and no conspecific females were present. Whether these softshell turtles can discriminate conspecifics from nonconspecifics is therefore still unclear.

Turtles present a somewhat confusing picture of species identification. While the ability of species to maintain a consistent gene pool seems evident from morphological and taxonomic data, many of the tests indicate responses to nonconspecifics which would be expected only for conspecifics. Several factors may be important in this apparent anomaly. Many of the studies were conducted in the laboratory, and species identification may not be clearly revealed in such artificial situations. One artifact is the juxtaposition in the laboratory of species that are not sympatric in the field. In nature, species confusion is only possible between sympatric species. Also, previous experience with nonconspecifics is limited to sympatric species members. If isolating mechanisms are to develop, they would surely develop between closely related, sympatric species. The questionable species discrimination in some of the turtles cited previously certainly points to the importance of sympatry in the development of species identification mechanisms.

Mechanisms of Species Identification

Visual cues in species identification may be employed by turtles. Species-typical head-bobs occur in several tortoise species (Auffenberg, 1965, 1966) and may be used in species identification. Typically, however, only males head-bob. Auffenberg (1966) suggested that a male gopher tortoise visually recognized a female even though she gave no obvious visual signal. More studies on many different species need to be conducted before the role of vision in species identification in turtles can be determined.

Olfactory cues may be important in tortoise species discrimination ability (Auffenberg, 1964). Chin glands in North American tortoises (genus *Gopherus*) produce secretions which elicit conspecific interactions (Rose et al., 1969; Rose, 1970). However, interactions among nonconspecifics have been described (Weaver, 1970) indicating that olfactory information does not work alone to produce species identification. The role of these odorous secretions in species discrimination needs further experimental confirmation.

Biting responses by males during courtship have been reported in many turtles. The most probable function of this biting, however, is to immobilize the female (Auffenberg, 1965), rather than to aid in species identification. Other tactile cues used especially by aquatic turtles may play a role in species identification. Jackson and Davis (1972a) described in detail the vibratory tactile courtship of the red-eared turtle. The courtship pattern differs for some closely related slider turtles (*C. s. taylori*, Davis and Jackson, 1973; and *Chrysemys concinna suwanniensis*, Jackson and Davis, 1972b). This tactile behavior could serve as an isolating mechanism (Jackson and Davis, 1972a). Cagle (1950), in fact, observed that female slider turtles (*C. s. troostii*) avoided improper courtship by different species males. This report is similar to Davis and Jackson's (1973) account of attempted cross-species courtship by a male slider. The ability of turtles to use tactile cues to discriminate conspecifics from nonconspecifics has not been demonstrated, however, with proper control procedures. It may be that the tactile cues form a small part of the overall identification process. Cagle (1955) observed that a juvenile slider turtle (*Chrysemys floridana suwanniensis*) performed its vibration response to a snail which it later ate. Whether this response was related to the conditions of confinement or whether it suggests a broader function of the vibration response is not clear. The tactile response, however, seems important in the courtship of some turtles.

Species identification mechanisms in turtles have been studied only to a limited extent. Most of the discussion is speculative, and species discrimination is often confounded with sex discrimination. Olfactory, visual, and tactile cues may mediate species identification in turtles. (See Auffenberg, 1977, for a review of possible communication cues.) Male turtles have often been observed attempting to court nonconspecifics, but some isolating mechanism, such as female rejection or morphological incompatibility, generally keeps species separate. The fact that males respond to several species in a courtship fashion does not mean that these males lack species-identification processes. The discriminations required in laboratory tests simply may not parallel the turtle's life-history requirements. If proper species identification requires interaction between two organisms, the reciprocal communicative cues in laboratory tests may not be those required to discontinue interaction, especially when passivity on the part of the female is one of the cues eliciting continuation of courtship.

The origin and development of identification processes have not been systematically studied in turtles. Data on discrimination ability of hatchling turtles in relation to conspecific and nonconspecific cues are needed. In addition, manipulating rearing experiences might provide information on the development of the identification process. Turtles present several unique problems for such research. Many are aquatic, and turtles generally do not reach sexual maturity as rapidly as other reptile groups. Nevertheless, given the adaptive radiation of several groups of turtles, studying them may provide confirmation or extension of principles of species identification found in other reptiles.

Crocodilians

Few reports regarding species identification in crocodilians exist. Some authors have reported adult protection in captive and free-living animals when the young give distress calls (Alvarez del

Toro, 1969; Hunt, 1975; Kushlan, 1973; McIlhenny, 1935). Whether these calls permit species identification, however, is questionable, since one report indicated that several captive crocodilian species responded to the distress calls of young animals (Hunt, 1977).

Garrick (1975) studied the roars of the Chinese alligator (*Alligator sinensis*) and concluded that these roars might contain information about sex and individual identity. Neill (1971) suggested that bellowing served a territorial rather than a species-identification function. Garrick and Lang (1977)observed differences in territory defense, courtship, and precopulatory behavior between the American alligator (*Alligator mississippiensis*) and the American crocodile (*Crocodylus acutus*). Visual and acoustic communication signals were employed by both species, but the relative frequencies of groups of signals were related to habitat type. Garrick and Lang suggested that courtship differences could have originated as species-isolating mechanisms. Several mechanisms must be involved in crocodilian species identification. Further empirical investigations are certainly required.

Levels of Analysis of Species Identification

Various levels of differences exist in mechanisms of species identification in reptiles. At the broadest level, differences occur among major groups of reptiles. Lizards seem to identify conspecifics by visual characteristics. Snakes seem to depend more on chemical stimuli for discrimination among closely related species. Turtles use a variety of mechanisms, including tactile stimuli, but it would be hard to specify one as predominant.

Within each of these major groups one also finds variability in the usual way species identify conspecifics. Some lizards are sensitive to body form while others are sensitive to head-bob displays or dewlap color. Some turtles rely on head-bob patterns, while others rely on olfactory or tactile cues. These differences within each of the major groups depend on the number of closely related sympatric species. The character displacement observed by Ferguson (1973) in spiny lizards (genus *Sceloporus*) is a prime example of this principle. Differences are also related to life history specializations for each species.

At the individual level, we again find variability in mechanisms of species identification. An animal may respond to another potential conspecific in a different manner depending on several situational variables. Distance between two animals in itself influences some aspects of lizard displays (Hover and Jenssen, 1976; Milstead, 1957; Stamps and Barlow, 1973). The response of an approached animal is another situational factor which introduces variability in the mechanisms used for species identification. Behaviors may be chained so that the ultimate response depends on a series of reciprocal signals among animals (e.g., Auffenberg, 1965). These behavioral chains may be complex, suggesting that species identification is not a simple matter of finding one proper releaser which then determines the nature of all future interactions.

For a complete understanding of species identification mechanisms in reptiles, all levels of variability need to be explored. Statements about broad phyletic trends give us little information in view of all of the other potential levels of variability. Each potential mechanism needs to be evaluated in many species for its contribution to the overall process of species identification.

Summary

Many reptiles behave in appropriate ways to conspecifics and in different ways to nonconspecifics. In certain cases, responses are indiscriminate, but these may be artifacts of laboratory observations. If the individual was given appropriate conspecific or sympatric nonconspecific reactions to its behavior, it might respond in a more proper fashion. In lizards, snakes, and turtles, specific mechanisms to segregate species seem to be found in cases where species confusion is most likely. These isolating mechanisms involve different sensory modalities among and within the major reptile groups. It should be clear from the previous presentation that looking to one single source for the identification process is simplistic. If, in fact, many mechanisms are used in species identification, and the mechanisms employed depend on situational variables, it would certainly be presumptive to explain the identification process by saying it was innate. However, many of the species-typical characteristics, both morphological and behavioral, which are employed in species identification may be evaluated for innate components. Even the discovery of such components, however, should not preclude the analysis of the ontogeny of the behavior patterns.

In this perspective, the analysis of species identification processes in reptiles is an enormous task. The data to date are very limited, especially in terms of the number of species studied and in terms of the life-history information required to make knowledgeable statements about the variability found at all levels of analysis. A model which begins with a concept of species identification as something animals do, rather than something they have, at least places the task in the realm of the empirical.

Experience and the Identification Process

In this final section some of the speculative information on the effects of experience on the identification process will be summarized. A concern for the effects of early experience is understandable given the profound influence of the socialization process on later behavior found in some mammals and the demonstrable effects of imprinting shown so clearly in some birds. That reptiles exhibit complex social organization in the absence of a socialization period with parents has implications for our understanding of the phylogeny of social organization in birds and mammals (see Burghardt, Greene, and Rand, 1977; Burghardt, 1977). While different experiences which might influence species identification have been described, whether they actually do have such an influence is unclear. Three types of experiences will be described: 1) parental behavior, 2) experience with nestmates and littermates, and 3) the laboratory environment.

Parental Behavior

While some snakes and lizards brood eggs (Bellairs, 1970, pp 429–430), the parental involvement usually ends with hatching. In the live-bearing Australian skink (*Sphenomorphus kosciuskoi*), Done and Heatwole (1977) observed no overt responses to the young by the mother. Whatever parental care might occur in lizards, snakes, or turtles is relatively short-lived since the young disperse after emerging.

There are consistent reports of parental care in crocodilians (McIlhenny, 1935) which range from nest opening (e.g., Herzog, 1975) to transportation and protection of young (e.g., Alvarez del Toro, 1969; Hunt, 1969, 1975, 1977; Kushlan, 1973). Some interaction between parents and offspring often occurs, but the importance of that interaction for species identification is unclear. No longitudinal studies of the effects of separation from time of hatching have been reported for any of those few species which appear to show parental behavior.

Nestmates and Littermates

In many reptiles, several young emerge from the nest around the same time. This provides an opportunity for early experience with conspecifics. In turtles, there is often a delay between time of hatching and emergence from the nest (e.g., Hendrickson, 1958). Carr and Ogren (1959) reported that even within the nest, turtles were stimulated to activity by conspecifics. Carr and Hirth (1961) found that green-turtle hatchlings (*Chelonia mydas*) had a better chance of emerging from their nests if they were in larger groups. These larger groups seemed to work together to dig through the roof of their nest. Communal emergence, however, is not universal in turtles. Burger (1976) described emergence of the diamondback terrapin (*Malaclemmys terrapin*) as separated by as much as eleven days in one nest.

Lizards and snakes also have a time for potential conspecific interaction when they hatch or are born. Broadley (1974) described communal nest sites in the African flat lizard (*Platysaurus intermedius rhodesianus*). Communal nests would increase the likelihood of conspecific interactions in early life, but the significance of early conspecific experiences for species identification has not been established.

Laboratory Environments

While reptiles are often reared in isolation from conspecifics in the laboratory, no systematic reports on how the isolation influences species identification are available. The laboratory may also produce behavior which would not be observed in the field. While several researchers observed no differences in head-bob displays of lizards in the laboratory and field (Done and Heatwole, 1977; Jenssen, 1971), Ferguson (1969) reported differences in conspecific discriminability in laboratory- versus field-tested Texas side-blotched lizards. Hunsaker (1962) described how isolating mechanisms in field-studied lizards depend on territorial behavior. In the laboratory, the possibility for such territorial establishment and defense is often eliminated, thereby destroying part of the normal species-identification process. While the mechanisms may be different in different species, other reports of indiscriminate courtship in laboratory animals need to be interpreted cautiously. A rapprochement between laboratory and field studies is needed. The field studies may serve as a source of hypotheses to be tested in the laboratory or they may serve to validate or invalidate hypotheses derived from laboratory studies. (See Mason, 1968, for a further discussion of the reciprocal relationship between laboratory and field studies.)

If the laboratory environment in itself can produce observable changes in the species-identification process, it seems reasonable that other experiences may have some effects on those processes. Burghardt (1967) and Burghardt and Hess (1966) have suggested the possibility of food imprinting in snapping turtles. There is no evidence, however, that such a system is employed in species identification. One reason, in fact, why the Burghardt and Hess studies were performed was that the possibility of parental (social) imprinting was absent in turtles. It is hard to imagine that imprinting to conspecifics would be an efficient way to establish species identity given that there is much variability in potential stimuli to which the young could be exposed when they emerge.

Summary and Conclusions

It is clear that reptiles are generally capable of making proper discriminations between conspecifics and nonconspecifics in their natural environments. Various kinds of isolating mechanisms have evolved to insure separation among closely related sympatric species. Some of these mechanisms involve morphological differences which are recognized by conspecifics or discriminated against by nonconspecifics. Other mechanisms include behavioral signals which directly release conspecific social behavior or act in concert with reciprocal signals from conspecifics. Many mechanisms may contribute to the identification process, and the sources of these mechanisms, their ontogeny, and how they work all need to be investigated to a much greater extent. The role of experience in the identification process in reptiles is virtually unknown. The influence of typical experiences on identification needs to be investigated in longitudinal studies. This will be an extensive task, especially in reptiles as turtles that have long intergeneration times and long life spans. Most reptiles are conveniently studied, however (Burghardt, 1977), and the research on them will prove interesting and informative.

Various research paradigms have been used in the past, and some of these could be extended and employed in future research. Quantification of variability in display behaviors, as performed already with some lizards and turtles, should clarify the extent of variability among and within species. Further discrimination studies with additional species-typical cues will clarify the extent of redundancy in species identification. Observed redundancies could, in turn, be evaluated in relation to ecological factors, as illustrated by Williams and Rand (1977).

The physiological tools suggested by Greenberg (1977) could be used to eliminate behavioral cues to species identity selectively. Responses to behaviorally impoverished conspecifics could provide information on the reciprocity of cues in the identification process. In addition, localizing neural control of conspecific recognition could provide another tool to investigate ontogeny of recognition processes.

Manipulating early rearing conditions of reptiles might provide especially interesting information. In addition to the effects of early experience on conspecific attractiveness, the effects of different experiences on predator–prey relationships, especially in snakes, could be established. Whether the king snake–rattlesnake interactions would be affected by early cross-fostering (e.g., rearing the species together from hatching) is only one example of the possibilities of this type of research.

One final research possibility with reptiles involves the influence of asexual reproductive strategies on species identification processes. Some lizards are parthenogenetic (see Cole, 1978), which changes the nature of selection pressures on the species. In addition, parthenogenetically produced siblings are genetically identical, making them ideal subjects for studies of variability in behavior. The unique life-history characteristics of reptiles and the previously cited research possibilities certainly suggest that the question of species identity in reptiles ought to receive much additional attention.

Acknowledgments

Financial support for this study and release from teaching responsibilities during January, 1978, were provided by Sterling College. The author thanks Gordon Burghardt, Harry Greene, and Thomas Jenssen for valuable comments on an early draft of the manuscript. Carol Froese encouraged the author to work on this project and assisted him in many other ways.

Note

1. For a revealing attack on this undeserved bias, read Burghardt's (1977) review of social behavior in neonate reptiles.

References

Alvarez del Toro, M. 1969. Breeding the spectacled caiman *Caiman crocodylus* at the Tuxtla Gutierrez Zoo. *Int. Zoo Yearb.* 9, 35–36.

Auffenberg, W. 1964. Notes on the courtship of the land tortoise *Geochelone travancorica* (Boulenger). *J. Bombay Nat. Hist. Soc. 61,* 247–253.

Auffenberg, W. 1965. Sex and species discrimination in two sympatric South American tortoises. *Copeia* 335–342.

Auffenberg, W. 1966. On the courtship of *Gopherus polyphemus*. *Herpetologica 22,* 113–117.

Auffenberg, W. 1977. Display behavior in tortoises. *Am. Zool. 17,* 241–250.

Beer, C. G. 1977. What is a display? *Am. Zool. 17,* 155–165.

Bellairs, A. 1970. *The life of reptiles.* Vol. 2. New York: Universe Books, pp 283–590.

Bogert, C. M. 1941. Sensory cues used by rattlesnakes in their recognition of ophidian enemies. *Ann. N.Y. Acad. Sc. 41,* 329–344.

Brattstrom, B. H. 1974. The evolution of reptilian social behavior. *Am. Zool. 14,* 35–49.

Broadley, D. G. 1974. Reproduction in the genus *Platysaurus* (Sauria: Cordylidae). *Herpetologica 30,* 379–380.

Brown, W. S., W. S. Parker, and J. A. Elder. 1974. Thermal and spatial relationships of two species of colubrid snakes during hibernation. *Herpetologica 30,* 32–38.

Burger, J. 1976. Behavior of hatchling diamondback terrapins (*Malaclemys terrapin*) in the field. *Copeia* 742–748.

Burghardt, G. M. 1967. The primacy effect of the first feeding experience in the snapping turtle. *Psychonomic Sci. 7,* 383.

Burghardt, G. M. 1977. Of iguanas and dinosaurs: Social behavior and communication in neonate reptiles. *Am. Zool. 17,* 177–190.

Burghardt, G. M., H. W. Greene, and A. S. Rand. 1977. Social behavior in hatchling green iguanas: Life at a reptile rookery. *Science 195,* 689–691.

Burghardt, G. M. and E. H. Hess. 1966. Food imprinting in the snapping turtle, *Chelydra serpentina*. *Science 151,* 108–109.

Cagle, F. R. 1950. The life history of the slider turtle, *Pseudemys scripta troostii* (Holbrook). *Ecol. Monogr. 20,* 31–54.

Cagle, F. R. 1955. Courtship behavior in juvenile turtles. *Copeia* 307.

Carpenter, C. C. 1967. Aggression and social structure in iguanid lizards. In W. W. Milstead (Ed.), *Lizard ecology, a symposium*. Columbia, Mo.: Univ. Missouri Pr. 87–105.

Carpenter, C. C. 1977. The aggressive displays of three species of South American iguanid lizards of the genus *Tropidurus*. *Herpetologica 33,* 285–289. (a)

Carpenter, C. C. 1977. Communication and displays of snakes. *Am. Zool. 17,* 217–223. (b)

Carpenter, C. C. and J. C. Gillingham. 1975. Postural responses to kingsnakes by crotaline snakes. *Herpetologica 31,* 293–302.

Carpenter, C. C., J. C. Gillingham, and J. B. Murphy. 1976. The combat ritual of the rock rattlesnake (*Crotalus lepidus*). *Copeia* 764–780.

Carr, A. and H. Hirth. 1961. Social facilitation in green turtle siblings. *Anim. Behav. 9,* 68–70.

Carr, A. and L. Ogren. 1959. The ecology and migrations of sea turtles, 3. *Dermochelys* in Costa Rica. *Am. Mus. Novit. 1958,* 1–29.

Clark, D. R. and R. J. Hall. 1970. Function of the blue tail-coloration of the five-lined skink (*Eumeces fasciatus*). *Herpetologica 26,* 271–274.

Cole, C. J. 1978. The value of virgin birth. *Nat. Hist. 87*(1), 56–63.

Crews, D. 1975. Inter- and intraindividual variation in display patterns in the lizard, *Anolis carolinensis*. *Herpetologica 31,* 37–47.

Davis, J. D., and C. G. Jackson. 1973. Notes on the courtship of a captive male *Chrysemys scripta taylori*. *Herpetologica 29,* 62–64.

De Fazio, A., C. A. Simon, G. A. Middendorf, and D. Romano. 1977. Iguanid substrate licking: A response to novel situations in *Sceloporus jarrovi*. *Copeia* 706–709.

Done, B. S., and H. Heatwole. 1977. Social behavior of some Australian skinks. *Copeia* 419–430.

Evans, L. T. 1956. The use of models and mirrors in a study of *Terrapene c. carolina*. *Anat. Rec. 125,* 610.

Ferguson, G. W. 1966. Releasers of courtship and territorial behaviour in the side-blotched lizard *Uta stansburiana*. *Anim. Behav. 14,* 89–92.

Ferguson, G. W. 1969. Interracial discrimination in male side-blotched lizards, *Uta stansburiana*. *Copeia* 188–189.

Ferguson, G. W. 1971. Observations on the behavior and interactions of two sympatric *Sceloporus* in Utah. *Am. Midl. Nat. 86,* 190–196.

Ferguson, G. W. 1972. Species discrimination by male side-blotched lizards *Uta stansburiana* in Colorado. *Am. Midl. Nat. 87,* 523–524.

Ferguson, G. W. 1973. Character displacement of the push-up displays of two partially sympatric species of spiny lizards, *Sceloporus* (Sauria: Iguanidae). *Herpetologica 29,* 281–284.

Frankenberg, E. 1974. Vocalization of males of three geographical forms of *Ptyodactylus* from Israel (Reptilia: Sauria: Gekkoninae). *J. Herpetol. 8,* 59–70.

Garrick, L. D. 1975. Structure and pattern of the roars of Chinese alligators (*Alligator sinensis* Fauvel). *Herpetologica 31,* 26–31.

Garrick, L. D. and J. W. Lang. 1977. Social signals and behaviors of adult alligators and crocodiles. *Am. Zool. 17*, 225–239.

Gillingham, J. C. 1974. Reproductive behavior of the western fox snake, *Elaphe v. vulpina* (Baird and Girard). *Herpetologica 30*, 309–313.

Gorman, G. C. 1969. Intermediate territorial display of a hybrid *Anolis* lizard (Sauria: Iguanidae). *Z. Tierpsychol. 26*, 390–393.

Gorman, G. C. and L. Atkins. 1968. Natural hybridization between two sibling species of *Anolis* lizards: Chromosome cytology. *Science 159*, 1358–1360.

Gorman, G. C., P. Licht, H. C. Dessauer, and J. Boos. 1971. Reproductive failure among the hybridizing *Anolis* lizards of Trinidad. *Syst. Zool. 20*, 1–18.

Greenberg, N. 1977. A neuroethological study of display behavior in the lizard *Anolis carolinensis* (Reptilia, Lacertilia, Iguanidae). *Am. Zool. 17*, 191–201.

Gregory, P. T. 1975. Aggregations of gravid snakes in Manitoba, Canada. *Copeia* 185–186.

Harless, M. D. and C. W. Lambiotte. 1971. Behavior of captive ornate box turtles. *J. Biol. Psychol. 13*, 17–23.

Hawley, A. W. L. and M. Aleksiuk. 1976. Sexual receptivity in the female red-sided garter snake (*Thamnophis sirtalis parietalis*). *Copeia* 401–404.

Henderson, R. W. 1974. Aspects of the ecology of the juvenile common iguana (*Iguana iguana*). *Herpetologica 30*, 327–332.

Hendrickson, J. R. 1958. The green sea turtle, *Chelonia mydas* (Linn.) in Malaya and Sarawak. *Proc. Zool. Soc. London, 130*, 455–535.

Herzog, H. A., Jr. 1975. An observation of nest opening by an American alligator *Alligator mississippiensis*. *Herpetologica 31*, 446–447.

Hover, E. L. and T. A. Jenssen. 1976. Descriptive analysis and social correlates of agonistic displays of *Anolis limifrons* (Sauria, Iguanidae). *Behaviour 58*, 173–191.

Hunsaker, D. 1962. Ethological isolating mechanisms in the *Sceloporus torquatus* group of lizards. *Evolution 16*, 62–74.

Hunt, R. H. 1969. Breeding of spectacled caiman *Caiman c. crocodylus* at Atlanta Zoo. *Int. Zoo Yearb. 9*, 36–37.

Hunt, R. H. 1975. Maternal behavior in the morelet's crocodile, *Crocodylus moreleti*. *Copeia* 763–764.

Hunt, R. H. 1977. Aggressive behavior by adult morelet's crocodiles *Crocodylus moreleti* toward young. *Herpetologica 33*, 195–201.

Jackson, C. G. 1977. Courtship observations on *Chrysemys nelsoni* (Reptilia, Testudines, Testudinidae). *J. Herpetol. 11*, 221–222.

Jackson, C. G. and J. D. Davis. 1972. A quantitative study of the courtship display of the red-eared turtle, *Chrysemys scripta elegans* (Wied). *Herpetologica 28*, 58–64. (a)

Jackson, C. G. and J. D. Davis. 1972. Courtship display behavior of *Chrysemys concinna suwanniensis*. *Copeia* 385–387. (b)

Jenssen, T. A. 1970. Female response to filmed displays of *Anolis nebulosus* (Sauria, Iguanidae). *Anim. Behav. 18,* 640–647.

Jenssen, T. A. 1971. Display analysis of *Anolis nebulosus. Copeia* 197–208.

Jenssen, T. A. 1977. Morphological, behavioral and electrophoretic evidence of hybridization between the lizards, *Anolis grahami* and *Anolis lineatopus neckeri,* on Jamaica. *Copeia* 270–276.

Jenssen, T. A. 1977. Evolution of anoline lizard display behavior. *Am. Zool. 17,* 203–215.

Jenssen, T. A. and E. L. Hover. 1976. Display analysis of the signature display of *Anolis limifrons* (Sauria: Iguanidae). *Behaviour 57,* 227–240.

Kushlan, J. A. 1973. Observations on maternal behavior in the American alligator, *Alligator mississippiensis. Herpetologica 29,* 256–257.

Legler, J. M. 1955. Observations on the sexual behavior of captive turtles. *Lloydia 18,* 95–99.

Mahmoud, I. Y. 1967. Courtship behavior and sexual maturity in four species of kinosternid turtles. *Copeia* 314–319.

Marcellini, D. L. 1974. Acoustic behavior of the gekkonid lizard, *Hemidactylus frenatus. Herpetologica 30,* 44–52.

Marcellini, D. L. 1977. The function of a vocal display of the lizard *Hemidactylus frenatus* (Sauria: Gekkonidae). *Anim. Behav. 25,* 414–417. (a)

Marcellini, D. L. 1977. Acoustic and visual display behavior of Gekkonid lizards. *Am. Zool. 17,* 251–260. (b)

Maslin, T. P. and J. M. Walker. 1973. Variation, distribution and behavior of the lizard, *Chemidophorus parvisocius Zweifel* (Lacertilia: Teiidae). *Herpetologica 29,* 128–143.

Mason, W. A. 1968. Naturalistic and experimental investigations of the social behavior of monkeys and apes. In P. Jay (Ed.), *Primates: Studies in adaptation and variability.* New York: Holt, Rinehart and Winston, 398–419.

McIlhenny, E. A. 1935. *The alligator's life history.* Boston: Christopher.

Meltzoff, A. N. and M. K. Moore. 1977. Imitation of facial and manual gestures by human neonates. *Science 198,* 75–78.

Milstead, W. W. 1957. Some aspects of competition in natural populations of whiptail lizards (Genus *Cnemidophorus*). *Tex. J. Sci. 9,* 410–447.

Milstead, W. W. 1967. *Lizard ecology, a symposium.* Columbia, Mo.: Univ. Missouri Pr.

Neill, W. T. 1971. *The last of the ruling reptiles.* New York: Columbia Univ. Pr.

Noble, G. K. 1937. The sense organs involved in the courtship of *Storeria, Thamnophis* and other snakes. *Bull. Am. Mus. Nat. Hist. 73,* 673–726.

Noble, G. K. and H. T. Bradley. 1933. The mating behavior of lizards; Its bearing on the theory of sexual selection. *Ann. N.Y. Acad. Sci. 35,* 25–100.

Noble, G. K. and H. J. Clausen. 1936. The aggregation behavior of *Storeria dekayi* and other snakes, with especial reference to the sense organs involved. *Ecol. Monogr. 6,* 269–316.

Porter, R. H. and J. A. Czaplicki. 1974. Responses of water snakes (*Natrix r. rhombifera*) and garter snakes (*Thamnophis sirtalis*) to chemical cues. *Anim. Learn. Behav. 2,* 129–132.

Pyburn, W. F. 1955. Species discrimination in two sympatric lizards, *Sceloporus olivaceus* and *S. poinsetti. Tex. Journ. Sci. 7,* 312–315.

Rand, A. S. and E. E. Williams. 1970. An estimation of redundancy and information content of anole dewlaps. *Am. Nat. 104,* 99–103.

Rose, F. L. 1970. Tortoise chin gland fatty acid composition: Behavioral significance. *Comp. Biochem. Physiol. 32,* 577–580.

Rose, F. L., R. Drotman, and W. G. Weaver. 1969. Electrophoresis of chin gland extracts of *Gopherus* (Tortoises). *Comp. Biochem. Physiol. 29,* 847–851.

Ruibal, R. 1967. Evolution and behavior in West Indian anoles. In W. W. Milstead (Ed.), *Lizard ecology, a symposium.* Columbia, Mo.: Univ. Missouri Pr.

Stamps, J. A. and G. W. Barlow. 1973. Variation and stereotypy in the displays of *Anolis aeneus* (Sauria: Iguanidae). *Behaviour 47,* 67–94.

Weaver, W. G., Jr. 1970. Courtship and combat behavior in *Gopherus berlandieri. Bull. Fl. State Mus. 15,* 1–43.

Wickler, W. 1976. Evolution-oriented ethology, kin selection, and altruistic parasites. *Z. Tierpsychol. 42,* 206–214.

Williams, E. E. and A. S. Rand. 1977. Species recognition, dewlap function and faunal size. *Am. Zool. 17,* 261–270.

Woodbury, A. M. 1952. Hybrids of *Gopherus berlandieri* and *G. agassizii. Herpetologica 8,* 33–36.

Chapter 4

Species Identification in Birds: A Review and Synthesis

L. James Shapiro

This chapter is concerned with the manner in which birds form attachments to their species. As will be seen shortly, the degree to which a bird has to learn its species identity varies from one species to another. The diversity among this class of organisms makes it difficult to speak of a typical method.

The composition of the family group differs among the species of birds, as do the developmental stages of the young and their dependence upon parents to provide for their physiological needs. One of the important points to be made in this chapter is that no discussion of species identification in birds can be presented without an appreciation of the great diversity that does exist within this class of organisms and without appreciating the survival requirements of the birds and the adaptive value of their behavior. Most important of all, we must realize that we are talking about a continuum of behaviors in a class of organisms that varies widely among its component orders and families with respect to almost every aspect of its anatomical, physiological, and geographical requirements. Each expressed behavior in this continuum of behaviors has a particular advantage for the species concerned. To speak of "species identification" in birds as if there

Funds from Canadian National Research Council grant A0697 contributed to the cost of producing this chapter.

were only one way in which this process can occur is to deny the evolutionary history of this class of organisms and the adaptive radiation that has occurred during its evolutionary history.

The manner in which a young bird eventually identifies a mate and reproduces ranges from mostly innate to mostly learned, but it is never either one or the other. The *source* of the contribution to the nature–nuture interaction can be either purely innate or purely learned, but their *product* cannot be solely attributed to either innate factors or to learned factors. This point will be stressed throughout this chapter. The arguments concerning this issue are well known to the scientific community (Jensen, 1961; Lehrman, 1953, 1970; Schneirla, 1956) and need not be elaborated here. The consequence of such a point of view, however, is that one has to appreciate the diversity of factors that can potentially affect birds as a class of organisms and the wide range of adaptations that the various species have adopted in forming an attachment to their species. The major part of this chapter will review some of these factors and the remainder of it will attempt to synthesize the information in an orderly way.

The Review

Not all birds form an attachment to their species in the same manner, at the same time, or in the same habitat. The particular manner in which an attachment to one's species occurs in a particular bird must be viewed from the only biological perspective that makes any sense to the species involved, i.e., it must confer upon the organism involved some potential reproductive advantage to ensure that its genes are passed on to the next generation. We must realize that genes only give an organism a *potential* to do something. The environment *requires* that the organism perform some task. Failure to accomplish the task will result in the death of the organism or its inability to reproduce successfully, in which case it is genetically dead.

There is more than one kind of an attachment that can be formed in a bird's lifetime. Several questions must be asked concerning these attachments and the answer to each question is dependent upon the kind of bird that we are talking about, the type of parental rearing that it has received, and its developmental stage. The questions are (1) What kind of attachments are formed at what developmental stages? (2) How long do these attachments last? (3) What mechanisms mediate each of them? and, most important of all, (4) Do any of these attachments affect the ability of an individual bird to accurately identify its own species? The following sections discuss most of these topics.

Different Kinds of Birds

Birds may be classified into five different categories according to their condition when they hatch (Skutch, 1976). Altricial birds such as a robin are hatched naked, cannot thermoregulate efficiently immediately after hatching, cannot locomote, and must be fed by their parents. Ordinarily they cannot leave the nest for at least a week and are wholly dependent upon their parents for food and protection during this time.

Semialtricial birds, such as the herring gull, are able to leave the nest within a day or two, but they remain near it and are still dependent upon their parents for food, protection, and warmth. To offset the hazards of hatching on the ground instead of in a tree, these birds hatch in a less helpless state than altricial birds. At hatching their eyes are open, they have down on them, and they can locomote fairly soon after this event. They cannot forage far from the nest, however, nor can they survive long without the assistance of their parents for most of the necessities of life.

Subprecocial birds, such as the American coot, are able to leave their nest as soon as they hatch or very shortly thereafter, can follow their parents while they forage for food, and are fed from the bills of the parent birds. The newly hatched birds have their eyes open and a covering of down enveloping them, but they cannot feed themselves. Their parents deliver food to them as they find it. The food is not carried to the newly hatched birds at or near the nest. This is the chief difference between the subprecocial bird and the semialtricial bird.

Precocial birds, such as a duck or a chicken, are hatched with a covering of down, can thermoregulate shortly after hatching, can locomote shortly after hatching, and do not have to be fed by their parents. They can either find their own food or their parents help them to locate it. An example of the former would be most ducklings, while an example of the latter would be a domestic chick. A duck is the most precocial bird (Nice, 1962).

Superprecocial birds are wholly independent of their parents as soon as they hatch or at least from the time that they are dry. There are only two known examples of this type of bird. They are the megapodes and the parasitic black-headed ducks of southern South America. Megapodes are gallinaceous birds found in eastern Indonesia, Polynesia, New Guinea, and Australia. They dig pits in the soil or build mounds of rotting vegetable matter in which they lay their eggs and then cover them up. The heat of the sun and the heat generated by the decomposing vegetable matter warm the eggs and allow them to incubate (Pettingill, 1970).

Frith (1962) has found that within one hour after emerging from its mound the mallee-fowl chick (a megapode) can run firmly. After two hours it can run quite efficiently and can even support itself above the ground for up to 40 feet. Twenty-four hours after hatching it can fly strongly. Even if it sees its parents tending the mound they have built, neither the parents nor the chick pay any attention to each other. The parents do not assist the chick in any way.

The black-headed duck of South America is a parasitic egg layer. It will deposit its eggs in the nest of almost any large bird that nests in the appropriate vegetation. Upon hatching the duckling does not depend upon any of its hosts for food, but may accept brooding for the first day or two. After that it swims off by itself and is completely capable of finding its own food, avoiding its enemies, and keeping warm without being brooded. Parents of this species show no interest in their offspring (Weller, 1968).

The important thing to remember about the superprecocial birds is that they represent one end of a continuum of hatching states found in young birds. From altricial birds to superprecocial birds we find that the young hatch in increasingly mature states with correspondingly less dependence upon their parents.

Reproductive Isolating Mechanisms

Like all other animals, birds form pairs with members of the opposit sex of their own species. They also avoid mispairings with opposite-sexed members of the wrong species. As stated elsewhere (Chapter 1, this volume), there is behavioral order rather than behavioral chaos in the natural environment. There are numerous factors that account for this phenomenon. All of them are considered to be reproductive isolating mechanisms, and they refer to those characteristics possessed by or associated with a species that prevent it from mating with members of another species (Alcock, 1975; Brown, 1975; Dobzhansky, 1937, 1970; Littlejohn, 1969; Mayr, 1963). The factors may be behavioral or they may be physical. In either case, the different species are reproductively isolated from each other, and this factor limits the number of mismatings that can occur among the many avian species that exist today.

Alcock (1975) and Brown (1975), following Mayr (1963), classify isolating mechanisms into those that prevent inappropriate matings (premating reproductive isolating mechanisms) and those that operate following an attempt to copulate with a member of the wrong species (postmating reproductive isolating mechanisms).

Premating reproductive isolating mechanisms reduce the probability that a member of one species will mate inappropriately with a member of another species. Included in this category are mechanisms which actually prevent encounters between species and mechanisms which prevent copulation when opportunities to interbreed do exist. Examples of the former are geographical isolation, ecological isolation, seasonal isolation, and temporal isolation. Postmating reproductive isolating mechanisms include mechanical isolation, gametic isolation, inviability of hybrids, and sterility of hybrids. These mechanisms interfere with the success of an attempted or actual hybrid mating.

Sexual isolation, also called behavioral or ethological isolation (Brown, 1975), is an example of a premating reproductive isolating mechanism designed to prevent copulation between members of noncongeneric species when opportunities to breed do exist. When birds live and breed in the same habitat at the same time, interbreeding is a possibility. Birds that live in proximity to each other, however, almost always have elaborate courtship activities which differ strikingly from species to species. Only when females are courted by males of their own species and perceive the proper courtship ritual will they become receptive and permit copulation to occur.

The importance of these isolating mechanisms is that they ensure that avian species do not dilute their unique adaptations to their ecological niches through hybridization. Hence, their reproductive advantage over other species coexisting in the same area is preserved. Ordinarily, reproducing with one's species is rigidly controlled by the organism's accessibility to other species in its environment and the degree to which it is genetically predisposed to react to those species. The existence of isolating mechanisms tremendously reduces the potential confusion that could exist if a great variety of species choices existed for an organism.

Whereas reproductive isolating mechanisms affect the choice of potential mates in many species and limit the choice of such mates, such regulating and limiting forces are removed when experimenters expose birds to inappropriate or atypical stimulus objects in their laboratories. The birds being used as experimental subjects frequently will never encounter some of the choices they have been exposed to in our laboratories. Yet these same laboratory alternatives may possess components of the natural situation that the investigator does not even suspect exists. His cause–effect statements concerning his results are thereby very misleading.

Ordinarily, in the natural environment we do not find atypical rearing conditions. Any atypical condition is likely to be specific to a particular situation and result in the death of the young bird or in some physical deformity, should it survive. Such a deformity would probably result in a lower probability that this bird would be able to mate successfully, thereby eliminating its genes from the gene pool. If some laboratory experiments have atypical conditions and the attending confounding variables associated with them, it does not mean that all laboratory experiments investigating the formation of an attachment in birds are meaningless. To evaluate laboratory experiments in a better perspective, however, it is a good idea to look to the natural situation to see what factors seem to be important there. When that situation is understood, then laboratory experiments can shed further light on the mechanisms mediating the responses demonstrated in the birds' natural environment.

Patterns of Incubation

The following information is taken from Skutch's (1976) excellent coverage of this topic. It is presented as a background to the possible experiential avenues that certain variables may travel in influencing the formation of a species identity in birds.

Incubation involves the application of heat to promote the embryo's development. Without the external application of heat, bird eggs fail to hatch. The only birds to hatch without the warmth provided them by some kind of parent, natural or foster, are those warmed by man-made incubators or the more ancient ones constructed by the megapodes.

Skutch (1976) outlines 17 well-defined types of incubation. The three major categories into which the different types fall involve the cases where both parents incubate, only one parent incubates, or more than two birds, not all of which are the biological parents, incubate the same nest. The three categories provide three major avenues along which environmental variables may potentially influence the developing embryo. Table 1 outlines the possible combinations between the incubating parent(s) and the parent(s) caring for the young. Innate factors would exert a greater influence on the formation of an attachment to one's species if there were no parents to serve as a model for the newly hatched offspring as is the case with the megapodes. The same situation would prevail for the case where the parents that are present are of the wrong species, as is the case with the offspring of parasitic egg layers.

The potential influence of a single parent depends upon the sex of that parent and, among other factors, upon whether that parent cares for the offspring or if both parents care for them. If we assume that only the incubating parent cares for the offspring, the "other factors" alluded to above that may be important in helping the offspring form an attachment to its species are (1) the degree to which the incubating parent vocalizes during incubation and hatching, (2) its visual characteristics, (3) whether the parent is representative of a sexually monomorphic ("looks alike") species or of a sexually dimorphic ("looks different") species, and (4) the kind of bird that we are talking about (altricial through super-precocial).

It should be emphasized that the genetic contribution to a proper species identification may be sufficient in any case to ensure that the correct attachment is made or it may need guidelines imposed upon it by the environment. As indicated previously, it will never be the case that either the genetic factor or the environmental factor will be the sole contributor to the formation of an attachment to one's species. Either factor, however, may play a proportionally greater role than the other factor in the formation of an attachment to one's species. In the megapodes, for example, where no parent incubates the eggs or cares for the young, it is obvious that innate factors will play a large component in assisting these birds later to mate with the proper individual. But this fact does not mean that learning plays no role. The same thing may be said for the offspring of parasitic egg layers. On the other hand, birds incubated and cared for by both parents (or even by several pairs of parents or several members of the flock representing both sexes) probably have less dependence upon the innate factors and rely more heavily on the experiential aspects of their

Table 1. Possible Combinations Between the Incubating Parent(s) and the Parent(s) Caring for the Young

	Natural or Foster Parents							
Incubates the eggs	Male and female	Female	Female	Female	Male	Male	Male	Neither
Cares for young	Male and female	Male and female	Male	Female	Male and female	Female	Male	Neither

upbringing, with respect to the proper selection of a future mate. The innate predisposition to respond in a particular way to a particular species is intimately involved with the opportunities the newly hatched bird has to interact with its environment. Research in this area has frequently overlooked or totally ignored this interactive interpretation of attachment behavior in birds.

To separate the relative contributions of genetic and environmental factors is no easy task. Given the number of species of birds known to exist today, the various patterns of incubating and rearing these species, and the diverse ecological systems supportive of avian existence, each of which interacts uniquely with the morphological characteristics of each species, it is not likely that researchers are going to be able to gain a complete understanding of the process of forming an attachment to one's species in each species of bird in the near future. Some clues to likely avenues of investigation may be obtained by looking at selected examples of avian research which indicate that certain variables may be important in this process. In the following sections this research has been divided into pre-hatching affects and post-hatching affects; but this is not to imply the lack of a connection between them. I have also tried to interpret these studies from a naturalistic point of view. Hence, the research has been classified into a framework of situations where a parent is either present or not, and in either case, certain variables may be influencing the offspring's later choice of mate.

As long as there is the possibility that the incubating parent can influence the developing embryo's future attachment behavior, then the possible mode of interaction must be found and the specific details of the interaction must be investigated. We must not think that we know all the possible direct and indirect influences that can affect this interaction. The formation of an attachment to one's species may be directly affected by the parent's incubating activities or it may be indirectly affected by the effect such activities have on some other form of attachment which must precede the formation of an attachment to one's species.

The following descriptions suggest a few possibilities and provide some evidence for the plausibility of such interactions. The generality of the following statements is not great because of the wide discrepancies among the studies concerning almost all pertinent aspects of the variables involved. Tantalizing possibilities for studying attachment behavior are suggested, however, by these results.

Characteristics of Incubation

Prehatching Effects with a Parent Present

The presence of a parent or any adult bird of the same species incubating a clutch of eggs provides a source of stimulation for the developing embryos. Conversely, the sounds and movements of the embryos provide a source of stimulation for the incubating parent bird and for each other.

A few days before hatching both chicks and ducklings begin to hear (Gottlieb, 1965; Gottlieb and Vandenbergh, 1968). The presence of a vocalizing parent incubating eggs could provide a source of stimulation that may enhance a later attachment to the parent and, perhaps, to the species. Kuo (1932) and Gottlieb and Kuo (1965) have demonstrated that fetuses of both chickens and ducks begin to vocalize a few days before hatching. Such activity from within the shell provides for a two-way channel of communication between parent and young. For the embryo there are three auditory inputs: one from itself, one from its peers in adjacent eggs, and one from the bird above it. Audiospectrographic analysis (Gottlieb and Vandenbergh, 1968) has revealed that the calls of the embryo prior to hatching are similar to distress, contentment, and broodinglike calls that they make after hatching.

Prenatal auditory discrimination is possible in the white Peking duck (Gottlieb, 1971) and in the wood duck (Heaton, 1972). Exposure to prenatal auditory stimulation has affected later approach and following responses in guillemots (Tschanz, 1968), domestic chicks (Grier et al., 1967), and ducklings (Gottlieb, 1971).

Gottlieb (1965, 1971) has shown that conspecific maternal calls reinforce certain embryonic movements made just prior to hatching. Impekoven (1973) has shown that when these calls are made contingent upon a response of the embryo, the responses increase significantly. The functional significance of this observation is that it might speed up hatching and serve to synchronize the hatch. Hess's (1972) experiments with mallards support the idea that vocal interactions between the hen and her brood can synchronize hatching. Such interactions may also help the young to recognize the individual voice of their parent. Several studies indicate that young precocial and semiprecocial birds can discriminate between their own parents and other adults (Beer, 1969, 1970; Evans, 1970a, 1970b; McBride et al., 1969; Ramsay, 1951; Stevenson et al., 1970; White, 1971).

Impekoven (1976) has shown that parent-incubated laughing-gull chicks respond positively to certain calls of their parents but incubator-reared chicks of the same species do not respond to recordings of these calls. Developing embryos respond selectively to their own sounds, also. Vince's (1964, 1966) work indicates that these sounds help to synchronize the hatch. Two kinds of sound seem to be important. One is a low-frequency sound from 20–40 Hz which is emitted by an embryo prior to penetrating the air cell membrane. A second sound, called "clicking," occurs following lung ventilation and is associated with respiration in the embryo. It is a repetitive sound composed of brief notes of several frequencies up to 8000 Hz. McCoshen and Thompson (1968) found these "clicking" sounds to be similar in 12 avian species. According to these authors the clicking sounds continue after hatching and, in one species, lasted for up to five days post-hatch. These sounds may be potentially important in maintaining the cohesiveness of the brood or in facilitating a parent–brood interaction.

Rajecki (1974) has found that when chicks are stimulated with sound prenatally, there is a preferential response to these sounds after hatching. This finding is in agreement with those reported by Hess (1973), which indicated that prenatal stimulation of young in naturalistic settings contributes to the formation and maintenance of attachments between parent and young.

As mentioned previously, Vince (1964) has shown that when eggs are placed in contact with each other they seem to synchronize their hatching better. In bobwhite quail she has found that this can occur by having the less advanced eggs speed up their development to match that of the more advanced eggs. Not only can eggs be accelerated in development, but the opposite can also occur. If one places normally developing Japanese quail eggs next to eggs at a more primitive stage of development, then their hatch time is retarded (Vince and Cheng, 1970). This phenomenon is apparently more widespread than previously suspected. Vince (1966) has found these sounds and movements in several species of birds. They may be a means of communicating with each other and the parent as well as a means of becoming sensitive to such sounds. They also seem to affect the rate of occurrence of some behaviors, such as standing and walking (Vince and Chinn, 1971, 1972). Woolf et al. (1976) have shown that the accelerating affect can result from a stimulus duration of as little as two hours any time during the last three days of incubation.

Thus it may be said that auditory stimuli remain a very viable option for studying the formation of a bird's attachment to its species. It should be noted that in all probability the role of audition is limited by the nature of the stimuli, its source (embryo or incubating bird), the pattern of incubation involved, and the environmental requirements of the species concerned. Other stimuli are also important in forming an attachment to one's species, however. These stimuli may act in addition to sound, in place of it, or by interacting with it.

Visual stimulation of incubating eggs can affect later post-hatch behavioral responses (Dimond, 1968; Gold, 1969; Rajecki, 1974). It can also reduce incubation periods (Lauber and Shutze, 1964; Siegel et al., 1969; Shutze et al., 1962) and apparently enhance development in the growing embryo (Harth and Heaton, 1973). Some of these effects are apparent after only ten hours of incubation. Heaton's work (1971, 1973, 1976) demonstrates that in some species visual-system competence occurs before the final quarter of incubation.

Other systems that are mature and activated prior to hatching in birds are taste (Vince, 1977) and olfaction (Tolhurst and Vince, 1976). Work reviewed by Gottlieb (1968) and Freeman and Vince (1974) shows that tactile, vestibular, and proprioceptive systems all begin to function during incubation. All of the above systems may interact with each other in a stimulatory sense to facilitate a post-hatching attachment between the developing embryo and the bird above it. It is also possible that the presence of one or more of these sensory systems activates or primes other systems, all of which may be necessary to form an attachment to one's species.

Prehatching Effects Without a Parent Present

The only birds to hatch without the warmth provided them by some kind of parent are those warmed by man-made incubators or those (e.g., the megapodes) residing within bird-made incubators. Without an adult sitting above it, how does the developing embryo communicate with the outside world? Indeed, does it communicate with the outside world? In attempting to answer these questions there are two points that should be made. First, as mentioned earlier, the megapodes are superprecocial birds and as such represent the advanced end of a continuum of hatching states found in young birds. Examples of their extreme state of development were given previously. Second, in the preceding section there were some systems mentioned that could interact with each other or which could act independently and simultaneously to facilitate communication between the developing embryo and the bird over it. It is conceivable that the more advanced states of the megapodes allow these sensory systems to monitor their external, parentless surroundings in a more sensitive manner and for a longer prehatch time than in other birds. In birds hatched in such an advanced developmental state, there is no need to interact with a parent. Hence those mechanisms which in other birds would be used to mediate a bond with a parent could be used in the megapodes to mediate an attachment to a particular environment. Such a statement does not imply what has been termed "locality imprinting" (Thorpe, 1944, 1945), "environmental imprinting" (Hess, 1973; Kaufman, 1973; Lamprecht, 1977; Mulder, 1975), or "habitat imprinting" (Hess, 1964). What it suggests is that the evolving sensory systems of the megapode could assist it in forming an attachment to a particular kind of environment that it may later prefer if given a choice of several environments in which to live. For example, the decaying vegetation used to create the mound in which the megapode egg is laid may have a particular kind of odor associated with it that is dependent upon the vegetation that the adults can gather in that locality. The megapode adults continue to bring fresh vegetation to the mound to maintain its internal temperature. Consequently, bits and pieces of recognizable flora are present. If the megapode olfactory system is operative before it hatches, then the developing embryo should be able to smell the decaying vegetation before it hatches and to see some of the vegetation used to construct the mound after it hatches. Presumably there is some adaptive advantage to be conferred upon the newly hatched megapodes by staying in or around such vegetation. Such a system or one similar to it seems plausible because megapode eggs will not hatch if they are exposed to light (Batlin, 1969). Hence the developing visual system of the megapode embryo probably plays little of a prehatch role in mediating a bond to a particular post-hatch environment.

The reproductive isolating mechanisms discussed earlier assume an important role in a situation such as the one in which the megapodes are found. Without a parent with which they can communicate prior to hatching or interact with after hatching, they must remain in a location or geographical region where there is a high probability that they will later encounter an adult of their own species. Innate factors and reproductive isolating mechanisms will play a larger role in this situation than in avian species having a parent above them before they hatch and around them after they hatch. However, this is not to imply that there is no genetic basis to the formation of a species attachment in this situation either. It is only being assumed that the genetic basis underlying the formation of the attachment plays a correspondingly less important role under the latter circumstances. Similar comments can be made about the black-headed duck of South America.

Figure 1 illustrates some of the factors that attend the formation of an attachment with and without a parent present. It is assumed that most of the time the adult bird attending the eggs is the parent of those eggs; but that condition need not be true, as Skutch (1976) has pointed out. It may be seen from this figure that with a parent/adult bird present the offspring can be more dependent in a less favorable environment. With the parent absent, however, the offspring must be more independent and hatch in a more favorable environment. In addition, genetic components probably play a greater role in the latter case than in the former case. Although the different types of birds are arranged in a linear order from lower left to upper right, one need not impose such order to the diversity of species and the many ways in which

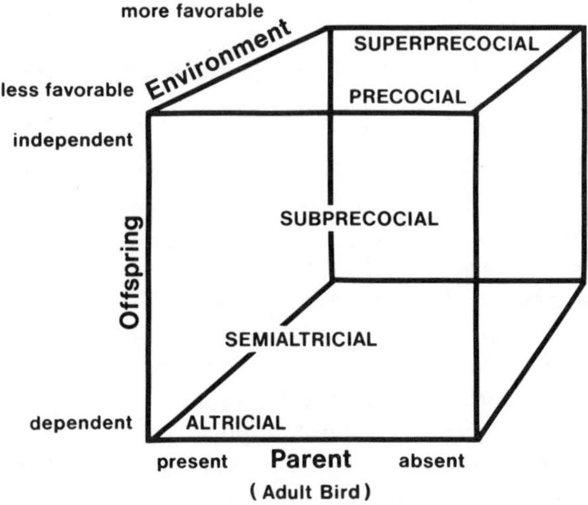

Figure 1. Hypothetical representation of some of the factors attending the formation of an attachment with and without a parent present.

attachments can be formed. For convenience's sake the different kinds of birds have been arranged in this manner.

Hence it may be concluded that birds which hatch without an adult incubating them probably use sensory systems commensurate with the stage of development at and just preceding hatching to help them form attachments to peers and preferred mates later in their lives. Reproductive isolating mechanisms, which can include genetic factors, probably play a greater role in forming a species identity in megapodes than in any other kind of birds.

Post-hatching Effects

In proceeding directly to a discussion of post-hatching effects associated with patterns of incubation we are omitting a very dynamic aspect of the life of the young bird. I am referring to the actual processes involved in hatching. What possible role these processes play in the formation of an attachment is not known and space does not allow a detailed description of this extremely interesting phenomenon.

There is another body of literature, however, that ostensibly deals with post-hatching affects on the newly hatched bird. This literature could generally be classified as pertaining to the phenomenon known as "imprinting." Most of this research has been concerned with the potential post-hatching effects of treatment conditions administered under well controlled artificial conditions usually unrelated to those found in the natural environment. The results of these experiments are undoubtedly valid under the conditions they were obtained. The problem with them is that they are hard to interpret with respect to those factors actually affecting the behavior of the newly hatched bird. In addition, these studies do not take into consideration the pattern of incubation that is involved and generally assume most avian species can be treated as if they were equal.

Besides the imprinting literature, there are studies that have investigated potential post-hatching affects on birds. In order to evaluate these affects, one must know what kind of bird one is working with. By "kind of bird" I do not mean the species involved, but where the bird lies on the altricial–superprecocial continuum. In terms of forming an attachment to one's own species, the presence of a parent would seem to be very important. But if one is working with a superprecocial bird (e.g., a megapode) how is the same thing accomplished?

In order to understand the formation of a species identity in a bird adequately, one should know what the newly hatched bird's relationship is to its parent or parent surrogate. Is a natural parent of the same species in attendance at hatching or is an adult surrogate of another species in attendance (as would be the case if the biological parent was a parasitic egg layer)? If some adult bird was present for a newly hatched bird to interact with, then there may very well be innate predispositions causing the newly hatched bird to interact with those aspects of its environment that generalize to the concept of "live adult bird." If the newly hatched bird is a superprecocial bird, the species identity may be so governed by innate schema that the absence of a parent does it no harm. Its advanced state of maturity at hatching and lack of any dependence on a parent renders it very capable to exist on its own. This latter state of affairs does not mean that no learning takes place in forming an attachment to one's species. In megapode chicks, for example, there may be quite a bit of learning that occurs through perception of their feet or through their perception of the shape of their own bodies. They may even see their own reflection in drinking water. The color of their own plumage and the sounds that they emit may also guide their eventual choice of a mate resembling their own self-perception. This situation would be like "self-imprinting" (Salzen and Cornell, 1968) and, along with reproductive isolating mechanisms, could interact with genetic factors to assist the superprecocial megapode in eventually making the correct attachment to a sexually mature individual of the same species. Even if the interaction between the learned components of the megapode species identity on one hand and the genetic components governing the same thing on the other was not totally controlling, nonfatal trial-and-error interactions between megapodes and other avian species in the environment may result in their attention being directed to those birds offering less resistance to their advances, i.e., to conspecific adults of the opposite sex.

In those species of birds that hatch in the presence of a grown bird, there may be common features associated with most "adult birds" that mediate an attachment between the newly hatched bird and the adult tending it. Some of these factors could be tactile stimuli, temperature, smell, sound, and visual stimuli.

Driver (1960, 1962) has proposed that a "brooding reflex" exists in waterfowl and has hypothesized that it may be "the mechanism whereby the newly hatched duckling orientates itself to the position of optimum warmth and mechanical protection beneath the female: It involves an active search by the duckling for a feeling of enclosure around the head or part of the head" (p 416). Taylor, Sluckin, Hewitt, and Guiton (1967) seem to support Driver's suggestion when they state that chicks may be capable of becoming "imprinted" to objects of specific textures. Sluckin, Taylor, and Taylor (1966) and Taylor and Sluckin (1964) provide additional support for this suggestion. In a normal pattern of incubation involving the presence of at least one of the biological parents, such a tactile sensitivity could mediate an attachment between parent and young and, indeed, even among the young. Hence, in addition to, in place of, or in interaction with a genetic component, a young bird may begin to learn a species identity adequately almost as soon as it is hatched. Indeed, there is no reason to believe that this learning did not begin long before hatching took place. Thus, post-hatching learning may only continue the prehatch experiences or perhaps just refine them.

Acting at the same time as a tactile sensitivity and even complementing it, the young bird's desire for warmth may cause it to seek out the adult above it or alongside of it. Such a mechanism could mediate an attachment already formed or facilitate an attachment being formed. Kaufman and Hinde (1961) have shown that cold adversely affects the behavior of chicks and Fischer (1970) and Salzen and Tomlin (1963) have shown that attenuation to cold temperatures impairs following. Fischer (1970), however, has shown that experienced cold (i.e., being warm but feeling cold) enhances following. Young birds being brooded by a parent may follow the parent more if, in addition to other attractive factors associated with the parent, it is also keeping the young bird warm. Again, such a finding is dependent upon the kind of bird we are referring to. Heat is an effective reinforcer in young chicks (Zolman, 1968), and they can certainly learn at an early age (Marley and Morse, 1966).

Smell, as previously mentioned, is a viable medium which could serve as a factor mediating an attachment between parent and young, as auditory and visual stimuli do.

Fischer (1976) and Gottlieb (1971) have adequately demonstrated a species-specific preference for a maternal call in chicks and in ducklings. Shapiro (1971) has confirmed these findings for the first seven days of the duckling's life. Shapiro (1977) has also shown that white Peking ducklings have an apparently innate visual preference for the ancestral form of the mallard, as opposed to a preference for a live female white Peking duck.

There are many indications that innate visual preferences could play a role in mediating an attachment between parent and young. Fantz (1957) has shown that innate form preferences exist in chicks, as have Goodwin and Hess (1969), and innate depth discrimination also exists (Fantz, 1958). Hess (1956) and Hess and Gogel (1954) have shown that chicks have natural preferences for objects of different colors. All of these demonstrated preferences in young birds seem to make sense if they are fit into a model that takes into consideration the kind of bird involved, its level of dependency upon a parent or adult bird, and the requirements of the environment into which it hatches. Viewed in this manner, the post-hatching environment of the young bird should contribute quite a bit to the formation of a species identity. As mentioned previously, however, the genetic contribution to a proper species identification may be sufficient in any case to ensure that the correct attachment is eventually made. It may also need guidelines imposed upon it by the environment. Such an environment is provided by the pattern of incubation requiring a parent to be present after hatching occurs. The more independent the newly hatched bird is, the less structure needs to be imposed upon the genetic components.

One factor which has not been mentioned until now is the possible effect that broodmates might have upon the formation of a proper species identification. Since we will return to this topic later in this chapter, let it only be said here that the stimuli peers provide for a newly hatched bird may be sufficient in themselves to ensure that an adequate species identity is formed in young birds. Such a system of cues may be independent of those provided by the parent, used in addition to them, or complement them. If used independently of the cues provided by the parent, such a system would be useful in the case where a set of monogamous parents rear more than one young and split up after leaving the nest. The father takes charge of some of the young and the mother keeps the rest. According to Skutch (1976) this practice is widespread among the families of birds. If broodmates provide sufficient cues to form a species identity, then there would be no advantage to being with the male or the female. In addition, some parents drive their offspring away as soon as they become self-supporting. Depending upon the level of the altricial–

superprecocial continuum that one is dealing with, broodmates may provide themselves with sufficient cues to continue forming a species identity or they may generate sufficient cues to complete the process begun earlier.

Early experiences are thought to be important in forming a proper species identity in young birds. Indeed, anyone reading Immelmann's (1972) excellent article on this topic can hardly come to any other conclusion. Early experiences *are* important in later mate selection. Warriner, Lemmon, and Ray (1963) have shown that male pigeons will usually select a female of the same color as its parents. Schutz (1965, 1971) has shown that male mallards will mate with nonconspecific females of the species they were reared with but, in the reverse situation, that female mallards usually will not. Brosset (1971) working with doves and Immelmann (1969, 1970) working with finches have also found that birds reared with foster parents of a different species would later prefer that foster species when it was time to mate.

The problem with such research is that there are too many exceptions to the results for one to have much faith in them. For instance, Fabricius and Fält (1969) found that female mallards *will* mate with males of a foster species if you confine them with that different species for a longer period of time than Schutz did. Brown (1975) reports the following: (1) mallards reared in visual and auditory isolation for up to nine weeks of age will later mate normally with others of their own species; (2) male and female mallards can be "imprinted" to other objects, using standard criteria of "imprinting," and they will still mate with members of their own species; (3) some birds "imprinted" to another species will mate with both the foster species and their own species.

Given the small number of species worked with to date, the large number of avian species yet to be studied systematically in the same way, and the exceptions to the results obtained to date, judgment concerning the consequences of atypical early rearing experiences in birds should probably be suspended, at least for now. The potential for early experiences to affect later behavior certainly cannot be denied. In birds, however, the potential consequences do not seem to be as damaging as in other species. For example, the owners of ornamental collections of waterfowl routinely rear their ducklings under bantam hens. These ducklings do not seem to have any problems in mating later on. Parasitic egg layers also have foster parents, and they do not seem to have any problems in mating later on, either. Such exceptions to the rule may be a function of our ignorance of the correct conditions necessary to form the proper species identification in the first place. It may also be that early experiences are not as important as we would like to think and that the genetic contribution to a proper species identification may be sufficient to ensure that the correct attachment is formed. An alternative view is that the early experiences investigated to date, along with the procedures used in these investigations, have not been appropriate to the natural circumstances of the birds in question, and the resulting information has been difficult to assimilate into a natural view of things. Certainly Smith and Bird (1963a,b) provide evidence that the more stimuli one uses in investigating species identification, the more successful one will be at observing the formation of an appropriate bond.

Kinds of Attachment

We really do not know how many different kinds of attachments a young bird forms in its lifetime. Forming an attachment to one's species is only one kind of an attachment that the young bird can establish. Other kinds of attachment that can be formed include an attachment to one's parent or parents and, later, an attachment to a particular mate.

What role these attachments play in the formation of an attachment to one's species is not known, and the exact relationship among these three different forms of attachment is not clear. In fact, I am not sure that they *are* different forms of attachment. They may all be formed at the same time or at least concomitantly and involve similar mechanisms; but because of the natural course of development, either the proper stimulus to which a response can be directed is not available or the response has not matured and cannot be directed to an available stimulus. The latter interpretation could account for the behavior of subjects used in studies where immature organisms were injected with male or female sex hormones. In these studies the experimenters were able to elicit mature sexual responses from immature subjects (Bambridge, 1962; Guiton, 1962; Schein and Hale, 1959).

There is another way to view the potential relationship that exists among the three forms of attachment previously mentioned. Each attachment may be dependent upon the successful completion of a preceding attachment before a subsequent attachment can begin to be formed. Each attachment could involve a different process and have a different consequence. It is just as likely, however, that the three forms of attachment may involve similar processes but have different consequences at different developmental stages.

Different kinds of attachment may be formed at different developmental stages, and the processes involved in the formation of these attachments may or may not be similar. The immediate consequences of such attachments certainly do differ, however.

Assume, for a moment, that there are three stages of post-hatch development in a bird's life. The first stage of development we will identify as the first week or two. The second stage we will consider to be that period of time when the bird is becoming less dependent upon the adult or the family unit and more capable of operating on its own. The third stage we will assume is adulthood and sexual maturity. The exact timing of these stages would depend upon the species of bird that is being talked about.

Immediately after hatching and during the first stage of post-hatch development, a young bird may form an attachment to its parent or parents, regardless of what its "parent" may be. That is, a duckling may consider its "parent" to be a hen, a human, or a hot-water bottle. The immediate consequence of this attachment at this developmental stage is that the newly hatched bird has its biological needs taken care of. As long as this is the case, anything may substitute for the normal biological caretaker. Hence we have biological parents caring for their own offspring, foster parents caring for the young of parasitic egg layers, and a host of inanimate, biologically inappropriate stimulus objects used in "imprinting" experiments functioning successfully as maternal surrogates.

Another consequence of this first developmental stage is that the parent or parents and the young are in proximity to each other. During this period of time four kinds of relationships may be formed within the family unit. They are relationships between mother and hatchling, father and hatchling, hatchling and father, and hatchling and mother. Parenthetically, only the latter relationship, i.e., the relationship between the hatchling and the *mother* has traditionally been referred to as "imprinting."

Another kind of attachment that can be formed during this initial developmental stage is one that evolves among the hatchlings themselves. This may ultimately be the most important attachment that the newly hatched bird forms. This attachment will be discussed in later sections.

During the second stage of development the young bird can potentially form an attachment to its species. As we have seen, in general it appears that such an attachment can be genetically programmed but modified by prenatal and postnatal experiences. It can also be mostly programmed and fairly independent of environmental experiences or mostly influenced by environmental experiences but guided to some extent by genetic capabilities. Again, it depends upon the kind of bird we are talking about. At any rate, this bond may be formed at a later stage of development than the one at which the bird forms a bond to its own parents and, as a consequence, selects a class of organisms with which to associate in some capacity.

During the third developmental stage, the bird can form an attachment to a particular individual taken as a mate. The consequence of this bond would be that the pair reproduces and contributes to the gene pool.

It does not follow that the same mechanisms mediate each of the above forms of attachment at different developmental stages. Nor does it follow that each of the attachments lasts for the same period of time and has similar consequences. Unfortunately, much research concerning avian attachment behavior has assumed that an experiment concerning a brief period of time early in the life of a particular bird can form the basis for generalizing to all forms of attachment during most developmental stages in many kinds of birds. Such is not the case at all. In fact, such may not even be the case for male and female young of the same age and of the same species (see Roy, Chapter 1, this volume). Hence different kinds of attachment can potentially be formed at different developmental stages, and they can potentially be mediated by different mechanisms. Each attachment may have a different consequence. Investigators should be aware of these possibilities when they plan experiments and subsequently interpret their results.

Research on Species Identification in Birds

Imprinting has often been cited as the explanation for species identification in birds. The young bird is supposed to form an attachment to the first moving object that it sees, and the formation of this attachment has been cited as the mechanism mediating the later choice of a mate. In point of fact such a description is merely a label, not an explanation. Nevertheless, much research concerning species identification in birds has been done under the name of "imprinting."

Since the literature describing this phenomenon is now extensive, the reader is referred to several sources for a history of the topic and a review of the literature (i.e., Bateson, 1966; Hess, 1973; Sluckin, 1965, 1970; Smith, 1969). Certainly the studies reported in these sources are valid within the context of the experimental situation in which they were conducted. In most cases the results are also reliable. There *is* a problem when these results are generalized to more natural situations involving the full range of organisms known as birds and incorporating all developmental stages within each species of bird. At the very least the same term should not be used to describe so many different forms of attachment. Consider the following phrases, all of which use the same term to describe what is presumably a different form of attachment: "filial imprinting" (Burghardt, 1973), "maternal imprinting" (Klopfer, Adams, and Klopfer, 1964), "parental imprinting" (Burghardt, 1973), "self-imprinting" (Salzen and Cornell, 1968), "sibling imprinting" (Klint, 1973), "olfactory imprinting" (Carter and Marr, 1970; Muller-Schwarze and Muller-Schwarze, 1971), "color imprinting" (Immelmann, 1972), "sexual imprinting" (Fabricius and Fält, 1969; Gallagher, 1976; Immelmann, 1972, 1975; Klint, 1973; Schutz, 1971; Walter, 1973), "food imprinting" (Burghardt and Hess, 1966; Frank and Meyer, 1970; Hess, 1962, 1964; Meyer and Frank, 1970), "environmental imprinting" (Kaufman, 1973; Lamprecht, 1977), "habitat imprinting" (Hess, 1964), and "host imprinting" (Immelmann, 1975), among others. The use of the same basic word "imprinting" to describe these different behaviors implies that similar mechanisms mediate them. It is difficult to understand how one word can mean so many different things to so many different people. The glibness with which investigators generalize across different species of similar ages or across different ages within similar species is alarming.

Another problem associated with the use of the term "imprinting" is that virtually all of the characteristics originally associated with the term have been questioned by experimenters who have failed to support them empirically. It remains to be determined whether the original attributes of imprinting are actually wrong or only partially right, or whether the dissenting voices are only reflecting opinions which are themselves based on inappropriate procedures, oversimplified assumptions, or a lack of a suitable framework. Perhaps it would be a good idea if the term "imprinting" were laid to a well deserved rest, at least until the researchers involved decide on what they want to label the formation of an attachment in a bird. My own nomination for such a label would be the phrase "avian attachment behavior" with a modifier before the phrase identifying both the organism and the developmental stage of the organism in question.

By taking an abstract word like "imprinting" and treating it like a real thing, one is doing two things: (1) reifying the concept and (2) committing a nominal fallacy. In the latter case one names the phenomenon but does not explain it. The result of this procedure is counterproductive in that one discourages further research from being undertaken in the mistaken belief that the question originally asked has been answered.

What we should be doing in our investigations of attachments formed in birds is to concentrate on the mechanism or mechanisms mediating these attachments in a particular species at a specified developmental age or ages and then to limit our conclusions to the specifics of the experiment, rather than generalizing to many different species and many different developmental stages. Investigators should also learn more about the natural environment in which their subjects are normally reared. We seem to forget that the rearing conditions imposed upon subjects in the laboratory occur in an artificial situation. It is one thing to find out that an organism possesses a mechanism that allows it to mediate a particular situation or exhibits a preference given a particular choice. It is another thing to know that the same organism actually uses that mechanism or exhibits that preference in its natural habitat. If the mechanism *is* used, it may not be used in the bird's natural environment in the way the investigator hypothesized its use in the laboratory.

To date, very few of the approximately 8600 (Dorst, 1971; Rutgers and Norris, 1970) species of birds have been used in behavioral research. This fact, by itself, precludes any valid generalizations from being made about species identification in birds. Sluckin (1970) has outlined the various procedures used in many of the behavioral tests that have been done to assess the strength of an attachment in some birds. One of the most common indications of the formation of an attachment is whether the bird will approach or follow or stay near the stimulus object previously or currently presented to it. In using this indication of an attachment many investigators have confused following with attachment. Following is not necessary for an attachment to be formed (Collins, 1965; Klopfer and Hailman, 1964). Such a statement is perfectly obvious to anyone who has studied birds that nest on cliffs, such as the kittiwake. If the young of this species attempted to leave their nest they would almost certainly fall to their immediate death; yet they form the proper attachment to their species. The question can then be asked: In those species where it is present, does following occur because an attachment has already been formed or does the subject follow in order to form an attachment? The answer may be affirmative to both parts of the question, depending upon the kind of bird involved and its stage of development. Because many of the studies using the following response have not taken these questions into consideration, the following response may not be the best indicator or the most complete indicator with which to assess the formation of an attachment.

To paraphrase a passage from Beach (1955), it is logically indefensible to categorize any rearing condition as normal unless the characteristics of typical or normal rearing conditions have been thoroughly investigated and are well known. Such is not the case with imprinting research. Much more research will have to be done on many more avian species in their natural habitat before we can use the current literature on imprinting to answer the question of how a bird forms an attachment to its species. In addition, one must be cautious in equating the term "imprinting" with many different kinds of avian attachments. Imprinting seems to be a term that has incorporated all other terms relating to this topic. Among these other terms are species identification, species orientation, species identity, species awareness, and species affiliation. In each case it matters little how the bird is socially attached to or forms bonds with another organism or object. Nor does it seem to matter at what developmental stage these attachments occur. In all cases the bird is said to be "imprinted." This is an inappropriate use of the term "imprinting."

The Synthesis

When I taught an introductory psychology course, I began the section on child psychology with this quotation: "Understanding the atom is child's play, relative to understanding child's play."

The same may be said for understanding the formation of a bird's attachment to its species. Table 2 lists some of the factors which must be considered. The amount of information required to understand the process in a single species of bird is probably a lifetime occupation for a single person if, in fact, it can be done at all. To begin to comprehend the processes applicable to a single family of birds is mind boggling. To even attempt to think that we can understand the same processes for all species of birds is almost inconceivable.

To compound the problem, the literature seems to be confusing in some cases. For instance, Lorenz (1957) states that "young birds of most species, if reared in isolation, do not recognize other birds of their species when they are brought together" (p 102). Research from the Avian Behaviour Laboratory at the University of Manitoba questions whether this statement applies to the precocial birds used as subjects in this laboratory. It has been found that when mallard ducklings are hatched and reared in auditory and visual isolation from other mallard ducklings and are then exposed individually to a small group of mallard ducklings (usu-

Table 2. Some of the Factors Contributing to the Complexity of Species Identification in Birds

1. Number of species of birds (approximately 8600)
2. Environmental requirements of each species
   a. Reproductive isolating mechanisms and evolutionary processes leading to adaptations to these environments
3. Pattern of incubation (altricial–superprecocial continuum)
4. Degree of genetic involvement
5. Degree of experiential involvement
   a. Prehatching effects
   b. Posthatching effects
6. Interaction between genetic and environmental factors
7. Developmental stages of each species
8. Kinds of attachment formed at each stage
9. Length of attachment
10. Mechanism mediating each attachment
11. Function of each form of attachment at each developmental stage
    a. Same attachments formed at different developmental stages having different functions
    b. Different attachments formed at the same developmental stages having similar functions
12. Possible interactions among all of the above

ally three), six parakeets, and nine chicks, the ducklings will always prefer the group of ducklings (Szkraba, 1978). The same result can be obtained if the ducklings are reared in visual but not auditory isolation (Sallows, 1977) or if they are reared communally (Hoag, 1977). Hoag's and Sallow's experiments have been replicated with chicks. In Szkraba's experiment individual duck eggs were removed from a larger incubator two or three days before the embryos broke the inner shell membrane and were placed into individual glass-enclosed hatchers, which were in turn placed in commercially manufactured sound-attenuated chambers, each of which was located on a different floor of the building housing the laboratory facilities of the Avian Behaviour Laboratory. The subjects were subsequently reared in these chambers.

In another study Shapiro (1977) found that white Peking ducklings apparently have an unlearned preference for a live adult female duck that looks like a mallard (a rouen duck) as opposed to a live adult female Peking duck or a chicken. These results also question Lorenz's statement without, one may add, implying that his statement is wrong. The point to be made is that such results make the topic of species identification in birds very complex.

The studies cited above also question whether a statement Roy made in Chapter 1 applies to the kind of waterfowl used in the Avian Behaviour Laboratory. He said that "an innate mechanism for 'knowing' what species one belongs to does not exist in many species and that early experiences, notably exposure to a particular species, are crucial for determining each organism's species identity." Neverthless, he also states that "behavioral order rather than behavioral chaos" seems to be the rule in the animal world. It is the unravelling of this mystery in birds to which the study of species identification in birds is directed.

Flexibility in Studying Species Identification in Birds

Because the behavior of any living organism is so complex and varied, it is foolish to think that there is only one way that a class of organisms such as birds can form an identity with its species. Every species living in a different habitat is going to have different pressures exerted on it. Consequently, the mechanisms involved in forming an attachment to one's species may be different under different circumstances. Alternatively, under different circumstances the mechanisms involved may be the same, but their utilization may be different in different species. In other words, birds use a variety of processes to form an attachment to their species. It may be mostly innate or mostly learned; but whatever the process or processes involved, there is a *continuum of strategies* that have been adopted to ensure that the best process or combination of processes is used with each species.

Unfortunately, many experiments studying avian attachments have used inanimate stimulus models to study what is essentially an interactive process. The use of inanimate models should not be neglected under the proper experimental conditions. Nevertheless, if an inanimate stimulus object and a live subject are used, it makes no sense to speak of the formation of an attachment if, by attachment, one means a form of dyadic stimulation, or stimulation where the behaviors of one animal become the cues for the response of the other. An inanimate cube or a decoy cannot respond to a living organism.

Shapiro and Bruce (1977) have adopted the position that one can form a continuum on which to order the appropriateness of stimulus objects used as models in research involving avian attachments. Beginning from the least desirable end of the scale and working towards the more favorable end the list is as follows:

1. Inanimate and biologically inappropriate
2. Inanimate but resembling the biologically appropriate
3. Live and biologically inappropriate
4. Live and biologically appropriate

By biologically appropriate, Shapiro and Bruce mean a stimulus object that a subject normally encounters in its natural environment as part of its normal development. They feel that the only meaningful results that one can obtain in this area of investigation at this point in our knowledge of the phenomenon are results that are the product of a live, biologically appropriate stimulus object interacting with a subject of the same species.

Just as the subjects under study can use a continuum of strategies to form an attachment to their species, so too can investigators employ a continuum of strategies to study their subjects. Some investigative strategies might involve looking at a continuum composed of attachments formed at different stages of development in the same species and the relationships among the various forms of attachments, a continuum involving different functions of similar attachments, a genetically learned continuum, or simply similar attachments in different species. The list is only limited by the originality of the investigator. The important point to remember is that the behavior under investigation is a complex interaction of many factors, all of which occur in a class of organism that itself varies widely with respect to almost every aspect of its anatomical, physiological, and environmental requirements. Hence perhaps the best strategy is to learn as much about the organism as possible, first, and then let the bird speak for itself with respect to the most likely variables that the investigator should begin to study experimentally. Thus erroneously conceived ideas that find experimental support in the laboratory will not be unduly advocated until some enterprising investigator decides to see if this viewpoint really holds up when viewed in its natural environment.

The Effects of the Brood

With rare exceptions (Immelmann, 1972), the effects of the brood on the formation of a species identity in birds are never mentioned in the literature concerning this topic. Two factors which would seem to be important in determining the role of the brood in forming a species identity would be the independence of the young at hatching (i.e., its placement on the altricial–superprecocial continuum) and the rearing practices of the parents after hatching. (That is, do the parents split up the brood and each take one of perhaps two offspring and rear them separately?) It may very well be that the influence of the brood is different in different kinds of birds. In the absence of much empirical data, we can only speculate on the possible role of the brood in the natural environment and see if it is plausible to continue to consider the brood as a possible mechanism assisting in the formation of an attachment to one's species. Immelmann (1972) seems to feel that siblings can mediate the formation of an attachment to one's species only when the object of the normal attachment, i.e., the parent, is absent. In other words, in laboratory experiments or in cross-fostering experiments in the field, the brood may be an important factor, but what its influence is under normal circumstances is at present unknown.

The precocial duckling, in its natural habitat, is usually exposed simultaneously to both its mother and its siblings. Normally the hen and her brood remain in close proximity to one another for the first few weeks of the duckling's life. Many "imprinting" experiments assume that during this early stage of development, a duckling forms an attachment to its mother (see Bateson, 1966; Hess, 1973; Sluckin, 1965; Smith, 1969 for a review of this literature). Shapiro (1972), however, has postulated that the bond uniting the family is not exclusively between each duckling and the hen but is due, in part, to a bond among the ducklings. It is also possible that in addition to the ducklings following the mother hen, the female also stays near or follows the ducklings. Indeed, Beard (1964) has reported that "the ducklings . . . usually preceded the hen, which followed slowly, very much on the alert . . ." (p 409). Similarly, Collias and Collias (1956) have stated that "broods show a strong tendency to keep together regardless of whether the hen is present" (p 498).

The effectiveness of the brood in eliciting an approach response has been investigated by me and by my students. The attractiveness of a brood of ducklings used as a stimulus object was demonstrated by Storey and Shapiro (1972), when a brood of ducklings was simultaneously presented to individual ducklings along with a live female mallard and a stuffed model of a female mallard. The ducklings were overwhelmingly preferred. Storey (1976) has further demonstrated that individual ducklings will prefer a brood of ducklings significantly more than a live nonbrooding mallard female or the maternal call of the female mallard. Using white Peking ducklings (*Anas platyrhynchos*), Kirvan and Shapiro (1972) and Rogan and Shapiro (1972) both demonstrated that a brood of nine ducklings is preferred significantly more than a brood of six ducklings or a brood of three ducklings. This preference exists in mallard ducklings (Rogan and Shapiro, 1974) and with subjects that have been reared communally (Rogan and Shapiro, 1973) or reared in isolation (Shapiro and Rogan, 1973). Additional research (Smith and Shapiro, 1975) made the brood of nine, which was formerly the largest brood, the smallest brood in a simultaneous test of a duckling's preference for a brood of nine, 12, or 15 ducklings. The larger broods were preferred significantly more than the brood of nine or an empty quadrant.

Bruce and Shapiro (1977) have more recently shown that mallard ducklings cannot differentiate their own brood from strange broods, but that they prefer larger broods significantly more than they do smaller broods, even if the smaller brood is their own brood. Shapiro and Bruce (1977), in addition, have also found that mallard ducklings can differentiate their own mother from strange mothers, but that they prefer a brood of ducklings significantly more than they do their own mother. These experiments demonstrate the importance of the brood to individual ducklings. The brood seems to be preferred more so than one's own mother, and larger broods seem to be preferred more so than smaller broods.

As the brood is growing, they are providing each other with visual, auditory, tactile, and olfactory stimuli that may be very important in helping them later form an attachment to their species. There are, in addition, probably both male and female ducklings in the brood, so that the problem associated with understanding how both male and female offspring identify with a female parent, but later mate with an appropriate member of the opposite sex, is potentially understandable.

Although superprecocial birds do not need a parent present, the presence of broodmates, at least in the vicinity, might serve to assist them in forming an attachment to their own species, as opposed to another species. Reproductive isolating mechanisms will isolate these birds from many other species and also increase the density of their own species with respect to other birds present.

Ordinarily the brood is not separated from its parent or parents. The presence of the brood along with the parents provides the potential for a very complicated series of interactions, within which a single organism could probably form a very strong attachment to one's species. The parents probably form one model for the offspring. The offspring can potentially form another model for themselves. Should it be necessary, the brood can serve as a backup model for the formation of an attachment to one's species. Interacting with each duckling's receptivity to the stimuli being provided for it by the environment is its genetic predisposition to respond to certain stimuli. In those species, for example, that are not so isolated and perhaps not cared for as well by their parents, genetic components probably play a larger role in members forming the proper attachment to their species.

Forming an attachment to one's brood may assist individual birds in forming an attachment to their species. Certain situations arise normally in the lives of many young birds that make an attachment to one's brood very important and lend more support to the hypothesis that in some birds the most important bond that a young bird can make is to its broodmates and not to its parents.

In some ducks the hen alone takes care of the ducklings. When the mother's postnuptial molt occurs she loses all of her flight feathers simultaneously and remains flightless until they are renewed. At this time the hen deserts her brood and remains in hiding until she can fly again. The abandonment of still flightless ducklings is widespread among diving ducks (Skutch, 1976). The hen's desertion occurs at widely different stages in the development of the young. In late-hatching broods the ducklings may only be two or three weeks old. It is incumbent upon the ducklings to survive on their own. The advantages associated with larger broods, mentioned above, may play a role in assisting such a brood to survive.

Abandoned ducklings sometimes congregate in large broods of different ages. Such gang broods have the advantages associated with larger broods discussed above, but they also require that the young birds have some idea of what to congregate with in order to derive the benefits associated with such behavior.

Hence the brood may facilitate the formation of an attachment to one's species in the presence of parents, or it may be sufficient to form such an attachment in the absence of the parents. Add to this a genetic predisposition to respond to certain characteristics of one's species, a long period of parent–brood interactions in some species, and an ability to learn to recognize members of one's species, and we have the basis for a strong role for the brood to play in helping individual birds to identify with their species.

Current Interpretations of the Way in which Birds Form Attachments to Their Species

As was pointed out earlier, there is no "one way" in which birds form an attachment to their species. This statement is as it should be, since the selection pressures and genetic variations associated with each bird differ from species to species. The method of forming an attachment to one's species will differ with each subtle change in selection pressure and expressed advantageous genetic mutation. This situation means that one must view the formation of an attachment to one's species as a continuum in which one can find many variations, though all have the same function. One continuum, for example, that must be examined is the continuum ordering the kinds of birds from altricial to superprecocial. The young of these birds may be ordered in terms of their increasing maturity at hatching and their decreasing dependence on their parents. These young birds will accordingly form an attachment to their species in different ways, depending upon the degree of dependence that exists between them and their parents.

Another continuum that must be examined is the genetically learned continuum. Ordered along this continuum are mechanisms which can mediate an attachment to one's species on the basis of mostly learned experiences or mostly genetic predispositions, but never exclusively one or the other. Hess and Hess (1969) suggest that in wild mallard ducklings there is a greater reliance on an "innate schema" in forming an attachment to one's parent than was previously thought.

Acting concomitantly with the above continuua is the fact that several reproductive isolating mechanisms may prevent a particular species from associating with many other species when it is forming a particular attachment. Hence, depending upon the number of other species around at the time of hatching, a young bird may be able to form an attachment to several different kinds of parents. Given normal maturation, however, the same bird may be in a very different physiological state and even in a different geographical location when it is ready to form a different kind of attachment, say to its species or to a particular mate. At this time the innate predisposition to respond only to certain courtship displays and not to others may result in an appropriate attachment being made to another of its species.

Given the preceding discussion, it does not make sense to use one term, i.e., "imprinting," to represent the many different kinds of attachments that can be formed by birds. By not favoring the use of this term, however, one does not deny any special characteristics that may be associated with attachment processes nor does one view an attachment as a fixed entity that can only occur once in an organism's life. The same type of attachment may be formed several times within a developmental stage or different forms of attachment may occur at different stages of one's existence.

Just as there are several reasons why use of the term "imprinting" is restricted, there are also several reasons why one does not rely exclusively on the following response as an indication that an attachment has been formed. The primary reason is that one does not feel that one knows what the appropriate cues are that will elicit the subsequent response. If there are innate predispositions to react to certain stimuli associated with a mallard female, and an investigator uses a mallard decoy in his experiment, then he will have an extraneous variable present which will confound his results.

Given the complexity of the processes involved, the number of species that can form attachments, and the kinds of attachments that can be formed, investigators would be well advised to refrain from making sweeping generalizations about the formation of an attachment to one's species in birds. Many interpretations can be made for similar data in different birds. To conclude that they are all the same would be erroneous.

Conclusions

In Chapter 1 of this volume some statements were made that require a response with respect to the formation of an attachment to one's species in birds. Those statements are intended to apply to a wider phylogenetic range than do the statements in this chapter. It is the narrower perspective involved here, however, that prompts these concluding remarks.

A central statement in Chapter 1 is that an innate mechanism for "knowing" what species one belongs to does not exist in many species and that early experiences are crucial for determining each organism's species identity. In response to this statement one must point out that no simple statement is available to cover adequately the range of birds to which the statement can apply. The continuum ordering birds from altricial to superprecocial provides examples that can both support the statement and refute it at the same time. Superprecocial birds are not exposed to any model that can provide them with "early experiences" crucial to forming an attachment to their species. These birds must have a very well developed, innate response to stimulus objects that will later represent a potential mate for them. These birds apparently "know" what species they belong to. Some altricial birds, on the other hand, do not seem to have the same knowledge and must be cared for and fed by their parents. These birds do not seem to "know" who their parents are.

Some precocial birds seem to be able to form an attachment to any adult, indeed any object, that can confer upon them those necessities required to survive. Later they seem to be able to mate appropriately with a member of their own species. Collectors of ornamental precocial waterfowl routinely hatch these birds under chickens. The hatched waterfowl have no apparent problems later on when they mate with members of their own species. Parasitic egg layers leave their eggs in the nest of another species. The hatched young later mate successfully, even though they had the wrong model on which to form an attachment to their species.

Unpublished research from my own laboratory (Hoag, 1977; Sallows, 1977; Szkraba, 1978) indicates that if naive chicks or ducklings are given a choice of approaching and staying near a small group of their own species, six parakeets, or a large group of another avian species, they will always prefer the smaller group of their own species. The response is almost immediate, lasts for at least the first seven days of their lives, and occurs regardless of whether the subjects have been reared in visual isolation, visual and auditory isolation, or communally.

Hence with respect to the statement referred to above and as applied to birds, it is suggested that the reader take a more liberal approach to the suggestion in Chapter 1 and examine each species individually with respect to the several continuua mentioned earlier.

Two additional statements are made in Chapter 1 (numbers two and three) which imply (1) that learned attachments to one's species can have innate components and (2) that innate predispositions to respond to one's species do not ensure that a proper attachment will be formed to one's species. The latter statement seems to imply that some learning takes place to make the attachment complete. Both statements fit in nicely with the learned–genetic continuum mentioned earlier. In the megapodes, genetic components probably play a far greater role in forming an attachment to one's species than they do in altricial birds. In no case, however, is the formation of the attachment purely innate or purely learned.

Statement number 4 made in Chapter 1 also agrees with the content of this chapter. It states that not all species use the same mechanisms to form an attachment to their species. As I have previously stated, different mechanisms can mediate different forms of attachment in the same bird, and different mechanisms can also mediate the same attachments in different species.

I also agree with statement number 5 in Chapter 1, which says, in essence, that different forms of attachment can be formed at different stages of development. Similarly, statement number 6 in Chapter 1, that an attachment does not have to be exclusively with one species, raises no objections either. Ordinarily, however, just the reverse would be true, as pointed out in Chapter 1. If the exception to the rule raises the probability of the organism's survival, however, then the statement is more correct than incorrect. Certainly, many of the "imprinting" experiments using inanimate, biologically inappropriate stimulus objects would seem to support this statement.

Finally, two more statements in Chapter 1 seem fitting to conclude this chapter. First, one should not assume that both sexes form attachments in the same manner, at the same time, or for the same duration of time. Second, one must realize that much of the knowledge about birds has come from work done on a very small proportion of the total number of species of birds and was done during a small time sample of the evolutionary process at that. Hence, to generalize the information too widely would be inappropriate.

Acknowledgment

This chapter is dedicated to the memory of my father, Herman Bernard Shapiro, a courageous man who had many qualities that made it easy for me to identify with the human species.

References

Alcock, J. 1975. *Animal behavior: An evolutionary approach.* Sunderland, Mass.: Sinauer Assoc.

Baltin, S. 1969. Zur Biologie und Ethologie des Talegalla-Huhns (*Alectura lathami* Gray) unter besonderer Berücksichtigung des Verhaltens während der Brutperiode. *Z. Tierpsychol. 26,* 524–572. Cited in R. W. Oppenheim. 1972. Prehatching and hatching behaviour in birds: A comparative study of altricial and precocial species. *Anim. Behav. 20,* 644–655.

Bambridge, R. 1962. Early experience and sexual behaviour in the domestic chicken. *Science 136,* 259–260.

Bateson, P. P. G. 1966. The characteristics and context of imprinting. *Biol. Rev. 41,* 177–220.

Beach, F. A. 1955. The descent of instinct. *Psychol. Rev. 62,* 401–410.

Beard, E. B. 1964. Duck brood behavior at the Seney National Wildlife Refuge. *J. Wildl. Manage. 28,* 497–521.

Beer, C. G. 1969. Laughing gull chicks: Recognition of their parents' voices. *Science 166,* 1030–1032.

Beer, C. G. 1970. Individual recognition of voice in the social behavior of birds. In D. S. Lehrman, R. A. Hinde, and E. Shaw (Eds.), *Advances in the study of behavior.* Vol. 3. New York: Academic Pr.

Brosset, A. 1971. L'"imprinting," chez les Colombidés—étude des modifications comportementales aucours de vieillisement. *Z. Tierpsychol. 29,* 279–300.

Brown, J. L. 1975. *The evolution of behavior.* New York: Norton.

Bruce, M. and L. J. Shapiro. 1977. *Brood recognition in ducklings.* Paper presented at the meeting of the Animal Behavior Society, University Park, Pennsylvania, June.

Burghardt, G. M. 1973. Instinct and innate behavior: Toward an ethological psychology. In J. A. Neven (Ed.), *The study of behavior: Learning, motivation, emotion, and instinct.* Glenview, Ill.: Scott, Foresman.

Burghardt, G. M. and E. H. Hess. 1966. Food imprinting in the snapping turtle. *Chelydra serpentina. Science 151,* 108–109.

Carter, C. S. and J. N. Marr. 1970. Olfactory imprinting and age variables in the guinea pig, *Cavia porcellus. Anim. Behav. 18,* 238–244.

Collias, N. E. and E. C. Collias. 1956. Some mechanisms of family integration in ducks. *Auk 73,* 378–400.

Collins, T. B., Jr. 1965. Strength of the following response in the chick in relation to degree of "parent" contact. *J. Comp. Physiol. Psychol. 60,* 192–195.

Dimond, S. J. 1968. Effects of photic stimulation before hatching on the development of fear in chicks. *J. Comp. Physiol. Psychol. 65,* 320–324.

Dobzhansky, T. 1937. *Genetics and the origin of species.* New York: Columbia Univ. Pr.

Dobzhansky, T. 1970. *Genetics of the evolutionary processes.* New York: Columbia Univ. Pr.

Dorst, J. 1971. *The life of birds.* Vol. 1. London, England: Weidenfeld and Nicolson.

Driver, P. M. 1960. A possible fundamental in the behavior of young nidifugous birds. *Nature 186,* 416.

Driver, P. M. 1962. Parent/young relationships in the ontogeny of ducklings. (Abstr.) *Anim. Behav. 10,* 388–389.

Evans, R. M. 1970. Parental recognition and the "mew call" in black-billed gulls (*Larus bulleri*). *Auk 87,* 503–513.(a)

Evans, R. M. 1970. Imprinting and mobility in young ring-billed gulls, *Larus delawarensis. Anim. Behav. Monogr. 3*(3), 195–248.(b)

Fabricius, E. and O. Fält. 1969. Sexuell prägling hos grässandhonor. *Zool. Revy 31,* 83–88.

Fantz, R. L. 1957. Form preferences in newly hatched chicks. *J. Comp. Physiol. Psychol. 50,* 422–430.

Fantz, R. L. 1958. Depth discrimination in dark-hatched chicks. *Percep. Mot. Skills 8,* 47–50.

Fischer, G. J. 1970. Arousal and impairment: Temperature effects on following during imprinting. *J. Comp. Physiol. Psychol. 73,* 412–420.

Fischer, G. J. 1976. Chick (*Gallus domesticus*) approach preferences for natural and artificial sound stimuli. *Dev. Psychol. 12,* 39–46.

Frank, L. H. and M. E. Meyer. 1970. Food imprinting in domestic chicks as a function of social contact and number of companions. *Psychonomic Sci. 19*, 293–295.

Freeman, B. M. and M. A. Vince. 1974. *Development of the avian embryo*. London: Chapman and Hall.

Frith, H. J. 1962. *The Mallee-Fowl: The bird that builds an incubator*. Sydney, Australia: Angus and Robertson.

Gallagher, J. 1976. Sexual imprinting: Effects of various regimens of social experience on mate preference in Japanese quail. *Coturnix coturnix japonica. Behaviour 57*, 91–115.

Gold, P. 1969. Effects of prehatching stimulation on growth and behavior in a domestic chick. (Abstr.) *Am. Zool. 9*, 1074.

Goodwin, E. B. and E. H. Hess. 1969. Innate visual form preferences in the pecking behavior of young chicks. *Behaviour 34*, 223–237.

Gottlieb, G. 1965. Prenatal auditory sensitivity in chickens and ducks. *Science 147*, 1596–1598.

Gottlieb, G. 1968. Prenatal behavior of birds. *Q. Rev. Biol. 43*, 148–174.

Gottlieb, G. 1971. *Development of species identification in birds: An inquiry into the prenatal determinants of perception*. Chicago: Univ. Chicago Pr.

Gottlieb, G. and Z.-Y. Kuo. 1965. Development of behavior in the duck embryo. *J. Comp. Physiol. Psychol. 59*, 183–188.

Gottlieb, G. and J. G. Vandenbergh. 1968. Ontogeny of vocalization in duck and chick embryos. *J. Exp. Zool. 168*, 307–326.

Grier, J. B., S. A. Counter, and W. M. Shearer. 1967. Prenatal auditory imprinting in chickens. *Science 155*, 1692–1693.

Guiton, P. 1962. The development of sexual responses in the domestic fowl in relation to the concept of imprinting. *Symp. Zool. Soc. London 8*, 227–234.

Harth, M. S. and M. B. Heaton. 1973. Nonvisual photic responsiveness in newly hatched pigeons (*Columba livia*). *Science 180*, 753–755.

Heaton, M. B. 1971. Ontogeny of vision in the peking duck (*Anas platyrhynchos*): The pupillary light reflex as a means for investigating visual onset and development in avian embryos. *Dev. Psychobiol. 4*, 313–332.

Heaton, M. B. 1972. Prenatal auditory discrimination in the wood duck (*Aix sponsa*). *Anim. Behav. 20*, 421–424.

Heaton, M. B. 1973. Early visual function in bobwhite and Japanese quail embryos as reflected by pupillary reflex. *J. Comp. Physiol. Psychol. 84*, 134–139.

Heaton, M. B. 1976. Developing visual function in the red jungle fowl embryo. *J. Comp. Physiol. Psychol. 90*, 53–56.

Hess, E. H. 1956. Natural preferences of chicks and ducklings for objects of different colors. *Psychol. Rep. 2*, 477–483.

Hess, E. H. 1962. Imprinting and the critical period concept. In E. L. Bliss (Ed.), *Roots of behavior*. New York: Hoeber.

Hess, E. H. 1964. Imprinting in birds. *Science 146,* 1128–1139.

Hess, E. H. 1972. "Imprinting" in a natural laboratory. *Sci. Am.* August, 24–31.

Hess, E. H. 1973. *Imprinting: Early experience and the developmental psychobiology of attachment.* New York: Van Nostrand.

Hess, E. H. and W. C. Gogel. 1954. Natural preferences of the chick for objects of different colors. *J. Psychol. 38,* 483–493.

Hess, E. H. and D. B. Hess. 1969. Innate factors in imprinting. *Psychonomic Sci. 14,* 129–130.

Hoag, P. J. 1977. *Effects of species on brood size: 3 ducklings, 6 parakeets, 9 chicks.* Unpublished manuscript. (Available from L. J. Shapiro, Avian Behaviour Laboratory, Department of Psychology, University of Manitoba, Winnipeg, Manitoba, Canada, R3T 2N2.)

Immelmann, K. 1969. Über den Einfluss frühkindlicher Erfahrungen auf die geschlechtliche Objektfixierung bei Estrildiden. *Z. Tierpsychol. 26,* 677–691.

Immelmann, K. 1970. *The ecological significance of isolating mechanisms based on imprinting.* Sonderdruck aus Verhandlungsbericht der Deutschen Zoologischen Gesellschaft, 64. Tagung.

Immelmann, K. 1972. Sexual and other long-term aspects of imprinting in birds and other species. *Adv. Study Behav. 4,* 147–174.

Immelmann, K. 1975. Ecological significance of imprinting and early learning. *Ann. Rev. Ecol. Syst. 6,* 15–37.

Impekoven, M. 1973. Response-contingent prenatal experience of maternal calls in the peking duck (*Anas platyrhynchos*). *Anim. Behav. 21,* 164–168.

Impekoven, M. 1976. Responses of laughing gull chicks (*Larus atricilla*) to parental attraction- and alarm-calls, and effects of prenatal auditory experience on the responsiveness to such calls. *Behaviour 56,* 250–278.

Jensen, D. D. 1961. Operationism and the question "Is this behaviour learned or innate?" *Behaviour 17,* 1–8.

Kaufman, L. 1973. *Environmental imprinting in chicks reared on the visual cliff.* Paper presented at the meeting of the Western Psychological Association, Anaheim, Cal. April.

Kaufmann, I. C. and R. A. Hinde. 1961. Factors influencing distress calling, with special reference to temperature change and social isolation. *Anim. Behav. 9,* 197–204.

Kirvan, B. A. and L. J. Shapiro. 1972. *The effect of brood size in eliciting an approach response in white peking ducklings.* Paper presented at the meeting of the Southern Society for Philosophy and Psychology, St. Louis, Mo., March.

Klint, T. 1973. Praktdräkten som "sexuell utlösare" hos gräsand. *Zool. Revy 35,* 11–21.

Klopfer, P. H., D. K. Adams, and M. S. Klopfer. 1964. Maternal "imprinting" in goats. *Proc. Nat. Acad. Sci. 52,* 911–914.

Klopfer, P. H. and J. P. Hailman. 1964. Basic parameters of following and imprinting in precocial birds. *Z. Tierpsychol. 21,* 755–762.

Kuo, Z. Y. 1932. Ontogeny of embryonic behavior in aves. IV. The influence of embryonic movements upon the behaviour after hatching. *J. Comp. Psychol. 14,* 109–121.

Lamprecht, J. 1977. Environment-dependent attachment behaviour of goslings (*Anser indicus*) due to environment-specific separation experience. *Z. Tierpsychol. 43,* 407–414.

Lauber, J. K. and J. V. Shutze. 1964. Accelerated growth of embryo chicks under the influence of light. *Growth 28,* 179–190.

Lehrman, D. S. 1953. A critique of Konrad Lorenz' theory of instinctive behavior. *Q. Rev. Biol. 28,* 337–363.

Lehrman, D. S. 1970. Semantic and conceptual issues in the nature–nurture problem. In L. R. Aronson, E. Tobach, D. S. Lehrman, and J. S. Rosenblatt (Eds.), *Development and evolution of behavior: Essays in memory of T. C. Schneirla.* San Francisco: Freeman.

Littlejohn, M. J. 1969. The systematic significance of isolating mechanisms. In International Conference on Systematic Biology, *Systematic biology.* Washington, D.C.: Nat. Acad. Sci. Publ. *1692.*

Lorenz, K. 1957. Companionship in bird life. Fellow members of the species as releasers of social behavior. In C. H. Schiller (Ed.), *Instinctive behavior.* New York: International Univ. Pr.

Marley, E. and W. H. Morse. 1966. Operant conditioning in the newly hatched chicken. *J. Exp. Anal. Behav. 9,* 95–103.

Mayr, E. 1963. *Animal species and evolution.* Cambridge, Mass.: Harvard Univ. Pr.

McBride, G., I. P. Parer, and F. Foenander. 1969. The social organization and behaviour of the feral domestic fowl. *Anim. Behav. Monogr. 2*(3), 127–181.

McCoshen, J. A. and R. P. Thompson. 1968. A study of clicking and its source in some avian species. *Can. J. Zool. 46,* 169–172.

Meyer, M. E. and L. H. Frank. 1970. Food imprinting in the domestic chick: A reconsideration. *Psychonomic Sci. 19,* 43–45.

Mulder, J. B. 1975. *Possible effects of imprinting on bedding preferences of rats and mice.* Paper presented at the meeting of the Animal Behavior Society, Wilmington, N.C., May.

Müller-Schwarze, D. and C. Müller-Schwarze. 1971. Olfactory imprinting in a precocial mammal. *Nature 229,* 55–56.

Nice, M. M. 1962. Development of behavior in precocial birds. *Trans. Linn. Soc. N.Y. 8,* 1–211.

Pettingill, O. S., Jr. 1970. *Ornithology in laboratory and field.* 4th ed. Minneapolis: Burgess.

Rajecki, D. W. 1974. Effects of prenatal exposure to auditory and visual stimuli on social responses in chickens. *Behav. Biol. 11,* 525–536.

Ramsay, A. O. 1951. Familial recognition in domestic birds. *Auk 68,* 1–16.

Rogan, J. C. and L. J. Shapiro. 1972. *The effects of brood size on eliciting an approach response in white peking ducklings (A replication).* Paper presented at the meeting of the American Association for the Advancement of Science, Washington, D.C., December.

Rogan, J. C. and L. J. Shapiro. 1973. *The effects of brood size on eliciting an approach response in white peking ducklings reared communally.* Unpublished manuscript. (Available from L. J. Shapiro, Avian Behaviour Laboratory, Department of Psychology, University of Manitoba, Winnipeg, Manitoba, R3T 2N2.)

Rogan, J. C. and L. J. Shapiro. 1974. *The effect of brood size on eliciting an approach response in mallard ducklings.* Paper presented at the meeting of the Canadian Psychological Association, Windsor, Ontario, June.

Roy, M. A. 1979. An introduction to the concept of species identification. In M. A. Roy (Ed.), *Species identity and attachment: A phylogenetic evaluation.* New York: Garland.

Rutgers, A. and K. A. Norris (Eds.). 1970. *Encyclopaedia of aviculture.* Vol. 1. London: Blanford Pr.

Sallows, T. E. 1977. *Effects of species on brood size preference when subjects are raised in visual isolation.* Unpublished manuscript. (Available from L. J. Shapiro, Avian Behaviour Laboratory, Department of Psychology, University of Manitoba, Winnipeg, Manitoba, R3T 2N2.)

Salzen, E. A. and J. M. Cornell. 1968. Self-perception and species recognition in birds. *Behaviour 30,* 44–65.

Salzen, E. A. and F. J. Tomlin. 1963. The effect of cold on the following response of domestic fowl. *Anim. Behav. 11,* 62–65.

Schein, M. W. and E. B. Hale. 1959. The effect of early social experience on male sexual behaviour of androgen injected turkeys. *Anim. Behav. 7,* 189–200.

Schneirla, T. C. 1956. Interrelationships of the "innate" and the "acquired" in instinctive behavior. In P. P. Grassé (Ed.), *L'Instinct dans le comportement des animaux et de l'homme.* Paris: Masson.

Schutz, F. 1965. Sexuelle Prägung bei Anatiden. *Z. Tierpsychol. 22,* 50–103.

Schutz, F. 1971. Imprinting of sexual behavior of ducks and geese through social experience during their juvenile period. *Journal of Neuro-Visceral Relations, 10,* 339–357.

Shapiro, L. J. 1971. *The effects of rearing conditions on reactions to maternal calls in white peking ducklings.* Paper presented at the meeting of the Animal Behaviour Society, Logan, Utah, June.

Shapiro, L. J. 1972. *Imprinting: Another look at the Cheshire cat.* Paper presented at the meeting of the Southern Society for Philosophy and Psychology, St. Louis, Mo., March.

Shapiro, L. J. 1977. Developing preferences for live female models of the same or other species in white peking ducklings. *Anim. Behav. 25,* 849–858.

Shapiro, L. J. and M. Bruce. 1977. *The use of biologically appropriate models to study avian attachment behavior.* Paper presented at the meeting of the American Ornithologists' Union, Berkeley, Cal., August.

Shapiro, L. J. and J. C. Rogan. 1973. *The effects of repeated testing on evaluating the development of a preference in white peking ducklings.* Paper presented at the meeting of the Canadian Psychological Association, Victoria, B. C., June.

Shutze, J. V., J. K. Lauber, M. Kato, and W. Wilson. 1962. Influence of incandescent and colored light on chick embryos during incubation. *Nature 96,* 594–595.

Siegel, P. B., S. T. Isakson, F. N. Coleman, and B. J. Huffman. 1969. Photoacceleration of development in chick embryos. *Comp. Biochem. Physiol. 28,* 753–758.

Skutch, A. F. 1976. *Parent birds and their young.* Austin, Tex.: Univ. Texas Pr.

Sluckin, W. 1965. *Imprinting and early learning.* Chicago: Aldine.

Sluckin, W. 1970. *Early learning in man and animal.* London: Allen and Unwin.

Sluckin, W., K. F. Taylor, and A. Taylor. 1966. Approach of domestic chicks to stationary objects of different texture. *Percep. Mot. Skills 22,* 699–702.

Smith, F. V. 1969. *Attachment of the young: Imprinting and other developments.* Edinburgh: Oliver and Boyd.

Smith, F. V. and M. W. Bird. 1963. Varying effectiveness of distant intermittent stimuli for the approach response in the domestic chick. *Anim. Behav. 11,* 57–61. (a)

Smith, F. V. and M. W. Bird. 1963. The relative attraction for the domestic chick of combinations of stimuli in different sensory modalities. *Anim. Behav. 11,* 300–305. (b)

Smith, W. and L. J. Shapiro. 1975. *The effects of brood size (15, 12, 9) on eliciting approach responses in white peking ducklings.* Unpublished manuscript. (Available from L. J. Shapiro, Avian Behaviour Laboratory, Department of Psychology, University of Manitoba, Winnipeg, Manitoba, R3T 2N2.)

Stevenson, J. G., R. E. Hutchinson, J. B. Hutchinson, B. C. R. Bertram, and W. H. Thorpe. 1970. Individual recognition by auditory cues in the common tern (*Sterna hirundo*). *Nature 226,* 562–563.

Storey, A. E. 1976. *The effects of live or inanimate models on the development of preferences in white peking ducklings.* Unpublished Master's thesis. Univ. Manitoba.

Storey, A. E. and L. J. Shapiro. 1972. *Preference of white peking ducklings for live or inanimate models.* Paper presented at the meeting of the Southern Society for Philosophy and Psychology, St. Louis, Mo. March.

Szkraba, P. 1978. *Effects of species on brood size preference when subjects are raised in visual and auditory isolation.* Unpublished manuscript. (Available from L. J. Shapiro, Avian Behaviour Laboratory, Department of Psychology, University of Manitoba, Winnipeg, Manitoba, R3T 2N2.)

Taylor, A., W. Sluckin, R. Hewitt, and P. Guiton. 1967. The formation of attachments by domestic chicks to two textures. *Anim. Behav. 15,* 514–519.

Taylor, K. F. and W. Sluckin. 1964. An experiment in tactile imprinting. *Bull. Br. Psychol. Soc. 17,* 10A.

Thorpe, W. H. 1944. Type of learning in insects and other arthopods. Part III. *Br. J. Psychol. 34,* 66–76.

Thorpe, W. H. 1945. The evolutionary significance of habitat selection. *J. Anim. Ecol. 14,* 67–70.

Tolhurst, B. E. and M. A. Vince. 1976. Sensitivity to odours in the embryo of the domestic fowl. *Anim. Behav. 24,* 772–779.

Tschanz, B. 1968. Trottellummen. Die Entstehung der persönlichen Beziehungen zwischen Jungvogel und Eltern. *Z. Tierpsychol. 4,* 51–100.

Vince, M. 1964. Social facilitation of hatching in the bobwhite quail. *Anim. Behav. 12,* 531–534.

Vince, M. 1966. Artificial acceleration of hatching quail embryos. *Anim. Behav. 14,* 389–394.

Vince, M. A. 1977. Taste sensitivity in the embryo of the domestic fowl. *Anim. Behav. 25,* 797–805.

Vince, M. A. and R. Cheng. 1970. The retardation of hatching in Japanese quail. *Anim. Behav. 18,* 210–214.

Vince, M. A. and S. Chinn. 1971. Effect of accelerated hatching on the initiation of standing and walking in the Japanese quail. *Animal Behaviour 19,* 62–66.

Vince, M. A. and S. Chinn. 1972. Effects of external stimulation on the domestic chick's capacity to stand and walk. *Br. J. Psychol. 63,* 89–99.

Walter, M. J. 1973. Effects of parental colouration on the mate preference of offspring in the zebra finch, *Taeniopygia guttata castanotis* Gould. *Behaviour 46,* 154–173.

Warriner, C. C., W. B. Lemmon, and T. S. Ray. 1963. Early experience as a variable in mate selection. *Anim. Behav. 11,* 221–224.

Weller, M. W. 1968. The breeding biology of the parasitic black-headed duck. *Living Bird, 7,* 169–207.

White, S. J. 1971. Selective responsiveness by the gannet (*Sula bassana*) to played-back calls. *Anim. Behav. 19,* 125–131.

Woolf, N. K., J. L. Bixby, and R. R. Capranica. 1976. Prenatal experience and avian development: Brief auditory stimulation accelerates the hatching of Japanese quail. *Science 194,* 959–960.

Zolman, J. F. 1968. Discrimination learning in young chicks with heat reinforcement. *Psychol. Rec. 18,* 303–309.

Chapter 5

Birds: The Role of Mate Choice

Fred Cooke

The theme of this book is species identity, a phrase which refers to the various mechanisms whereby individuals of a species associate, mate, and in other ways interact with members of their own species. This complex topic can perhaps be simplified somewhat by concentrating on those interactions which lead to the leaving of offspring.

Natural selection favors those organisms that can accurately assess the potential fecundity of a mate. This is probably one of the most powerful selective forces in developing a sense of species identity. A great blue heron who mates with a chickadee is likely to leave fewer offspring than one who mates with another great blue heron. Mating across species barriers is maladaptive in that offspring production is reduced or nonexistent. It is, however, probably incorrect to think of the distinction between intraspecific mating and interspecific mating as a clear-cut one. In reality the limits of a species are often blurred, and within species there will be varying degrees of suitability of a mate. Bateson (1977) pointed out that sexual imprinting may act not as a mechanism for developing a sense of species identity, but more for allowing an optimal mate to be selected. Thus mate suitability should be viewed as a continuum associated with the degree of similarity between the individual and its potential mate. At one extreme two unrelated species have little genetic similarity, and no offspring production would result from a mating attempt. Less extremely, two sibling species may be closer genetically, and a few fertile offspring may result. Even closer genetically would be different strains of a species; in this case fertile offspring would be expected, and there may even be heterosis. If the genetic similarity is too close there may be inbreeding depression. Thus one might expect on theoretical grounds that to choose a mate too genetically different *or* too genetically similar will be maladaptive.

Species Identity and Mate Choice

The basis of choice may be innate or it may be learned during the premating stage of life. The extent to which learning plays a role in mate selection probably varies from species to species in birds, as evidenced in the variability of results from experiments on the role of prepairing experience in mate selection. Many interspecific cross-fostering experiments have been carried out between pairs of species of birds. These include, among others, experiments with ducks and geese (Schutz, 1965, 1970), with house and tree sparrows (*Passer domesticus* and *P. montanus*) (Cheke, 1969), with herring and lesser black-backed gulls (*Larus argentatus* and *L. fuscus*) (Harris, 1970), and with zebra and Bengalese finches (*Taeniopygia guttata* and *Lonchura striata*) (Immelmann, 1969). In most of these cases cross-fostered birds showed strong sexual attachment to members of the foster-parent species, and when given a choice between a member of their own species and that of their foster parents, preferred to mate with the foster species. The cases of the sparrows and gulls are interesting in that they occurred under natural conditions.

In contrast, some birds chose a mate of their own species despite a lack of previous intraspecific experience. Schutz (1965) found this to be the case among females of several species of water fowl, and Klint (1975), as a result of experiments with captive female mallards (*Anas platyrhynchos*), concluded that they probably recognize the males on an innate basis. Similar conclusions were drawn by Hess and Hess (1969) for wild mallards and for domestic fowl by Lill and Wood-Gush (1965).

Parasitic birds such as cuckoos, cowbirds, and viduine finches are raised with members of a host species and yet later in life associate and mate with members of their own species. King and West (1977) report an auditory mechanism whereby female brown-headed cowbirds (*Molothrus ater*) respond to members of their own species even when raised in visual and auditory isolation from conspecifics.

It seems likely that in most birds prepairing experience augments the process of choosing a mate of the same species. When early experience is manipulated to yield an environment dissimilar to that likely to be encountered in nature, in many cases the process of choosing a suitable mate is disrupted. Excellent summaries of this phenomenon have been written by Immelmann (1972, 1975).

In addition to interspecific pair formation, the subject of species identity has been investigated extensively through the use of intraspecific variation. Birds showing conspicuous plumage polymorphism either naturally or induced by selective breeding have been used to assess the effect of familial plumage color on mate selection. As with interspecific studies, most work has been carried out with pigeons (Goodwin 1958; Warriner, Lemmon, and Ray 1963; Kerfoot, 1964), with waterfowl (Schutz, 1965; Klint, 1975; Cooke, Finney, and Rockwell, 1976) and with estrildine finches (Immelmann, 1969; Walter, 1973; Immelmann, Kalberlah, Rausch, and Stahnke, 1977). In most of these cases birds, when raised with a particular plumaged family, would choose a mate of that plumage when given a choice of mates. In some cases (e.g., zebra finches) parental color seems to be the major familial influence; in others (e.g., bullfinches; Nicolai, 1956) the role of the siblings may be of major importance, and in yet others (e.g., snow geese) both parental and sibling color were shown to be important.

The importance of the intraspecific and interspecific studies to the topic of species identity is in showing that birds use familial plumage color as a means of recognizing members of their own species and, even within a species, of recognizing familial and nonfamilial plumage. From these examples one can conclude that the young birds have a genetic mechanism which allows them to respond to familial color and to use this cue as a basis for later mate selection.

Lorenz (1935) was the first to point out that some element of generalization is necessary when he referred to the development of supra-individual species-specific cues. The young bird must be able to generalize from the particular appearance of members of their own family to the abstract species-specific characters. It is clear from intraspecific studies that the generalization does not necessarily extend to the boundary of the species, and if the species has plumage polymorphism and if individuals have a preference for a mate with the same plumage as their parents, one might expect nonrandom mating of the morphs, with like morphs pairing preferentially with one another. The evolutionary consequences of this have been investigated theoretically by Mainardi (1964), and more recently with computer simulation by Kalmus and Maynard Smith (1966), and Seiger (1967). They showed that in a species with distinct plumage dimorphism where the dimorphism was genetically based and the two morphs differed by a single allele, if offspring chose a mate of the same plumage color as one of the parents, then after a few generations matings between the morphs would no longer occur, and the two morphs would be reproductively isolated from one another. The sense of "species identity" in an organism will, by defining for it those individuals that are suitable as mates, delineate the gene pool which comprises the population or species to which that organism belongs.

For these reasons it is important to investigate the subject of species identity through intraspecific variation. It must be stressed that the models of Kalmus and Maynard Smith (1966) and Seiger (1967) were theoretical, and no example of the phenomenon in nature was educed by the authors.

Mate Choice in Lesser Snow Geese

Since 1968 we have been investigating the topic of "species identity" in lesser snow geese (*Anser caerulescens caerulescens*), both in the field and under experimental conditions. The basic questions asked were (1) Do snow geese use familial plumage color as a guide in choosing their mates? If so, this might suggest that they are generalizing from family attributes to "morph" rather than species characteristics. (2) If the first question is answered affirmatively, what are the population and evolutionary consequences of the mate selection? (3) If a bird chooses a mate with a plumage color different from that of its family, are there any penalties in terms of reduced reproductive fitness in so doing?

The lesser snow goose is an ideal study organism in many ways. The species nests in the Canadian Arctic and in the northern USSR. In the Hudson Bay and Foxe Basin colonies both color morphs, the blue (the "blue" goose) and the white (the snow goose), are common. Approximately 14 such colonies occur in the Canadian Arctic. The large eastern colonies on Baffin Island are predominantly (70–80 percent) blue phase. The colonies on the west coast of Hudson Bay are 25–30 percent blue phase. The southernmost colony at Cape Henrietta Maria is around 55 percent blue phase. At the La Pérouse Bay colony in northern Manitoba where the author works, 25–30 percent of the birds are blue phase.

The genetics of the polymorphism is known and is reasonably simple (Cooke and Mirsky, 1972). Blue is dominant over white. Heterozygotes which are blue in phase often have some white plumage on their bellies.

Pair bonds are usually lifelong, and family groups stay together for almost a year. Such bonds are formed even under captive conditions. Pairing is not random in terms of color. In the La Pérouse Bay colony approximately 14–18 percent of the pairs were mixed, one blue and one white, whereas if mating were random, 38–42 percent should be mixed.

An analysis of the pair bonds at Boas River by Cooke and Cooch (1968) showed that when heterozygous and homozygous blue birds were compared, the heterozygotes had chosen white mates much more frequently than had the homozygotes. Because of the inheritance of color in these birds, homozygous blue birds must have had two blue parents whereas a heterozygous blue bird could have had one white parent. It was postulated, therefore, that the assortative mating observed in the snow geese could be explained on the basis of a bird choosing a mate with a plumage color similar to that of one of its parents. Thus offspring from white-phase parents would choose white mates, offspring from blue-phase parents blue mates, and offspring from mixed parentage would choose either color (or if one of the parents had a greater effect than the other, its color may have more influence on offspring choice).

While this hypothesis fitted the facts known about assortative mating in snow geese, there was no direct proof. Unless the birds could be shown to be making a choice under natural field conditions, the relevance of the study to the question of species identity would remain suspect. Some questions, however, could only be answered by manipulation of families under captive conditions.

For an understanding of the field approach it is necessary to know something of the breeding biology of the species. At the La Pérouse Bay colony, there is a breeding colony of some 3000 pairs. Many of these birds are marked with colored leg bands for individual recognition. Birds arrive at the colony in mid-May. Breeding birds are already paired and commence nest initiation soon after arrival. Hatching which occurs in late June is fairly synchronous, and goslings remain in the nest for about 24 hours. The parents and goslings then leave for the nearby salt marshes, which are the post-hatch brood-rearing areas. The goslings grow rapidly, and the adults undergo a molt of remiges and rectrices at this time. The flightless period lasts approximately six weeks, and adults and young are able to fly by early August. The colony is abandoned in late August or early September when the fall migration begins. During their migration to the Gulf Coast and on the wintering grounds they mingle with birds from the other Hudson Bay and Foxe Basin colonies. The geese appear to be gregarious throughout the year.

Birds have delayed breeding, commencing from the second to the fourth summer (Finney and Cooke, 1978). Most birds pair for the first time in their second winter or spring. Pairing takes place when birds from many different breeding colonies are intermingled. Thus when birds pair, the male and female are usually from different colonies. The pair returns to the natal colony of the female (Cooke, MacInnes, and Prevett, 1975).

Young birds usually stay with their parents for almost a year until their parents breed again (Prevett, 1972). During the incubation period the yearlings and other nonbreeding birds form flocks at the periphery of the breeding colony (Cooch, 1958). At the La Pérouse Bay colony there is a molt migration of the nonbreeding birds (Abraham, in preparation). Whether this occurs at all colonies is not established.

Pairs remain together until the death of one partner (Cooke and Sulzbach, 1978). Re-pairing is a frequent phenomenon.

Field Approach

The aim of the field approach was to find out the mate choice of a bird in relation to the family in which it had been raised. In order to do this, several conditions are necessary.

1. The nest must be visited while the goslings are hatching and the parental and gosling plumage colors must be ascertained: At this time a small metal web tag is attached, coded to indicate family of origin. Each year 3000–4000 goslings are so tagged.
2. The family must be located four to five weeks later during banding, when the goslings have plastic color-coded leg bands attached; about 500–800 of the goslings caught at the time of

banding have web tags which allow one to identify them as to the nest from which they hatched and color of parents.

3. The geese must then survive to breeding age, usually at two or three years of age.
4. They must return to and then be sighted nesting at the breeding colony at which we work.

Since 1973 birds of known parentage which inhabited La Pérouse Bay have been returning and breeding at the colony. There have now been more than 300 sightings. These findings are summarized in Table 1, which tabulates mate choice in birds in relation to the plumage color of their parents. Because of the tendency of females, but not males, to return to the natal colony, almost all the data in the table refer to females.

One can see from the table that birds with white parents tend to choose white mates, and birds with blue parents, blue mates. Offspring of mixed pairs choose either blue or white at roughly the proportion in which the two color phases occur in the areas where pair formation occurs. In all years the null hypothesis, that parental color was unrelated to mate-choice color, was rejected.

Table 1. Mate Selection of Geese of Known Parentage, La Pérouse Bay, 1972–1977

		Parental Color					
Year	*Offspring Choice*	*White Pair*	*Mixed Pair*	*Blue Pair*	*Total*	$\chi^{2\,a}_2$	*Probability p*
1972	White	2	0	0	2		
	Blue	1	0	2	3		
1973	White	23	5	2	30		
	Blue	4	1	4	9	7.6	<0.025
1974	White	29	8	2	39		
	Blue	2	6	11	19	26.3	<0.0001
1975	White	43	4	1	48		
	Blue	8	8	9	25	30.7	<0.0001
1976	White	36	8	0	44		
	Blue	3	5	12	20	36.8	<0.0001
1977	White	58	12	4	74		
	Blue	4	7	19	30	48.1	<0.0001

[a] χ^2 tests null hypothesis that parental color does not influence offspring mate selection.

This established an association, but not necessarily a cause-and-effect relationship. First, birds with white parents are themselves usually white because of the genetic basis of color inheritance, and, second, they also usually have white siblings. Mate selection may reflect not parental color but self color or sibling color. Looking at the mate choices of offspring of mixed matings helps one decide whether self color is important in mate choice.

If birds choose mates of their own color, then the white offspring from mixed pairs should choose white mates, and the blue offspring, blue mates. Of 38 white offspring 26 chose white mates, and 12 chose blue. Of 26 blue offspring 15 chose blue mates, and 17 chose white. There is thus a slight tendency to choose self color, but not sufficient to account for the assortative mating. This result probably reflects the preference of the mates of the birds analyzed.

Another question may be answered by a look at the matings of the offspring of mixed pairs to determine if one parental color was having more influence upon offspring choice than the other. It was found, however, that there was no significant difference in mate choice by the offspring from the two types of mixed parentages. Since both parents play similar roles in the care of the offspring such a result is not surprising.

It has not been possible to assess the role of sibling color in mate selection from the field data. Since siblings may stay together even after the young birds leave their parents (Prevett, 1972), it is possible that they influence mate choice more than Cooch and I envisioned when we first proposed the hypothesis in 1968. The role of the siblings has been investigated under experimental conditions.

Experimental Approach

The experiments summarized here have been published earlier and in greater detail (Cooke, Mirsky, and Seiger 1972; Cooke and McNally, 1975; Cooke, Finney, and Rockwell, 1976). The ultimate aim of these experiments was to raise goslings in a variety of foster-family situations and determine the color preferences of these birds in terms of approach response, free-choice association, and pair formation. It was discovered that young birds placed in a free-choice situation with unfamiliar test birds had a significant preference for birds of parental color as measured by an approach response and free association. Sibling color also appeared to be important, however, in that goslings with siblings of color opposite to that of the parents chose parental color less frequently than did goslings from pure families.

It was also shown that in open field situations in the absence of parents, birds usually associated with their siblings at both one and two years of age. When birds associated with nonsiblings, they showed a distinct tendency to associate with birds of the same color as their siblings.

For the pair-formation experiments, we created artificial families. Incubating eggs were sent to southern Ontario. These were incubator hatched and then given to foster pairs (mated pairs with one exception). Twelve families were formed with four to 11 offspring. The families were of six types (two replicates of each), four of which are found in the field and two which are not. The families were (1) all white, (2) all blue, (3) blue male × white female with mixed offspring, (4) white male × blue female with mixed offspring, (5) white parents, blue offspring, and (6) blue parents, white offspring. The families were initially in separate pens in visual isolation from each other. At six weeks of age they were put together in a large field and remained in that situation until pair formation.

Table 2 shows the mate choices of these captive birds set out in a format similar to the field data. Although sample size is small, the data show a significant departure from randomness in the same direction as the field data. From the "pure" families 13 offspring chose mates of a color similar to parental color; only two did not. This is compared in Table 3 with those families where the parents were of one color and offspring were of the opposite color, a situation which occurs very rarely in nature. A contingency χ^2 shows a significant difference between the two cases. When offspring color is different from parent color, the offspring choose parental color much less frequently and, in fact, show a slight though not significant preference for nonparental color. This strongly suggests that sibling color plays a role in determining mate choice. The original hypothesis that parental color influences subsequent mate selection of the offspring should therefore be modified to say that familial color influences mate selection. In other words if a bird has siblings and/or parents of both colors, then it will accept birds of either color as a mate; if it is from a pure family, it will show a distinct preference for a mate which is

Table 2. Mate Selection of Geese from Known Families; Captive Flock; $\chi^2_2 = 8.9$; $p < 0.05$

Family Colors / *Offspring Choice*	*White Pair with White Offspring*	*Mixed Pair with White and Blue Offspring*	*Blue Pair with Blue Offspring*	*Total*
White	7	7	0	14
Blue	2	6	6	14

Table 3. The Effect of Parental and Sibling Color on Mate Selection; Contingency, $\chi_1^2 = 6.14$; $p < 0.05$

	Family Structure	
Offspring Choice	*Parents and Offspring of Same Color*	*Parents and Offspring of Opposite Color*
Parental Color	13	6
Nonparental Color	2	8

of the same color as the familial color. Occasionally a bird is the only representative of its color in the family. All its siblings and its parents are of opposite color. According to hypothesis these birds should choose familial color. From the field data, one only has two families of this type, where the unique colored bird has been sighted with a mate. In both cases they did indeed choose familial color. This is a further indication that self-color is not important.

There is one further piece of evidence pertaining to this question. If familial color is important in determining mate choice and if this is what explains the assortative mating observed in the wild, then should a group of birds be raised in captivity as a large mixed flock, no assortative mating in terms of color should occur when this flock forms pairs. Goslings hatched in 1969 were raised in this condition, and when they formed pairs there was no evidence of assortative mating. The assortative mating which occurs in the wild can be eliminated under appropriate prepairing conditions. There seems to be nothing inherent in the specific observed pattern of assortative mating in snow geese; rather, it appears to result solely from the prepairing experience.

Another interesting finding occurred during the mate selection experiments. None of the birds chose a foster sibling as a mate, even though according to our hypothesis such siblings would be suitable mates. Although the sample size is small, this finding may suggest that some barrier to incest may occur in snow geese.

Interpretations and Conclusions

These examples from the case of the snow goose show that these birds are making use of the plumage color and perhaps other characteristics of their family when making decisions about mate choice. This example is the first, as far as we are aware, showing the role of prepairing experience in influencing mate choice in a wild population of birds. The results show that in this dimorphic species, birds which have been raised in a family comprising only one of the morphs are able to discriminate among the morphs and show a sexual preference for the familial morph. If the species is monomorphic, then choosing a mate of the same species as the family is clearly adaptive and provides a mechanism for the development of species identity. In the dimorphic species one might ask the question: Is there any evolutionary advantage associated with choosing a mate of familial morph? Various measurements of reproductive fitness have been collected from the various pairs of geese at La Pérouse Bay since 1968. It is possible to attempt to answer this question by examining the relative reproductive fitness of mixed pairs versus pure pairs. Is there a loss of fitness associated with pairing with a mate of nonfamilial color? Not all mixed pairs, of course, have chosen nonfamilial mates. Many of the birds are themselves products of a mixed mating and according to hypothesis have no preference. Nevertheless, as many as 40 percent of all the mixed pairs have arisen as a result of choosing a mate of nonfamilial color. Various measures of fitness were assessed. These include clutch size, egg loss, brood size, gosling mortality to six weeks of age, first-year mortality, age of first breeding, and adult mortality. Although in some years there seems to be a lower fitness associated with the blue phase, in no case is there evidence of a lower fitness associated with mixed pairs (Cooke, unpublished). It is always possible that there is a lowered fitness associated with some other measure of survival (e.g., desertion rate or longevity) or that there has been lowered fitness associated with an earlier period in the evolutionary history of the species.

A possible explanation of these findings is that the birds are generalizing from the specific appearance of their family not to the limits of the species but to some narrower definition of species which excludes the nonfamilial morph. Birds often do not recognize the nonfamilial morph as a member of their own species for purposes of mate selection. They have a narrower definition of the boundaries of their species than the biologist has. This may reflect the application of "species specific" recognition cues at the morph level.

One may wonder why some birds, even though they have parents and siblings of one color, nevertheless choose a mate of the opposite color? Cooke (1978) presents a full analysis of these exceptions and concludes that three biological phenomena which result in the production of families different from those predicted on the basis of the genetics of plumage color are not adequate to explain the exceptions. These phenomena are (1) intraspecific nest parasitism (Finney, 1975), (2) rape (Mineau and Cooke, 1979), and (3) fostering (Prevett, 1972). All can result in an offspring of one color becoming part of a family of the opposite color. In six families at La Pérouse Bay where detailed family history was known we have strong evidence that none of these events occurred, and yet the birds chose a mate of the color opposite to the family color.

It is suspected that a combination of probabilistic and phenotypic variation explains the exceptions, but I will oversimplify and concentrate on the genetic component of the phenotypic variation in the selectivity. Presumably such genetic variation is continuous; but again, in order to simplify, the population will be dichotomized into those birds which have a genetically narrow definition of self, leading them to use family color as their guide in mate selection, and those birds which have a broad definition of self, leading them to use species-specific cues (common to both morphs) as their guide to mate selection. The latter would be indifferent to the color of their mate regardless of their family color.

These two types are referred to as narrowly selective and broadly selective. The degree of assortative mating in the population is then a function of the proportion of these two types in the population. Given a genetic basis for this difference in the choice mechanism, the population could evolve towards greater panmixis, if birds with broadly selective genes were favored, or to greater assortativeness and possibly sympatric speciation, if birds with narrowly selective genes were favored. Given the fact that there seems to be no evidence at the present time for either advantage or disadvantage associated with mixed mating, then the present level of assortative mating may be expected to be maintained.

It would seem that genetic variation in the degree of selectivity would be expected in outbreeding populations, particularly where environments are unpredictable. To be too narrowly selective, i.e., to be willing to mate only with an organism phenotypically very similar to one's family, could lead to inbreeding, whereas to be too indiscriminate could lead to interspecific mating. In some environments a narrow selectivity might be preferred, in others, a broad selectivity.

This genetic variation in selectivity might apply more widely than simply to mate selection. Habitat selection and food selection are two areas where it might be expected. An organism which is willing to select only that environment to which it was initially exposed may be at an advantage in times of environmental stability, whereas an organism which is willing to select more broadly may be at an advantage in times of change. Some genetic variation, again, in selectivity might be expected in the population.

As an example of variation in habitat selection in snow geese, there are two distinct and discontinuous feeding areas at the La Pérouse Bay colony. Goslings upon hatching are led to one of these areas by their parents. The area chosen by the parents is usually the one where the mother was taken by her parents when she was a gosling. One year later, when the goslings return to the colony with their parents, they are chased away from the nesting area at nest initiation and remain as nonbreeding birds in one of the two feeding areas. About 90 percent of the birds return to the feeding area to which they were taken as goslings, but the remaining 10 percent go to the other feeding areas. The 90 percent may be composed largely of birds with a narrow selectivity; the remaining 10 percent may consist of birds with a broad selectivity (more experimental and more willing to break traditions). There seems to be an analogy between this variation in behavior and that found in mate selection. It would be interesting to know if the birds which are broadly selective in terms of mates are also broadly selective in terms of feeding habitats. We should be in a position to answer this question as the data accumulate.

To summarize: in an analysis of the question of species identity, it is important to investigate the question both interspecifically and intraspecifically and particularly to examine cases where the boundary between intraspecific and interspecific variation is not distinct. Most examples of research into these topics with birds have been carried out under experimental conditions; but an example of study where experimental analysis has been integrated with field observations has been described in detail.

References

Abraham, K. F. Moult migration of lesser snow geese. (In preparation.)

Bateson, P. P. G. 1977. How do sensitive periods arise and what are they for? XVth International Ethological Conference, Bielefeld.

Cheke, A. S. 1969. Mechanism and consequences of hybridization in sparrows *Passer*. *Nature 222*, 179–180.

Cooch, F. G. 1958. The breeding biology and management of the blue goose. Ph.D. thesis. Ithaca, N.Y.: Cornell Univ.

Cooke, F. 1978. Early learning and its effect on population structure. Studies of a wild population of snow geese. *Z. Tierpsychol. 46*, 344–358.

Cooke, F. and F. G. Cooch. 1968. The genetics of polymorphism in the goose (*Anser caerulescens*). *Evolution 22*, 289–300.

Cooke, F., G. H. Finney, and R. F. Rockwell. 1976. Assortative mating in lesser snow geese (*Anser caerulescens*). *Behav. Genet. 6*, 127–139.

Cooke, F., C. D. MacInnes, and J. P. Prevett. 1975. Gene flow between breeding populations of lesser snow geese. *Auk 92*, 493–510.

Cooke, F. and C. M. McNally. 1975. Mate selection and colour preferences in lesser snow geese. *Behaviour 53*, 151–170.

Cooke, F. and P. J. Mirsky. 1972. A genetic analysis of lesser snow goose families. *Auk 89*, 863–871.

Cooke, F., P. J. Mirsky, and M. B. Seiger. 1972. Colour preferences in the lesser snow geese and their possible role in mate selection. *Can. J. Zool. 50*, 529–536.

Cooke, F. and D. S. Sulzbach. 1978. Mortality, emigration and separation of mated snow geese. *J. Wildl. Manage. 42*(2), 271–280.

Finney, G. H. 1975. Reproductive strategies of the lesser snow goose (*Anser caerulenscens*). Ph.D. thesis. Kingston, Ont.: Queen's Univ.

Finney, G. H. and F. Cooke. 1978. Reproductive habits in the snow goose—the influence of female age. *Condor 80*, 147–158.

Goodwin, D. 1958. The existence and causation of colour preferences in the pairing of feral and domestic pigeons. *Bull. Br. Ornithol. Club 78*, 136–139.

Harris, M. P. 1970. Abnormal migration and hybridization of *Larus argentatus* and *L. fuscus* after interspecific fostering experiments. *Ibis 112*, 488–498.

Hess, E. H. and D. E. Hess. 1969. Innate factors in imprinting. *Psychonomic Sci. 14*, 129–130.

Immelmann, K. 1969. Über der Einfluss frühkindlicher Erfahrungen auf die geschlechtliche Objectfixierung bei Estrildiden. *Z. Tierpsychol. 26*, 677–691.

Immelmann, K. 1972. Sexual and other long-term aspects of imprinting in birds and other species. *Adv. Study Behav. 4*, 147–174.

Immelmann, K. 1975. Ecological significance of imprinting and early learning. *Ann. Rev. Ecol. Syst. 6,* 15–37.

Immelmann, K., H. Kalberlah, H. P. Rausch, and A. Stahnke. 1978. Sexuelle Prägung als Faktor innerartlicher Selektion bei Zebrafinken. *J. Ornithol. 119,* 197–212.

Kalmus, M. and S. Maynard Smith. 1966. Some evolutionary consequences of pegmatypic mating systems. *Am. Nat. 100,* 619–636.

Kerfoot, E. M. 1964. Some aspects of mate selection in domestic pigeons. Ph.D. thesis. Norman, Okla.: Univ. Oklahoma.

King, A. P. and M. J. West. 1977. Species identification in the North American Cowbird; appropriate responses to abnormal song. *Science 195,* 1002–1004.

Klint, T. 1975. Sexual imprinting in the context of species recognition in female mallards. *Z. Tierpsychol. 38,* 385–392.

Lill, A. and D. G. M. Wood-Gush. 1965. Potential ethological isolating mechanisms and assortative mating in the domestic fowl. *Behaviour 30,* 16–44.

Lorenz, K. 1935. Der Kumpan in der Umwelt des Vogels. *J. Ornithol. 83,* 137–213, 289–413.

Mainardi, D. 1964. Effecto evolutivo della selezione sessuale basata su imprinting in *Columba livia. Riv. Ital. di Ornitol. 34,* 213–216.

Mineau, P. and F. Cooke. 1979. Rape in the lesser snow goose, *Behaviour.* (In press).

Nicholai, J. 1956. Zur Biologie und Ethologie des Gimpels (*Pyrrhula pyrrhula* L.). *Z. Tierpsychol. 13,* 93–132.

Prevett, J. P. 1972. Family behaviour and age-dependent breeding biology of the blue goose, *Anser caerulescens.* Ph.D. thesis. London, Ont.: Univ. Western Ontario.

Schutz, F. 1965. Sexuelle Prägung bei Anatiden. *Z. Tierpsychol. 22,* 50–103.

Schutz, F. 1970. Zur sexuelle Prägbarkeit und sensiblen Phase von Gänsen und der Bedeutung der Farbe des Prägungsobjektes. *Verh. Dsch. Zool. Ges. Würzburg,* 301–306.

Seiger, M. B. 1967. A computer simulation study of the influence of imprinting on population structure. *Am. Nat. 101,* 47–57.

Walter, M. J. 1973. Effects of parental colouration on the mate preference of offspring in the Zebra finch. *Tainiopygia guttata castanotis* (Gould). *Behaviour 46,* 154–173.

Warriner, C. C., W. B. Lemmon, and T. S. Ray. 1963. Early experience as a variable in mate selection. *Anim. Behav. 11,* 221–224.

Chapter 6

The Domestic Dog: A Case of Multiple Identities

J. P. Scott

The concept of identity is one that has arisen out of human experience and is particularly important in clinical practice. It is also a concept that can be derived inductively from the study of animal societies. The basis of vertebrate social organization is the individual recognition of other animals of the same species, as was first demonstrated by Schjelderup-Ebbe (1922) in his studies of the peck order in chickens. Each hen in a flock recognizes every other hen and reacts differently to each. Similar to this case, in other groups of social vertebrates, each animal recognizes every other one in contrast to the situation in the social insects, where recognition is based on caste and colony odor.

Both vertebrates and invertebrates differentiate between their own and other species. A chicken does not respond to humans in the same way that it does to other chickens, and an ant will not respond to a grasshopper in the same way that it does to other ants. The species identity of ants can be altered by early rearing, as normally occurs in the so-called slave-making ants. The larvae of one species are carried off by adults of another. Reared by their captors, the young ants become attached to them and rear young for them (Wheeler, 1910).

Paper originally delivered under the title, "Consequences of Atypical Rearing Experiences in Dogs," as part of an APA Symposium, "Early Experience and the Development of an Adequate Species Identity," San Francisco, CA, August 27, 1977.

Similarly, the species identity of vertebrate animals can be modified by rearing them with another species. This raises the question of how species identity is acquired. A social vertebrate obviously has both a species identity and individual identity, operationally defined as the existence of differential behavior by an individual toward other species and toward other individuals of the same species.

Some other questions raised in Chapter 1 are: Is the process of acquiring an identity the same in nonhuman vertebrates as it is in man, and are animals aware of their own identities as are humans? Still other questions arise out of such comparisons, as will be pointed out later in this chapter, but first it is important to review some of the evidence regarding the process by which dogs acquire species identity. The dog is a particularly interesting animal in this respect since he normally becomes a part of human society while retaining at least some vestige of membership in the canine species, and so acquires a double species identity.

Experiments on Atypical Rearing

Rearing in Social Isolation

The first experiments along these lines were performed by a group of experimenters associated with Hebb at McGill University (Thompson and Heron, 1954; Thompson and Melzack, 1956). They took Scottish terrier puppies, separated them from their mothers and littermates, and raised them in boxes which permitted no vision of the outside world. The boxes had three chambers which could be separated by opaque barriers, so that each puppy could be fed, watered, and have its pen cleaned without seeing the human caretakers. These puppies were maintained for six months in this fashion before being removed and tested. The experimenters were primarily interested in effects on the development of intelligence and on the cognitive aspects of behavior, but found a variety of other abnormalities, among which were a lack of response to pain and a kind of behavior that could be called diffuse excitement. The McGill scientists did not study the problem of social attachment. In evaluating their experiments, it should be remembered that Scottish terriers have been selected for their ability to fight and attack prey animals. Along with this, they have a high threshold with respect to pain.

A similar technique was used by Fisher (1955) in an experiment which varied the social experience of puppies. In a litter of four fox terriers, one puppy was reared in a complete isolation, another was regularly removed and given human contact, and the remaining two were reared together. Fisher isolated his puppies for a much shorter period than did the McGill group, stopping the experiment at 16 weeks. The principal results were that puppies that had been given human experiences of a negative sort, i.e., had been punished for approaching the experimenter, nevertheless became attached to him. As soon as he stopped the punishment they maintained closer contact with him than puppies that had been treated with uniform kindness. This leads to the conclusion that the attachment process will take place in spite of negative reinforcement. Harlow et al. (1963) obtained similar results with infant rhesus monkeys belonging to "motherless" mothers. These infants became attached in spite of mistreatment by their mothers.

The most striking results, however, were obtained with the fox terriers that had been kept in complete isolation. Fisher at first attempted to reunite the four animals in each experiment, but found that the isolated animal in every case not only became subordinate, but was so severely attacked as to become badly injured. He therefore put four of these isolated animals together in a large outdoor run.

There was no fighting and very little social contact of any kind. One puppy might be playing with a water bucket (buckets were one of the few objects in the isolation pens), but the others paid little attention. Each puppy played by itself, and there was no allelometic behavior or pack formation.

The resemblance of these puppies to autistic children was striking (Chapter 14, this volume), which implies that autism is a condition in which the attachment process does not function properly. The difference is that autistic children are not reared in isolation, but they are unable to attach themselves because of internal factors; in contrast to the situation in dogs, which is produced by external factors. Similar resemblances to autism have been noted in rhesus monkeys reared in social isolation (Harlow and McKinney, 1971; see Mitchell and Caine, Chapter 10, this volume).

Based on the previous results, Fuller and Clark (1966a,b) devised an improved technique for rearing puppies in isolation, with which they did an extensive series of experiments with fox terriers and beagles. In their technique the puppies were isolated at approximately 3 weeks and were allowed to emerge at 16 weeks. The puppies were observed in their individual kennels through one-way glass. They were very little disturbed by the initial isolation and later were different from normally reared puppies only in that they almost never vocalized.

Unusual behavior began only when the puppies were removed from their kennels. If a puppy had been reared in complete isolation from 3 to 16 weeks, it would exhibit a variety of bizarre behaviors when removed from its pen. Some puppies would react by wedging themselves into the corner of the room, standing on their hind legs and lifting one forepaw in the air. This could be interpreted as an attempt to escape from the area, which was of course completely novel to them.

Tested in a variety of situations, social and otherwise, the isolated puppies were inferior to normally reared animals. Their behavior can be explained as the result of two factors. Puppies normally develop a fear response to strange objects, beginning at about 7 weeks and reaching a maximum at about 14 weeks (Scott et al., 1974). Beginning somewhat earlier, at approximately 3 weeks, they also develop a capacity for emotional distress in response to separation (Elliot and Scott, 1961). These isolated puppies had no experience with either strange objects or separation, but had developed complete capacities to give strong emotional reactions to these situations. These puppies should have suffered from intense fear and at the same time an intense separation response.

Fuller and Clark (1966a,b; Fuller, 1967) then performed a variety of experiments concerned with the prevention and alleviation of those conditions. They found that a puppy could be removed from isolation at any time up to seven weeks of age and still develop normally. Also, those puppies that were regularly removed from isolation for as little as 2 to 20-minute periods per week throughout the experiment still developed in a normal fashion. Experiments with the alleviation of the condition using a major tranquilizer, chlorpromazine, showed that it was only effective at the actual time of emergence and was not completely effective even then. This is consistent with other experiments (Scott, 1974) which showed that chlorpromazine does not effectively reduce distress vocalization due to separation.

Alleviation of the symptoms was most effective if treatment was given at the time of emergence. One of the most effective therapies was to provide contact with a playful and friendly normal puppy. Like the isolated animals raised by Fisher, these isolated puppies did not play spontaneously, perhaps because they had passed a critical age.

We can conclude that the experience of these puppies in isolation allowed them to become attached only to the physical surroundings of the isolation kennel and such objects as it contained. Removal of a puppy from the cage has the effect of removing it from all objects to which it has become attached, producing a maximum emotional response to separation. Such a puppy has no social identity, although it seems to have a cage identity. The results thus raise the question of whether attachment to sites or areas produces another kind of identity which is as essential to normal behavior as social identity.

Rearing with Other Dogs without Human Contact

In our experiments at Bar Harbor, we had access to three one-acre fields surrounded with high board fences. These fences permitted us to raise dogs under seminatural conditions without interference from random environmental factors. In one of our early pilot experiments, we raised a litter of puppies in such a field. The experimenter had only visual contact with them, going in once a day to leave food and water and making observations from outside the gate, with only the upper half of her body visible to the puppies. As in Fuller's experiments, this minimal contact was nevertheless sufficient to socialize the puppies to humans (Scott and Fuller, 1965).

Freedman, King, and Elliot (1961) then devised an experiment in which puppies were raised in these fields without visual contact, with food and water introduced through a hole in the fence. They removed puppies at 2, 3, 5, 7, and 9 weeks of age, giving them individual contact with human beings for one week in the normal kennel environment and then returning them to the field. All puppies were removed at 14 weeks of age. Those that had had no previous contact with humans reacted like little wild animals, showing every sign of extreme fear—defecating, urinating, and biting when caught. Experimental analysis of their reactions to humans indicated that the early experience with people for one week was most effective at age 5 and 7 weeks, considerably less effective at 9 weeks, and with almost no effect at 14 weeks. Contact at 2 weeks of age likewise had little effect.

These experiments led us to conclude that dogs have a critical period for primary socialization, or social attachment, extending from approximately 3 to 12 weeks, with an optimum period at approximately 6 to 8 weeks. From other experiments we can also conclude that there is a coinciding critical period for site attachment (Scott et al., 1974).

Rearing with Humans Alone

In our Jackson Laboratory experiments with genetics and social behavior (Scott and Fuller, 1965) we wished to find out whether puppies reared under laboratory conditions were acting like normal dogs; for comparison, we reared three male puppies in private homes, two of which had no contact with other dogs. Silver, the Shetland sheepdog, was reared in a home with two small children and a cat. He became closely attached to humans but never to other dogs, reacting to them only with barking and threats. He did not respond positively even to females in estrus, but was sexually responsive to the male cat. George, the beagle, was reared by a young married couple without children and was kept closely confined to their home. Again, he became closely attached to people, was indifferent to other dogs, and scarcely ever left the yard in order to wander or hunt. He became sexually responsive to the bag on a vacuum cleaner. Finally Gyp, the basenji, was reared in a home which included both young children and another dog and had few restrictions. He wandered widely and frequently, fought with other dogs, and formed attachments to both dogs and people. In contrast to the first two dogs, the basenji had a place in canine as well as human society.

Most of the experiments with social attachment in dogs have concerned their attachment to humans. This raises the possibility that the attachment process may be something peculiar to the dog–human relationship and does not reflect a process that is normal to the species. An answer to this question is provided by reversing the usual experiment, i.e., rearing puppies only with humans from birth until eight weeks, the optimum age for transferring a pup from dog to human society, and then introducing the puppy to dog companions. Because of the amount of labor and difficulty involved in hand-rearing puppies, this experiment was done only twice, once at the Jackson Laboratory with a hybrid puppy and once at the Bowling Green Canine Laboratory with a beagle. The latter animal was raised in an apartment by hand from approximately one week of age until nine weeks and then brought out to the Canine Research Laboratory and placed with a group of other young dogs. The puppy became strongly attached to humans in its original home and extended this relationship to other humans at the laboratory. When left with the other dogs, it appeared somewhat depressed for a day or two; but it rapidly adjusted to them and has since shown a good adjustment to both dogs and humans. Compared to dogs that have been raised only in the kennel with routine contact with humans, this dog is more friendly towards human handlers and shows less fear. Essentially the same result was obtained with the original hybrid puppy (Scott and Fuller, 1965).

Puppies that have been reared with human beings alone can be described as "almost human" in that if they do not meet other dogs until they become adults, they confine their social responses to humans and such other animals that they may have met early in life. Their basic canine behavior patterns are not altered, except as they are inhibited or enhanced by human training; but their social relationships are developed entirely with humans. We suspect that adult dogs reared in this fashion could develop relationships with other dogs under the proper conditions, but that these relationships would be comparatively weak. Such animals have an identity in human society but not in canine society; in contrast to the puppies reared only with other dogs, whose identity lies only in canine society; and to the puppies reared in isolation, who have no social identity, but only a place identity.

Effects of Rearing in a Single Restricted Environment

From the previous experimental evidence, early rearing even in complete isolation is not harmful, provided the puppies are transferred to the environment that will be their future home by six to eight weeks of age. The harm arises from not transferring animals at the proper time. The consequences of maintaining a puppy without transferral are demonstrated by the following experiments.

The first of these experiments was performed (unwittingly) in the course of rearing puppies at Guide Dogs for the Blind in San Rafael, California (Pfaffenberger and Scott, 1959; Pfaffenberger et al., 1976). In usual conditions of rearing, puppies were kept in kennels until 8 weeks of age. From 8 to 12 weeks they were removed from their mothers and littermates and tested outside the kennel (by several testers). This gave the puppies experience with a variety of people and new situations. At 12 weeks they were supposed to be placed in homes where they were reared and trained as family dogs up to the age of one year.

In the early days of the program there were a large number of failures when the dogs were finally trained as guides at one year of age. Analysis of their previous records showed that a substantial number of the failed puppies had been left in the kennels beyond 12 weeks without special attention. Over the next 4 or 5 weeks, the percentage of dogs later failing as guides rose dramatically. Genetic factors were also involved, since puppies that had failed in their puppy tests showed a larger percentage of failures than those that had passed the test.

This experiment drew attention to a condition which we have called the kennel-dog syndrome or, more properly, separation syndrome. Any dog that has been left in a kennel for six months or more and is then removed to become a house pet is highly likely to show symptoms of extreme timidity, particularly toward strange objects and people. Krushinski (1962) had noted that dogs that had been reared in the Pavlovian Laboratories and then were trained as war dogs frequently performed very badly. The kinds of symptoms that may be expected are shown by the following case histories of dogs that were raised in a kennel under precisely controlled experimental conditions and then placed in private homes at the age of six months. All of these dogs were raised in a good kennel environment, but without any special training or experience.

The first dog was a male Shetland sheepdog that was transferred to a family in which there were young children, a cat, and a toy poodle. The family reported that the dog was afraid of the poodle, so they gave the poodle away. The owners also gave away the cat for similar reasons. They could never housebreak the animal, because if they threatened it with a rolled-up newspaper it would crouch on the floor and involuntarily urinate. They kept it for approximately six months and then begged us to take it back, because they could not stand having this unhappy animal in their home. Perhaps the most striking symptom which they reported was that they had never seen the dog wag its tail once in the whole time they had it.

We returned the dog to the kennel in which it had been reared and within a few days it was running around confidently, wagging its tail and behaving no differently from the other dogs. The symptoms of the kennel-dog syndrome are thus very severe and persistent, but, at least within the time limit given here, restoration to the original home environment brings about complete recovery.

The second case was also a Shetland sheepdog raised in a similar manner. However, this particular dog (Val) had been observed to be unusually timid while in the kennel. He was sold at six months of age, but the owners brought him back after five days because they were afraid he would die. The dog had not eaten, drunk, urinated, defecated, or even moved in the whole time that they had it. Back at the kennel it recovered within a few hours and acted like its usual self. The author consequently attempted to find out what would overcome this extreme case of the separation syndrome. He took the dog home and kept it overnight, bringing it back each morning to the kennel, where it spent the day. At night its usual reaction was immediately to run under a chair and stay there without moving or responding, in spite of attempts to tempt it out with food and petting. After two weeks, it showed the first positive response and took food from the author's hand. At that point the author took the dog home permanently and continued to work with it to achieve confident behavior. Among other things, the author brought home a lively Telomian puppy at the optimum age of eight weeks, thus giving Val a canine companion. Eventually, after several months, Val was sufficiently at home in the new area that he would stay in the yard while not on a leash, especially if the Telomian was tied in the same area. He even wagged his tail occasionally, especially while outdoors.

However, the case had an unhappy ending. One day Val was left unattended in the yard for a quarter of an hour, and disappeared. A neighbor reported seeing him some distance away, but he was never found. Because he was so timid, it is unlikely that he could have been carried off by another person. He had probably wandered off in an attempt to find his original home which was approximately three miles away, but he never got there.

The final case was that of a male Telomian reared in the same fashion and transferred to a family of dog lovers. The new owners bred and showed Chihuahuas and so were not ignorant about dogs. After several weeks they reported that the dog had severely bitten three people, inflicting puncture wounds that required medical treatment. Their veterinarian called him a "fear biter" and said that he would have to be destroyed. We therefore observed and tested him in his new environment. It was obvious that he was not happy, although he showed no fearful behavior. When given a handling test, which involved approaching him suddenly, he attempted to bite the tester and would have done so if the tester had not been prepared. Thus we confirmed the reported aggressive behavior.

Since the owners could not manage the dog, we returned him to the kennel. Within a week he was acting in a completely normal fashion and never again attempted to bite anyone. This case suggests that separation syndrome may result in unusual aggressiveness, as well as in unusually fearful behavior. We must conclude that separation from a familiar environment produces a long-lasting emotional response which is distressful to the animal. Depending on its genetic disposition, the separated dog will respond either with fearful or aggressive behavior. To the dog, it must appear that environmental factors are causing its distress, and it responds by either avoiding them or attempting to attack them. If a separated dog were left alone and free, these responses would undoubtedly cause it to wander about until it came back to its original home.

Dogs with the separation syndrome are difficult to train, apparently because of their emotional reactions. To test this hypothesis, we devised controlled experiments in which Shetland sheepdogs, Telomians, and their hybrids were subjected to three experimental conditions. Group I (to which the dogs in the previous three cases belonged) was simply raised in the kennel with good physical care for the first five months of life. Group II was raised in an identical fashion, except that its members were given elementary training such as going on a leash, sit training, and running to a reward, all in the home pen. Thus the puppies were given simple obedience training plus preliminary training for certain problem-solving tests. Group III was given the same experience as Group II, except that the puppies were taken in pairs to a different building for their training. Thus they experienced separation from the home site and, more briefly, from their littermates.

At five months all puppies were given a series of tests in a strange location—a different room in the building to which the Group III pups had gone. At the end of these tests they were taken to a completely strange house and yard, where we observed their behavior and led them on the leash over a simple course. In this new situation the proportion of animals that could be rated as well adjusted was highest in Group III and about the same in Groups I

Table 1. Mean Numbers of Balks in Five Days of Leash Training

Breed or Cross	Group I	Group II	Group III	Total
Telomian	130	59	25	71
Sheltie	153	84	53	96
F_1	76	34	25	45
Total	120	61	34	

and II. The experience of interacting with people in the home pen did not help the Group II dogs when transferred to a totally strange situation. Their average score for fearfulness was 30, only a little better than the score of 32 for Group I dogs, whereas the corresponding score for Group III was 21 ($p < 0.05$). Furthermore, the degree of fearfulness was significantly greater in Shetland sheepdogs than in either Telomians or F_1 hybrids, indicating that this breed is genetically predisposed toward expressing the symptoms of separation syndrome in the form of fearful behavior.

Of the various tests that we used, leash control is one of the best indicators of good adjustment. A well-socialized dog will readily follow a person into a strange situation, while one that is not attached to humans will resist and literally have to be dragged. The test consists of leading the dog along a strange course which includes stair climbing. The Group I dogs showed a maximum number of balks, Group II about half as many, and Group III about half as many as Group II (see Table 1). The results thus clearly reflected the effects of early separation experience.

In another test we trained the dogs to sit quietly. There were almost no differences between the groups, indicating that learning to do nothing is not interfered with by the emotions of separation. In the other two tests, the Group I dogs were clearly worse than either of the other groups. In the most difficult of these, performance in the Wisconsin General Testing Apparatus, the Group II dogs did somewhat better than any of the rest, just as they had on leash control (Scott et al., 1973).

Heredity was also a factor. As might be expected, Shetland sheepdogs (shelties) did better than Telomians on the leash test; but the clearest result was that the sheltie–Telomian hybrids ranked first in all test situations, being superior to their purebred parents in all tests except inhibition, where the shelties were as good as the hybrids.

Conclusions

From this review of research concerning attachment and separation in dogs we can draw the following conclusions:

1. Dogs normally form attachments both to humans and other dogs. As a result, dogs are predominantly members of human society, but their identity can be shifted either in the canine or the human direction, according to the conditions in which they have been reared during an early critical period.
2. Site attachment is just as important as social attachment for dogs. Part of a dog's individual identity is belonging in a particular place.
3. Change in both social and site attachment in the dog is brought about most readily at a particular age in early development. At this age change takes place without serious consequences.
4. Separation from the individuals and places to which an animal is attached (loss of identity) produces more and more serious results as the animals get older.
5. Adjustment to separation in later life is facilitated by the experience of early separation at the optimum period. The reasons for this are still not completely clear, but the evidence suggests that if an animal has previously undergone change, it becomes familiar with at least two environments and therefore a second experience of separation will not be so likely to place the animal in situations that are entirely new. The experience of successfully coping with separation facilitates similar successful coping with subsequent separations.

Discussion

There are at least three ways of looking at the phenomenon just described: the concept of attachment, the concept of primary socialization and the organization of systems, and the concept of identity. Is there any point in adding this last concept of identity, which after all has been primarily derived from human experience, or would it be better to develop research from the more objective viewpoints of attachment and systems organization? Most researchers, particularly those working with nonhuman animals, would probably prefer the simple concept of attachment.

These three conceptual frameworks are not necessarily mutually exclusive, but can be related to each other under a more general synthesizing concept based on the following line of thought. The processes of social and site attachment are such that their results can be readily distorted by early experience, leading to bizarre and even dangerous consequences. Why have not the dog and other vertebrate species evolved a foolproof, invariant process whereby an individual is born attached to a particular species, thus eliminating the possibility of errors and distortions? One answer is that these processes do work in a highly predictable and invariant fashion within a well organized social system. It is only when we tamper with the system that we find the process of attachment to be sensitive to modification by early experience.

Moreover, vertebrate social organization is based on individual recognition and differential responses to the behavior of other individuals. There is no way that a genetically fixed system could produce this effect. Genetics, in fact, enter into the process by providing individual variation in form and other characteristics that make individuals more easily recognizable. Therefore the attachment process is a basic and necessary part of vertebrate social organization.

Invertebrate species, on the other hand, have evolved along the line of stereotyped methods of species discrimination; for example, sexual behavior patterns that do not correspond to those of other species. Within a species they discriminate among individuals only on the basis of caste. In the simplest form, castes consist of males, females, and young, each of which is separable in terms of appearance and behavior. Even so, insects such as ants and bees depend on attachment processes to differentiate between members of their own group and those of different social systems; that is, they have developed colony attachment processes which are similar to the individual attachment processes seen in the vertebrate. Once an individual or colony attachment process has been evolved, it of course eliminates the problem of species discrimination and species identity under normal conditions.

This reasoning establishes attachment and the resultant phenomenon of identity as basic processes in vertebrate social organization. This is an evolutionary concept, but one that is based on the concept of the organization of systems.

The humanly derived concept of identity does contribute something new. It is true that problems of identity can usually be explained as the result of attachment and separation; but the psychodynamic school of researchers has described a phenomenon that is not obvious in nonhuman animal societies—this is the process of identification. It has been poorly named, because the ability to act and feel as if one were another person almost implies a loss of identity. This ability obviously contributes a great deal to the cooperative functioning of human society, since it can be extended far more widely than to the members of one's immediate family.

It has been amply demonstrated that identification takes place unconsciously as well as the result of verbal thinking. Is it, therefore, a process that also may have evolved in other animals? McBride et al. (1969) with chickens and Ginsburg (1965) with wolves have found that individuals are able to step into new social roles within a dominance hierarchy, even though their experience may have been with other roles. There is also a good deal of confirmatory evidence from the study of primate societies.

Thus we can not only conclude that attachment is a basic and necessary process for vertebrate social organization, but we can hypothesize that attendant processes such as identification contribute to the successful functioning of both individuals within a society and the society as a whole.

References

Elliot, O. and J. P. Scott. 1961. The development of emotional distress reactions to separation in puppies. *J. Genet. Psychol.* *99*, 3–22.

Fisher, A. E. 1955. *The effects of differential early treatment on the social and exploratory behavior of puppies.* Ph.D. thesis. Pennsylvania State Univ.

Freedman, D. G., J. A. King, and O. Elliot. 1961. Critical period in the social development of dogs. *Science* *133*, 1016–17.

Fuller, J. L. 1967. Experiential deprivation and later behavior. *Science* *158*, 1645–1652.

Fuller, J. L. and L. D. Clark. 1966. Genetic and treatment factors modifying the post-isolation syndrome in dogs. *J. Comp. Physiol. Psychol.* *61*, 251–257. (a)

Fuller, J. L. and L. D. Clark. 1966. Effects of rearing with specific stimuli upon post-isolation behavior in dogs. *J. Comp. Physiol. Psychol. 61,* 258–263. (b)

Ginsburg, B. E. 1965. Coaction of genetical and non-genetical factors influencing sexual behavior. In F. A. Beach (Ed.), *Sex and behavior.* New York: Wiley, 53–76.

Harlow, H. F., M. K. Harlow, and E. W. Hansen. 1963. The maternal affectional system of rhesus monkeys. In H. L. Rheingold (Ed.), *Maternal behavior in mammals.* New York: Wiley.

Harlow, H. and W. T. McKinney, Jr. 1971. Nonhuman primates and psychoses. *J. Autism Child. Schizophr. 1,* 368–375.

Krushinski, L. V. 1962. *Animal behaviour: Its normal and abnormal development.* New York: Consultants Bur.

McBride, G., I. P. Parer, and F. Foenander. 1969. The social organization and behaviour of the feral domestic fowl. *Anim. Behav. Monogr. 2,* 127–181.

Pfaffenberger, C. J. and J. P. Scott. 1959. The relationship between delayed socialization and trainability in guide dogs. *J. Genet. Psychol. 95,* 145–155.

Pfaffenberger, C. J., J. P. Scott, J. L. Fuller, B. E. Ginsburg, and S. W. Bielfelt. 1976. *Guide dogs for the blind: Their selection, development, and training.* New York: Elsevier.

Schjelderup-Ebbe, T. 1922. Beiträge zur Sozial-Psychologie des Haushuhns. *Z. Psychol. 88,* 225–252.

Scott, J. P. 1974. Effects of psychotropic drugs on separation distress in dogs. *Excerpta Med. Int. Congr. Ser. 359,* 735–745.

Scott, J. P. and J. L. Fuller. 1965. *Genetics and the social behavior of the dog.* Chicago, Univ. Chicago Pr.

Scott, J. P., J. M. Stewart, and V. DeGhett. 1973. Separation in infant dogs: Emotional and motivational aspects. In E. Senay and J. P. Scott (Eds.), *Separation and depression.* Washington, D.C. American Association for the Advancement of Science.

Scott, J. P., J. M. Stewart, and V. J. DeGhett. 1974. Critical periods in the organization of systems. *Dev. Psychobiol. 7,* 489–513.

Thompson, W. R. and W. Heron. 1954. The effects of restriction on activity in dogs. *J. Comp. Physiol. Psychol. 47,* 77–82.

Thompson, W. R. and R. Melzack. 1956. Early environment. *Sci. Am. 194,* 30–42.

Wheeler, W. M. 1910. *Ants: Their structure, development and behavior.* New York: Columbia Univ. Pr.

Chapter 7

The Domestic Cat

Gary W. Guyot
Henry A. Cross
Thomas L. Bennett

Nothing's more playful than a young cat,
nor more grave than an old one.
[*Thomas Fuller (1654–1734) "Gnomologia"*]

Species identity, or knowing to which species one belongs, has generally been discussed as a product of socializing an individual to members of its own species. For instance, Lehrman and Rosenblatt (1971) stated,

> The development of appropriate social behavior implies . . . the animal's ability to direct its social behavior selectively toward, and to respond selectively to stimuli from, members of its own species. A number of observations suggest that some sorts of experiential interaction between the developing young animal and other members of its species play a role in the development of the behavioral tendencies that we take as evidence that the animal "knows" to what species it belongs (p 11).

The primary socializing agents in the early life of the mammal are the primary caretakers (usually the mother) and agemates (siblings and/or peers). These socializing agents participate in two major socialization processes. The first process is an emotional attachment the infant forms to its primary caretaker. It is usually described as the mother–infant bond, although an infant may form an attachment to other rearing objects, e.g., surrogates, siblings, or peers. The second socialization process is usually described as social play, which involves playful interactions between littermates or agemates.

These socialization processes have a major role in shaping the adult's behavior. Disrupting the socialization processes by varying the social conditions of early rearing has been found to have profound influences on the behavior of the adult. This has led to the assumption that in order to understand adult behavior, it is imperative that we understand the ontogeny of that behavior (Bekoff, 1972).

That socialization may lead to the formation of a species identity suggests that if an organism is socialized to members of a different species, it may form an identity with the alien species rather than its own. Fox (1971), for instance, after rearing dogs with kitten litters and finding that the normally social dogs displayed marked intraspecies avoidance and a debilitated mirror-image response, concluded, "These observations have shown that the development of 'species identity' (or self-recognition) is dependent upon the species to which the animal is socialized, and this socialization effect influences later social preferences for the same species, or alien foster-species" (p 262). The implications of this statement become intriguing when applied to the solitary carnivorous cat.

In this chapter the adult cat's solitary behavior will first be briefly described. Intraspecies early experiences and species identity through the socialization processes of attachment and social play in the kitten will be discussed. Within this structure the effects of varied social rearing conditions on those processes will be assessed. Finally, the authors will discuss interspecies early experiences and species identity with potential prey and with potential predators.

Intraspecies Socialization

Domestic cats (*Felis domesticus*) differ from other mammals traditionally used to study socialization processes because the adult cat is generally thought to be a solitary animal, showing little species affinity or social organization. The kitten, on the other hand, has been described as being highly social in its relationship with both its mother and other kittens. The ontogeny from the social kitten to the solitary adult is not fully understood.

Adult Solitary Behavior

Feral cats live a relatively solitary life and avoid conspecifics (Wilson and Weston, 1947). There is considerable evidence that adult cats are nongregarious animals who display marked territorial behavior (Rosenblatt and Schneirla, 1962). Even territorial marking seems to facilitate conspecific avoidance, since it does not appear to scare or intimidate another cat, but may prevent an unexpected encounter or sudden clash by giving information as to *who* is ahead and *how* far (Leyhausen, 1965).

While dominance hierarchies in cat communities do exist, they do not generally develop into a rigid social hierarchy, since a superior animal will give way to an inferior animal apparently to avoid social contact or a possible fight. This conspecific avoidance apparently rests on a relative social hierarchy that is locality–priority dependent. A superior cat, however, will not avoid a newcomer or an upcoming adolescent, in which case superiority must be determined (Leyhausen, 1965; Leyhausen and Wolff, 1959).

These behaviors suggest that the cat "knows" to what species it belongs and can recognize individuals within that species, although it is doubtful that cats have the capacity for self-recognition, since they respond to a mirror-image as if it were a strange cat (Ewer, 1968; Fox, 1974). Having species identity and the ability to recognize individuals within that species would seem to be necessary prerequisites for solitary behavior. However, they are not sufficient conditions. The normal solitary behavior of cats is disrupted when they are enclosed in a small colony or when a cat of superior strength enters the scene. In these cases aggression and chaos result (Leyhausen, 1965). Hence avoidance of conspecifics within a social community seems to require some minimum amount of space (territory), a code of conduct, and some form of species and individual recognition.

Kitten Social Development

In several articles, three major phases of kitten socialization have been described (Rosenblatt, 1971, 1972; Rosenblatt and Schneirla, 1962; Rosenblatt, Turkewitz, and Schneirla, 1961, 1969; Schneirla, 1959, 1965; Schneirla and Rosenblatt, 1961, 1963; Schneirla, Rosenblatt, and Tobach, 1963). These socialization processes are generally based on the feeding relationships between the mother and her offspring.

During the first feeding phase (0–20 days) the mother initiates the feedings, and the kittens generally aggregate in the home nest, orienting primarily by odor cues. In the second phase (20–30 days) either the mother or kitten may initiate feeding, visual orientation develops, home-nest aggregation declines, and the kittens generally become more active both in their play and seeking the mother outside the home nest. Although the mother may play with the kittens during this phase, as the play becomes more vigorous and the kittens become more active, the mother begins leaving them. This results in a decrease in the social bond between mother and infant. Kitten play is so important to this change and to the initiation of the weaning process, that Rosenblatt (1972) concluded that when there was only one kitten in a litter, weaning tended to take longer because there was less play activity to disturb the mother. In the third phase (30 days to weaning), with the mother leaving, the kitten must seek out the mother and initiate the feedings. Nursing, which occupies as much as 90 percent of the mother's time at week one, declines to as low as 16 percent at week five (Rheingold and Eckerman, 1971; Schneirla et al., 1963). Through these phases of kitten development there are close correspondences among (1) the sucking relationship with the mother, (2) the learning of nipple and home orientation through this social interaction, and (3) the maturing perceptual–motor systems of the kitten (Rosenblatt, 1972). This correspondence appears to rest on the kitten's ability to form complex associations.

Several studies have indicated that a kitten can form complex associations very early in its life (some as early as the first day after parturition). Normally these associations are based on mother–littermate interactions and on the environment in which they occur. Olfactory, thermal, tactual, and visual stimuli have been implicated in these associations, depending on the sensory–motor maturation of the kitten (Freeman and Rosenblatt, 1978a; Rosenblatt, 1972; Rosenblatt et al., 1969). It appears, in addition, that a kitten can orient better to the odor cues of its own home cage (odors left by its own mother and littermates) than to a strange home cage, where the odors were left by a strange mother and litter (Freeman and Rosenblatt, 1978b).

These associations are crucial to the socialization as well as to the survival of the kitten and are utilized in such processes as location of the mother and the preferred nipple, home-nest location and aggregation, and individual recognition, all of which could provide the bases for attachment formation and species identity in the kitten.

Atypical Rearing Conditions

Since the kitten can form many complex associations based on social interrelationships soon after parturition, one would suspect that atypical rearing conditions would cause some form(s) of atypical development. Rosenblatt et al. (1961) demonstrated that continuous social contact through the phases of kitten development is important to maintain socialization and sucking patterns. When they isolated infants from the normal rearing conditions for varying periods of time, the sucking patterns of the kittens were disrupted upon their return to the mother. In addition, several of the kittens displayed overt withdrawal and signs of excessive disturbance (hissing, piloerection, and so on). This may indicate that recognition of the mother and/or littermates was also disrupted by the periods of isolation.

Other investigators have also reported some effects of atypical rearing conditions. Seitz (1959) found that kittens that were weaned and isolated at two weeks of age were suspicious, fearful, and aggressive in their behavior toward other cats when tested at nine months of age. He also found that novelty had an unusually large impact on early weaned cats, causing them considerable stress. This finding was supported by Konrad and Bagshaw (1970) in early socially isolated cats.

Adult sexual behavior may also be affected by varying the early social rearing conditions. Rosenblatt (1965) reported that two adult male cats, isolated from birth, did copulate after many tests, but only briefly. They did not mate thereafter.

Social preference, finally, appears to be affected by isolation. Candland and Milne (1966) totally isolated 45-day-old kittens for 20, 40, or 100 days. They found that the longer the period of isolation, the more a kitten would prefer another kitten rather than a rearing object. Communally reared kittens preferred the rearing object to another kitten. With longer isolation, however, isolated kittens may respond with hostility toward conspecifics (Kuo, 1960). Kuo found that cats who were isolated for 10 months were later hostile toward other cats. However, cats that had some early experience with strange cats later displayed friendly or indifferent behavior toward strange cats.

These studies certainly indicate that atypical rearing conditions will lead to atypical behaviors and could influence the socialization processes of attachment and social play in the kitten.

Attachment in Kittens

The basis for socialization and for the formation of social bonds in mammals has generally been attributed to the concept of attachment. The distress produced by separation from rearing stimuli has historically been used as an index of attachment to those stimuli (Gerwirtz, 1972). Among the many behavioral measures utilized to assess this distress, vocalizations or distress cries often have been used. Vocalizations from separated kittens reliably elicit retrieval and return to the home site by the mother (Haskins, 1977).

Rosenblatt (1971, 1972) reported that a kitten's movements slowed down and its distress cries became softer and gradually stopped when, after separation, it regained contact with social or nonsocial rearing objects (i.e., mother, littermates, home nest, or a brooder). This was called the "settling reaction." It begins to appear the first day after parturition and may be mediated by contact and by thermal and olfactory stimuli.

It was decided to assess the effects of various rearing conditions on the attachment processes in 32 infant kittens (Guyot, Cross, and Bennett, 1979). Specifically, four males and four females were raised in each of the following rearing conditions: (1) two kittens by their biological mother, (2) one kitten by its biological mother, (3) two kittens with a surrogate, and (4) one kitten with a surrogate. Thus behavioral comparisons could be made between each of the four groups: between kittens reared with one littermate (littermate-reared), or without a littermate (littermate-deprived); and between kittens reared by their biological mother (mother-reared), or with a surrogate (surrogate-reared). These possible comparisons are summarized in Table 1.

The surrogate provided milk, contact (shag carpet), and thermal stimuli (32°–38°C). The kittens were reared from birth to six weeks with their mother, or from two days until six weeks with the surrogate. At six weeks of age all kittens were weaned and housed in standard laboratory cages either alone (littermate-

Table 1. Factorial Design of Rearing Conditions and the Nomenclature Utilized to Describe the Primary Groups and Interactions

	Type of Mother	
	Mother-reared	*Surrogate-reared*
Littermate-reared (one sibling)	Mother-littermate (Group 1)	Surrogate-littermate (Group 3)
Littermate Exposure		
Littermate-deprived (no sibling)	Mother-only (Group 2)	Surrogate-only (Group 4)

deprived subjects) or with their littermate (littermate-reared subjects).

In order to assess the possible attachments the kittens formed to their social and nonsocial rearing stimuli, each kitten was separated from its respective rearing condition and placed alone in an open field for 15 minutes each week from the second through the fifth week of age. During this brief separation period, the frequency of their vocalizations (distress cries) was recorded and used as an index of attachment to their rearing stimuli. Consistent observations were also made on the developing kittens in order to determine changes in social conditions and behavior during rearing.

In this study it was found that the frequency of distress cries, produced by separating an infant from its rearing condition, was generally commensurate with the social conditions at the time of separation. The vocalizations of the littermate-reared groups did not significantly differ throughout testing. The vocalizations of both groups, however, showed a significant decline in frequency at five weeks of age, when separation occurred during the third feeding phase (Figure 1). Our observations revealed that by this time the mother of the mother–littermate-reared kittens was spending most of her time away from her litter, and her kittens appeared to be growing more independent of each other by exploring alone for increasingly long periods of time. The surrogate–littermate-reared kittens initially spent most of their time together, sucking on each other, sleeping together, exploring together, and so on. By five weeks of age, however, they rarely slept together or sucked on each other and were observed also to be spending more time alone. This result would seem to indicate

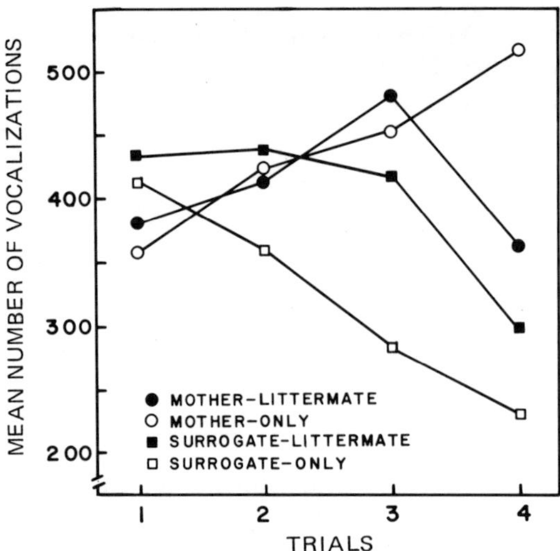

Figure 1. Distress cries produced by separation from the kitten's rearing condition.

that the mother per se is not necessary for the development of attachments, and it appears to be the kittens' growing independence (probably in conjunction with their sensorimotor maturation) that is responsible for the decline in attachments.

That the mother is important for facilitating the growth of independence by leaving her litter (in normal litters) is indicated by the distress cries produced by separating the mother-only reared kittens from their mother. Their cries increased with repeated separations and were significantly more frequent than all of the other groups at five weeks of age (Figure 1). In essence the mother did not wean her only infant, as she was rarely observed without being in close proximity to her kitten. Hence the kitten did not appear to gain its independence, and this apparently had the effect of increasing the infant–mother social bond. The result of this increase was also observed after weaning, when these kittens became hyperactive, lost approximately 20 percent of their body weight (even though they had learned to lap milk and eat soft food), and cried constantly for two weeks. The kittens in the other groups generally stopped crying within a week after weaning.

The frequency of distress cries produced by separating the surrogate-only reared kittens initially did not differ significantly from the other kittens. With repeated separations, however, the cries continued to decrease and were significantly less frequent than in all the other groups by four weeks of age (Figure 1). This result questions the ability of a soft, warm, nutrient-supplying nonsocial rearing object (a surrogate) to compensate adequately for social stimuli (mother and/or littermates) in the development of attachments. The effect these rearing conditions may have had on the formation of an adequate species identity will be assessed by looking at the continuance of the socialization process, i.e., social play.

Kitten Social Play

Although attachment is generally given as the basis for the formation of social bonds in infant mammals, social play is often seen as the "medium" through which an individual becomes socialized to conspecifics. Social play, for instance, has been empirically tied to the production of social skills necessary for functioning in elaborate primate social organizations (e.g., Dolhinow and Bishop, 1970; Harlow, 1969; Jolly, 1972). The exact function of social play in the domestic kitten remains unclear, since the adult cat is relatively solitary.

The emphasis for all play in kittens has been on its serving as practice for prey-catching behavior (e.g., Egan, 1976; Leyhausen, 1965/1973). The movements utilized in play and prey-catching behavior are so similar that Leyhausen described them as being instinctive. Support for Leyhausen comes from the fact that neither rearing kittens in total isolation (Kling, Kovach, and Tucker, 1969) nor inhibiting normal interactions with an active mother (Koepke and Pribram, 1971) seems to influence the development of play patterns. West (1974), however, noted that while the play patterns of the kitten seem suited to a carnivore, social play may also function to provide opportunities for the acquisition of social communication skills (which will be used in other social situations) and to maintain friendly social relations among members of a litter. While some excellent normative studies of kitten social play have recently appeared (Barrett and Bateson, 1978; West, 1974), the relative contribution of maternal and/or littermate interactions to the social play of juvenile cats has yet to be determined.

It was decided to assess this contribution by measuring the juvenile play of the same four groups used in the attachment study (Table 1). Each kitten was tested twice a week individually and twice a week with a normal laboratory-reared stimulus animal (generally of the same sex and age) from 8 to 28 weeks of age (Guyot, Bennett, and Cross, 1979). Each testing session lasted 20 minutes, and consisted of 80 15-second intervals. A behavior was scored if it occurred within a 15-second interval. All frequency data were transformed with $\sqrt{X + 1}$. While multiple categories of individual and social behavior were utilized, only the social play categories of approach, slap, wrestle, and bite (which showed similar group differences) will be discussed here.

In order to assess the effects of long-term social isolation, two male and two female kittens were reared in social isolation for five and a half months (termed "five-month isolates," Figure 2). These four kittens had rearing and separation testing identical to that of the surrogate-only reared kittens (Table 1). They were then housed individually in cages, handled four times a week, placed alone in an empty novel environment four times a week the month before testing, and tested in the playroom from 24 to 28 weeks of age. Their data are included for comparison (Figure 2). Also included for comparison are the data from four male and four female, individually housed, adult cats (obtained from an animal shelter) who were tested in the playroom for 8 weeks (Figure 2). These two groups were also tested twice a week individually and twice a week with a stimulus animal.

While Figure 2 shows that neither of these latter two groups exhibited much social play, there were drastic differences in their behavior. All the adults showed some social play behavior, although it was minor. Only one of the five-month isolates displayed any social behavior. They seldom moved and rarely initiated any social contact. The stimulus animal would often approach these subjects in a playful manner, but they seldom responded. None of these five-month isolates displayed any grooming, carpet scratching, or vocalizations during either individual or social testing, and they never went up on the shelves or ramp. This latter finding was also characteristic of the surrogate-reared kittens in our other groups, who seldom climbed on the shelves or ramps.

In comparison, play behavior was observed in all four of our groups previously tested in the attachment study (Table 1). However, important group differences emerged in these four groups. The social play of the mother-only reared kittens was the most frequent (as was their object play when tested individually). The surrogate-only reared kittens also displayed enhanced social play

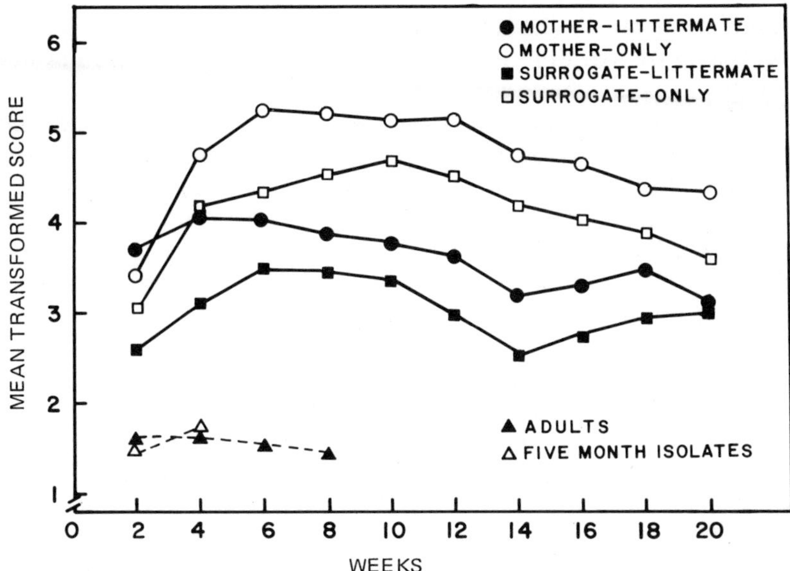

Figure 2. The mean frequency of social play for the combined categories of approach, slap, wrestle, and bite for each group.

(but not object play) when compared to the littermate-reared kittens. The littermate-reared groups displayed less frequent social play, and the surrogate–littermate reared kittens displayed the least amount of social play (Figure 2).

Our observations indicated that the littermate-deprived subjects did not learn the social communication skills necessary for appropriate social play. The one major characteristic that both littermate-deprived groups seemed to have in common was that they did not appear to be aware of social play signals. Both littermate-deprived groups displayed more biting and piloerections, for instance, than the littermate-reared groups. Normally in this study, if a kitten bit the stimulus animal too hard or fought too vigorously, the stimulus animal would often let out a high shriek and attempt to escape. Littermate-reared subjects, upon hearing the cry, would usually withdraw and either reapproach or wait for the stimulus animal to initiate another play bout. In the littermate-deprived groups, however, the cry and withdrawal only intensified biting and fighting behavior. Kittens in these two groups were also often observed in front of the stimulus animal, either slapping it vigorously several times with claws extended (mother-only reared kittens) or touching the stimulus animal several times (surrogate-only reared kittens).

Other observations also indicate a lack of appropriate play signals in these two groups. The surrogate-only reared kittens developed stereotyped postures in social approach. They would often approach the stimulus animal with an arched back, piloerected, and in a sideways posture described by Fox (1974) as an offensive–defensive threat. When they reached the stimulus animal (who was usually in a defensive crouch), they would stop and begin touching the stimulus animal. Wrestling in these surrogate-only reared subjects usually consisted of throwing their bodies on the stimulus animal or crawling all over it, rather than normal rough-and-tumble wrestling. The mother-only reared kittens were generally aggressive, often jumping on the stimulus animal with legs and claws extended. They would wrestle and bite furiously, then jump away and reattack. The stimulus animals would often try to avoid the littermate-deprived kittens, but these attempts again appeared to intensify the social behavior of the littermate-deprived kittens.

The social play of all the kittens eventually declined. This decline for the mother-littermate reared kittens at 16 weeks of age was identical to data from Koepke and Pribram (1971) and West (1974).

Our observations led to the conclusion that early experience with littermates is important for the learning of play signals in the domestic cat. Further, since the littermate-deprived subjects displayed enhanced social behavior when compared to the littermate-reared groups, these play signals may be necessary for the development of normal solitary behavior. Utilizing the animal's ability to direct its social behavior selectively toward, and to respond selectively to stimuli from, members of its own species (Lehrman and Rosenblatt, 1971) as a criteria for species identity, we would have to conclude that depriving kittens of littermate experience disrupts their species identity.

Interspecies Socialization

In the introduction it was noted that the concepts of socialization and species identity have not only been utilized to explain the results of intraspecies early experiences, but also have been generalized to early experience with "alien" species. In the carnivorous cat this generalization may be complicated because alien species for the cat may be potential prey (e.g., rodents, birds, etc.) or potential predators (e.g., dogs, humans, etc.). The term "potential" is explicitly used here on the basis of the cat's relationship to alien species, because there is little evidence that the predator–prey relationships are innate for the domestic cat (e.g., that kittens are born rat killers or that dogs are innate enemies of cats). The evidence weighs heavily, rather, on the fact that these are learned relationships.

Early Experience with Potential Prey

Most investigators agree that the motor patterns used in prey catching and killing are innate. These actions include a stalking run, crouch stalk, pounce, forepaw grab and pin, and a neck-oriented (killing) bite (Fox, 1975; Leyhausen, 1965, 1973). Whether or not these innate movements will be used to catch and kill prey, however, appears to rest on certain early experiences in the life of the kitten.

Kittens learn from observing the mother (Chesler, 1969; Wilson and Weston, 1947), and this is an important aspect of prey catching and killing. Leyhausen (1956) and Ewer (1969) reported at least three stages of the mother teaching her young how to apply their innate movements to prey. First she carries home prey which she has killed and eats it in the presence of her kittens. Later she leaves dead prey for her young to eat. Finally she brings home live prey and allows the kittens to kill it.

That observational learning is important in the acquisition of predator–prey relationships in kittens has been supported by a number of studies. Berry (1908) tested five-month-old kittens who had no previous experience with prey for their response to mice. He observed that the kittens even when food deprived did not kill mice or even behave agonistically toward them. Once they had observed their mother or another kitten kill and eat a mouse, however, they imitated this behavior.

Kuo (1930, 1938, 1960) reared kittens with a number of alien species, varying the amount of early experience with both alien species and conspecifics. Kuo (1930), for instance, observed the following relationships:

1. When 20 kittens were reared in total isolation from members of both alien and the same species, 9 killed prey later. That

figure increased to 18 if the isolates observed prey-catching and killing by another cat.

2. Of 21 cats reared with rat-killing mothers, 18 imitated this behavior.
3. None of the 18 kittens who were reared alone with rats or mice killed their cagemates or other strangers of the same species. Only 3 would kill any kind of rat or mouse. Observing another cat kill a rat or mouse did not change this relationship. Also, hunger did not change the number of killers.
4. Some of the kittens reared alone with rodents (a) played with them as they would have played with their littermates, (b) protected their rodent cagemates against intruders, and (c) displayed attachment responses (the number dependent on the type of rodent) toward their cagemates, e.g., meowing, searching, and restless behavior when their cagemates were removed.

Concerning this latter finding, when Kuo (1938) reared 17 kittens with *both* other kittens and a male or female albino rat, the kittens became attached to their littermates and not the rats. While they did not kill their rat cagemates (or similar strange adults), 12 of the kittens did kill and eat newborn "hairless" rats (the offspring of their rat cagemates), and killed shaved rats of the same size as their cagemates. Also, when they observed adult cats killing and eating rats, 6 imitated this behavior and attacked their rat cagemates.

The studies summarized in this section, taken as a whole, suggest the importance of a kitten being socialized to and forming a species identity with members of its own species. These processes would seem to be crucial for survival in the natural environment. The mother appears to be the best teacher of these survival skills. Littermates may also play a role, since in the normal rearing situation rivalry and competition among littermates encourages a prompt approach and kill (Berry, 1908; Egan, 1976; Leyhausen, 1965/1973). As Berry noted, however, this may be true only after the kittens have observed prey-catching and killing by another cat. Without these crucial conspecific early experiences, a kitten may become attached to, play with, and form a species identity with a potential prey species. The degree of these responses appears to be determined by the amount of early experience a kitten has with conspecifics.

Early Experience with Potential Predators

Throughout history humans have preyed upon the cat (Méry, 1972). However, humans also domesticated the cat, and strong affectional bonds have been observed between cats and humans (Ewer, 1973). Many people think that cats and dogs are natural enemies, but there have been many documented cases of cats and dogs developing strong bonds of affection (Beadle, 1977). Whether the cat views these alien species as predators or not may rest on early social interactions between the cat and these alien species.

Human handling has a pronounced effect on the socialization of kittens (Collard, 1967; Meier, 1961; Meier and Stuart, 1959; Wenzel, 1959; Wilson, Warren, and Abbott, 1965). In fact, if a kitten has no human contact the first two to three months of life, it will be almost impossible to handle thereafter (Collard, 1967; Fox, 1974). In addition, Collard found that if a kitten had only one handler, it became socialized only to that handler. It took more than one handler to socialize the kitten to the human species.

Although early experience with one human may not be enough to socialize it to the human species, early experience with one dog may be enough to socialize it to other dogs. Fox (1969) reared Chihuahua pups individually with kitten litters. He then tested the kittens for responses to their dog littermate and to strange Chihuahua dogs (who had no previous experience with cats). Although the kittens at first reacted negatively to the strange dogs, as soon as the dogs solicited play responses, the kittens played with all the dogs. Kittens without experience with dogs generally withdrew from them and assumed a defensive threat posture.

It appears that normally reared kittens without early experience with potential predators (humans or dogs) later avoid and become hostile toward them (Collard, 1967; Fox, 1969), whereas kittens who have had some experience with these potential predators appear to approach and act friendly toward them. It should be kept in mind, however, that in these studies of rearing kittens with potential predator species, the kittens also had some early experience with conspecifics. Citing a verbal communication with Thomas and his own observations, Leyhausen (1967/1973) concluded, "Cats of either sex which have been reared by humans and isolated at a very early age from all conspecifics later direct their sexual impulses irreversibly toward humans" (p 347).

It again appears that the degree of socialization responses and the formation of an appropriate species identity largely depend on whether the kitten has had early experience with conspecifics. After his extensive work on cross-species socialization, Kuo (1960) summarized:

> In other words, when one young animal grows up together with mates of its species as well as a different species, it is more attached to the mates of its own species than to the mates of another species whereas when one lone young animal of one species is reared together with one or more young of a different species, after the critical time (two to three months), its attachment is in most cases entirely fixed with the mates not belonging to its own species so that when it meets a stranger of its own species its attitude is either indifferent or hostile, depending on its past experience with strange animals (p 218).

Conclusion

In this chapter we have attempted to assess the processes of socialization that might lead to the formation of an appropriate species identity in the domestic cat. Normal kitten socialization appears to rest on the interplay of three socialization factors: the mother, the littermates, and the kitten itself. The mother, in correspondence to the kitten's sensorimotor maturation, provides an environment crucial to the survival of the kitten and to teaching the kitten how to survive on its own. The kitten apparently forms attachments to its mother and littermates. With the kitten's sensorimotor maturation and growing independence, attachments formed to the littermates will naturally decline with the increasing social distance between them. The mother, however, must leave her kittens (provide the social distance) in order for their attachment to her to decline. Littermate play would seem to be influential in providing the impetus for the mother to increase this social distance.

Littermate interactions (or social play) also seem important for the kitten to learn the social communication skills that are necessary for normal conspecific interactions. The effects of littermate deprivation seem to enhance the kittens' social behavior (compared to littermate-reared kittens) during the juvenile period. Thus it appears that littermate interactions may be important for solitary behavior (i.e., conspecific avoidance). When isolation extends beyond the two-to-three-month critical period for kitten socialization, however, conspecific social play and species identity may be totally debilitated.

If a kitten has interspecies as well as intraspecies early experience during the first two to three months of age, it displays positive approach and friendly responses to the alien species, but identifies with its own species. Without conspecific early experience the kitten appears to become attached to and form a species identity with the alien species, whether it be a potential predator or a potential prey. If a kitten has no early experience with any other organism during the critical period for socialization, however, it will avoid and become hostile to all species, including its own (Kuo, 1960). This would indicate a lack of any species identity.

In this chapter we have suggested that the early experiences during the first few months of the kitten's life are important for its later social behavior. The adult cat, however, is relatively solitary in its behavior. By focusing on those early experiences that lead to social avoidance as well as on those that lead to social approach, we may more fully appreciate and come to understand what these complex functional and adaptive processes of socialization and species identity in the domestic cat encompass.

Acknowledgments

The authors would like to thank M. Aaron Roy, Gary R. Byrd, and Sue Christenson Guyot, who made many suggestions that improved the quality of the presentation. We would also especially like to thank Sue Christenson Guyot, who typed several drafts of the chapter.

References

Barrett, P. and P. P. G. Bateson. 1978. The development of play in cats. *Behaviour 66*, 106–120.

Beadle, M. 1977. *The cat.* New York: Simon and Schuster.

Bekoff, M. 1972. The development of social interaction, play, and metacommunication in mammals: An ethological perspective. *Qu. Rev. Biol. 47*, 412–434.

Berry, C. S. 1908. An experimental study of imitation in cats. *J. Comp. Neurol. Psychol. 18*, 1–26.

Candland, D. K. and D. W. Milne. 1966. Species differences in approach-behavior as a function of developmental environment. *Anim. Behav. 14,* 539–545.

Chesler, P. 1969. Maternal influence in learning by observation in kittens. *Science 166,* 901–903.

Collard, R. R. 1967. Fear of strangers and play behavior in kittens with varied experience. *Child Dev. 38,* 877–891.

Dolhinow, P. J. and N. Bishop. 1970. The development of motor skills and social relationships among primates through play. In J. P. Hill (Ed.), *Minnesota symposium on child psychology 4,* 141–198.

Egan, J. 1976. Object play in cats. In J. S. Brunner, A. Jolly, and K. Sylva (Eds.), *Play–its role in development and evolution.* New York: Basic Books.

Ewer, R. F. 1968. *Ethology of mammals.* New York: Plenum Pr.

Ewer, R. F. 1969. The "instinct to teach." *Nature 222,* 698.

Ewer, R. F. 1973. *The carnivores.* Ithaca, N.Y.: Cornell Univ. Pr.

Fox, M. W. 1969. Behavioral effects of rearing dogs with cats during the "critical period of socialization." *Behavior 35,* 273–280.

Fox, M. W. 1971. *Integrative development of brain and behavior in the dog.* Chicago: Univ. Chicago Pr.

Fox, M. W. 1974. *Understanding your cat.* New York: Coward, McCann and Geoghegan.

Fox, M. W. 1975. The behavior of cats. In E. S. E. Hafez (Ed.), *The behavior of domestic animals.* 3rd Ed. London: Tindall.

Freeman, N. C. G. and J. Rosenblatt. 1978. The interrelationship between thermal and olfactory stimulation in the development of home orientation in newborn kittens. *Dev. Psychobiol. 11,* 377–457. (a)

Freeman, N. C. G. and J. Rosenblatt. 1978. Specificity of litter odors in the control of home orientation among kittens. *Dev. Psychobiol. 11,* 459–468. (b)

Gewirtz, J. L. 1972. On selecting attachment and dependence indicators. In J. L. Gewirtz (Ed.), *Attachment and dependency.* Washington, D.C.: Winston.

Guyot, G. W., T. L. Bennett, and H. A. Cross. In press. The effects of social isolation on the behavior of juvenile domestic cats. *Dev. Psychobiol.*

Guyot, G. W., H. A. Cross, and T. L. Bennett. In press. Early social isolation of the domestic cat: Responses to separation from social and nonsocial rearing stimuli. *Dev. Psychobiol.*

Harlow, H. F. 1969. Age mate or peer affectional system. In D. Lehrman, R. Hinde, and E. Shaw (Eds.), *Advances in the study of behavior.* Vol. 2. New York: Academic Pr.

Haskins, R. 1977. Effect of kitten vocalizations on maternal behavior. *J. Comp. Physiol. Psychol. 91,* 830–838.

Jolly, A. 1972. *The evolution of primate behavior.* New York: Macmillan.

Kling, A., J. K. Kovach, and T. J. Tucker. 1969. The behavior of cats. In E. S. E. Hafez (Ed.), *The behavior of domestic animals.* 2nd ed. Baltimore: Williams and Williams.

Koepke, J. E. and K. H. Pribram. 1971. Effect of milk on the maintenance of sucking behavior in kittens from birth to six months. *J. Comp. Physiol. Psychol. 75,* 363–377.

Konrad, K. W. and M. Bagshaw. 1970. Effect of novel stimuli on cats reared in a restricted environment. *J. Comp. Physiol. Psychol. 70,* 157–164.

Kuo, Z. Y. 1930. The genesis of the cat's response to the rat. *J. Comp. Psychol. 11,* 1–35.

Kuo, Z. Y. 1938. Further study of the behavior of the cat toward the rat. *J. Comp. Psychol. 25,* 1–8.

Kuo, Z. Y. 1960. Studies on the basic factors in animal fighting: VII. Inter-species coexistence in mammals. *J. Genet. Psychol. 97,* 211–225.

Lehrman, D. S. and J. S. Rosenblatt. 1971. The study of behavioral development. In H. Moltz (Ed.), *The ontogeny of vertebrate behavior.* New York: Academic Pr.

Leyhausen, P. 1956. *Verhaltensstudien an Katzen.* Berlin: P. Perey-Verlag.

Leyhausen, P. 1965. The communal organization of solitary mammals. *Symp. Zool. Soc. London 14,* 249–263.

Leyhausen, P. 1973. [The biology of expression and impression.] In K. Lorenz and P. Leyhausen (Eds.), *Motivation of human and animal behavior: An ethological view.* New York: Van Nostrand–Reinhold. (Originally published 1967.)

Leyhausen, P. 1973. [On the function of the relative hierarchy of moods.] In K. Lorenz and P. Leyhausen (Eds.), *Motivation of human and animal behavior: An ethological view.* New York: Van Nostrand–Reinhold. (Originally published 1965.)

Leyhausen, P. and R. Wolff. 1959. Das Revier einer Hauskatz. *Z. Tierpsychol. 16,* 666–670.

Meier, G. W. 1961. Infantile handling and development in Siamese kittens. *J. Comp. Physiol. Psychol. 54,* 284–286.

Meier, G. W. and J. C. Stuart. 1959. Effects of handling on the physical and behavioral development of Siamese kittens. *Psychol. Rep. 5,* 497–501.

Méry, F. 1972. *The life, history and magic of the cat.* New York: Grossett and Dunlap.

Rheingold, H. and C. Eckerman. 1971. Familiar social and nonsocial stimuli and the kitten's response to a strange environment. *Dev. Psychobiol. 4,* 71–89.

Rosenblatt, J. S. 1965. Effects of experience on sexual behavior in male cats. In F. Beach (Ed.), *Sex and behavior.* New York: Wiley.

Rosenblatt, J. S. 1971. Suckling and home orientation in the kitten: A comparative developmental study. In E. Tobach, L. Aronson, and E. Shaw (Eds.), *The biopsychology of development.* New York: Academic Pr.

Rosenblatt, J. S. 1972. Learning in newborn kittens. *Sci. Am. 227,* 18–25.

Rosenblatt, J. S. and T. C. Schneirla. 1962. The behavior of cats. In E. S. E. Hafez (Ed.), *The behavior of domestic animals.* Baltimore: Williams and Williams.

Rosenblatt, J. S., G. Turkewitz, and T. C. Schneirla. 1961. Early socialization in the domestic cat as based on feeding and other relationships between female and young. In B. F. Foss (Ed.), *Determinants of infant behavior II.* London: Methuen.

Rosenblatt, J. S., G. Turkewitz, and T. C. Schneirla. 1969. Development of home orientation in newly born kittens. *Trans. N.Y. Acad. Sci. 31,* 231–250.

Schneirla, T. C. 1959. An evolutionary and developmental theory of biphasic process underlying approach and withdrawal. *Nebr. Symp. Motivation 7,* 1–42.

Schneirla, T. C. 1965. Aspects of stimulation and organization in approach/withdrawal processes underlying vertebrate behavioral development. In D. S. Lerhman, R. A. Hinde, and E. Shaw (Eds.), *Advances in the study of behavior.* Vol. 1. New York: Academic Pr.

Schneirla, T. C. and J. S. Rosenblatt. 1961. Behavioral organization and genesis of the social bond in insects and mammals. *Am. J. Orthopsychiatry 31,* 223–253.

Schneirla, T. C. and J. S. Rosenblatt. 1963. "Critical periods" in the development of behavior. *Science 139,* 1110–1115.

Schneirla, T. C., J. S. Rosenblatt, and E. Tobach. 1963. Maternal behavior in the cat. In H. L. Rheingold (Ed.), *Maternal behavior in mammals.* New York: Wiley.

Seitz, P. F. D. 1959. Infantile experience and adult behavior in animal subjects: II. Age of separation from the mother and adult behavior in the cat. *Psychosom. Med. 21,* 353–378.

Wenzel, B. M. 1959. Tactile stimulation as reinforcement for cats and its relation to early feeding experience. *Psychol. Rep. 5,* 297–300.

West, M. 1974. Social play in the domestic cat. *Am. Zool. 14,* 427–436.

Wilson, M., J. M. Warren, and L. Abbott. 1965. Infantile stimulation, activity, and learning by cats. *Child Dev. 36,* 843–853.

Wilson, C. and E. Weston. 1947. *The cats of wildcat hill.* New York: Duell, Sloan and Pearce.

Part II

The Primates

Chapter 8

Atypical Behaviors and Social Preferences in Two New World Primate Species

M. Aaron Roy

New World primates are generally not as well understood as Old World primates. Only in the last 20 years have the former become used with any regularity as research subjects or as the object of species-typical behavioral observations. A discussion of the South and Latin American monkeys is nevertheless desirable. Analyses of the development of species-typical social preferences and behaviors in the New World types may offer considerable insight into the generalization of conclusions reached both with Old World[1] primates and with other nonprimate species discussed in this text.

The squirrel monkey and the marmoset (or tamarin[2]) are the two most frequently studied New World primates for which data relevant to species identification are available. Though there is much more to learn about atypical rearing and its effect on species identifications in both marmosets and squirrel monkeys, the diverse rearing styles and social systems which characterize these two species make them especially valuable to study. Since species-typical innate behaviors are found in both marmosets (Hampton, Hampton, and Landwehr, 1966) and squirrel monkeys (Ploog and MacLean, 1963), an insight into the species-identity processes of New World primates will occur by studying which organisms act as social releasers for these unlearned behaviors.

Squirrel Monkeys

Infant squirrel monkeys receive significantly less maternal care than do rhesus monkeys and chimpanzees. A neonate has a number of reflexes by which it maintains contact with the mother without requiring any maternal support (King, Forbes, and Forbes, 1974). As the mother moves about freely, the infant clings to her back, periodically moving laterally and ventrally to nurse. Mothers do not normally groom, cradle, or punish infants, though they may retrieve infants that stray or are in danger (Baldwin and Baldwin, 1971; Rumbaugh, 1965). Retrieval usually involves approaching the infant, establishing lateral contact, and nudging it with the whole body or with a lowered shoulder. Females without infants show "aunting" behavior by either letting infants cling to them or by retrieving infants (Dumond, 1968). An infant spends progressively less time with the mother, and by four months of age it is rarely on her; weaning is completed by six to seven months (Rosenblum, 1968; Vandenbergh, 1966). Males have no role in the care of the infant.

Breeding squirrel monkeys in the laboratory with any regularity has only come about in the last decade (Kaplan, 1977a). Thus, systematic observations on early developmental processes of both mother- and nursery-reared infants are just beginning to be reported for larger groups of subjects. Like other primates, infant *Saimiri* can discriminate their mothers from other females at an early age (Kaplan, Cubiciotti, and Redican, 1977; Kaplan and Schusterman, 1972; Rosenblum and Alpert, 1977). Infants will also develop strong attachments to surrogates when reared away from their mothers (Kaplan, 1977b; Kaplan and Russell, 1974).

Infant *Saimiri* as young as one day old thrive when nursery-reared. This occurs with hand-fed neonates (King and King, 1970), those reared in incubators with ad-lib food and no social contacts (Hinkle and Session, 1972), or those reared individually in wire cages with ad-lib food and visual–auditory contact with conspecifics (Kaplan and Russell, 1974). The many abnormal behaviors shown by some Old World types when reared in the nursery with reduced or no conspecific contacts are not usually reported for squirrel monkeys. There is nonnutritive sucking of digits (Kaplan, 1974; King and King, 1970), but not in all subjects, and to date, only one nursery-reared subject has been considered to show aberrant behavior (Kaplan, 1977c). These preliminary reports suggest that squirrel monkeys are relatively impervious to deprived rearing conditions.

The normal physical and near-normal behavioral development which is found in nursery-reared *Saimiri* had led to speculation that these primates are less dependent on contacts with conspecifics for species-typical behaviors to emerge (see Kaplan, 1977c). This position would seem to suggest that a species identity in squirrel monkeys is less, or not at all, dependent upon postnatal social experiences with conspecifics. Results from three projects with nursery-reared *Saimiri* which involve the author provide mixed support for this notion, i.e., squirrel monkeys may not be genetically insulated against atypical rearing experiences, and they may not inherit their species identity.

Nursery-reared Squirrel Monkeys

Our infant squirrel monkeys were separated from their mothers on the day of birth. All were hand-fed eight times per day with a nursing bottle and nipple until 34 days of age when the bottles were mounted on the cage during each feeding. Bottles fitted with a licking tube were substituted at 14 weeks of age. Weaning occurred at 38 weeks. Surrogates were made of rolled terry-cloth towels fastened with a rubber band. For the first month each neonate was placed in a solid-walled, floor-heated aluminum container which enabled the infant to hear and smell other squirrel monkeys. Beginning with the second month, each infant was placed individually in a wire cage which permitted it to see, hear, smell, and infrequently touch conspecifics.

STUDY 1. Sixteen of these surrogate-reared *Saimiri* are now living in a social group in an outbuilding (Roy, Wolf, Martin, Rangan, and Allen, 1978). Seven remained in the nursery for 12 months before social grouping, while nine were placed together when they were four months old. Observations were made on this group when the members ranged in age from three to four years. No gross behavioral abnormalities were noted. Conspecific interactions were normal in all areas observed: play, dominance, reproduction, and maternal care. Atypical behaviors, however, were evident: two subjects engaged in unusual drinking patterns involving fingers or a tail tip; many showed nonnutritive sucking of digits, arms, and tails; and subjects of both sexes would approach the front of the cage, usually at the beginning of an observation period, and make a genital display toward the observer. Fifty percent of the females gave birth following the first breeding season of maturity in the males, and all infants that lived received adequate maternal care. Aunting behavior was present. The near-normal behavior shown by these adult *Saimiri* (in spite of no contact with older, mother-reared conspecifics) lends support to the notion that *Saimiri* are less dependent upon those types of early social experiences with conspecifics which apparently facilitate normal development in rhesus monkeys and chimpanzees.

STUDY 2. Another group of squirrel monkeys, which was reared and maintained in single cages in a nursery, has developed various aberrant behaviors which are similar to those actions produced by nursery-reared rhesus monkeys and chimpanzees but not previously reported in *Saimiri* (Roy, in preparation). These 11 *Saimiri* received the standard nursery care, but their surrogates and towel bedding were removed at 14 weeks of age. Systematic observations were initiated when the subjects averaged 7 months of age. At this time the nursery-reared infants ate, drank, and manipulated their cages in a fashion similar to four mother-reared infants who served as controls. Thirteen activities, however, were noted in the nursery-reared squirrel monkeys which were never seen in the control subjects. These behaviors included variations of stereotypic posturing, agitation, and nonnutritive orality (see Table 1). The obvious differences in the type of abnormal behaviors as compared to those found in laboratory-reared rhesus monkeys and chimpanzees apparently reflect variations in physique and behavioral complexity between these Old and New World primate species. All subjects engaged in at least one form of nonnutritive orality and in the stereotypic behaviors of huddling and rolling-in-a-ball. Self-startling and despair occurred infrequently. Males tended to be more deviant than females, though there was considerable variability within each sex in the type and degree of aberrant behavior (see Figure 1). A follow-up series of observations was initiated when the subjects averaged 15½ months in age. Abnormal behaviors persisted in those monkeys who exhibited them during the earlier observations. It was as if the development of excessive orality and of holding the tail as a surrogate (as in the huddling variations) prevented the development of normal behaviors.[3]

Table 1. Abnormal Behaviors and Their Description

Nonnutritive Orality	
1. Sucking digits (SD)	Sucking on any digit, usually thumb or index finger.
2. Chewing on body (CH)	Repeated mouthing and/or licking any extremity without causing irritation; not hygienically related.
3. Sucking tail (ST)	Sucking tip of tail.
4. Sucking penis (SP)	Sucking nonerected penis.
Stereotypic Postures	
5. Huddling (H)	Grasping tail, which is wrapped ventrally between legs and arms, with hand(s) and forearm(s); motionless.
6. Huddling with a ball (HB)	Forearms holding large plastic ball ventrally, tail wrapped ventrally between legs and arms; motionless.
7. Huddle rocking (HR)	In huddle position, but rocking back and forth.
8. Huddle hopping (HH)	In huddle position, but moving up and down and hopping forward (similar to a kangaroo); mildly agitated.
9. Huddle walking (HW)	Walking on both legs and one arm while other arm holds tail ventrally; no agitation.
10. Rolling in a ball (RB)	Curled into a tight huddle position while rolling about the floor; irregular movements.
Agitated Actions	
11. Self-startle (SS)	Curled up quietly in a tight huddle position; then, without apparent stimulation, rapid jumping up and down with shrieking; short duration.
12. Despair (D)	Repeated grasping of face and/or top of head very tightly, frequently covering eyes.
13. Looping (L)	Vertical circling about the cage.

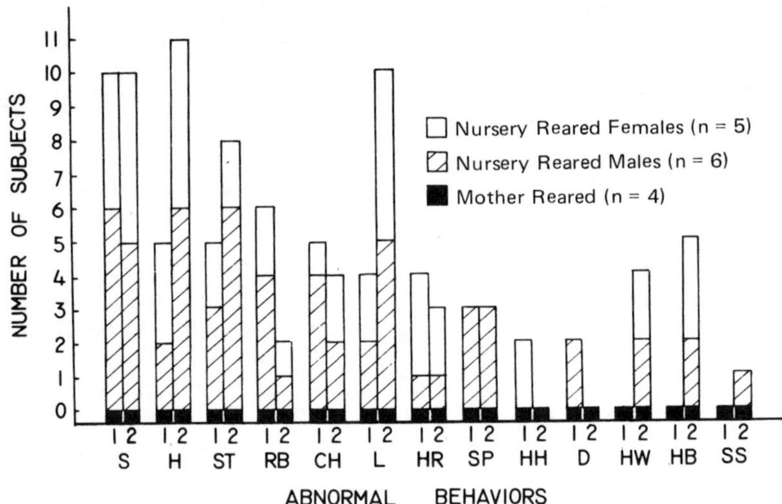

Figure 1. Percentage of subjects showing abnormal behavior patterns. (Numbers 1 and 2 represent the first and second observation periods, respectively, and the abnormal behaviors are described in Table 1.)

STUDY 3. The "genital display" in squirrel monkeys is an innate behavior made by both sexes during social greetings and during dominance interactions (Latta, Hopf, and Ploog, 1967). It is controlled by the caudal part of the globus pallidus (MacLean, 1972, 1975). One thigh is spread with a half-bent knee, there is supination of the foot with abduction of the big toe, and either penis erection or clitoral enlargement occurs. Vocalizations and some urination frequently occur simultaneously. The genital display occurs when one monkey is facing another or when viewing its mirror reflection (MacLean, 1964). Systematic observations were initiated when it became obvious that nursery-reared *Saimiri* regularly displayed toward humans (Roy et al., 1978) whereas feral-reared monkeys did not. This suggested that early social experiences influence which organisms will act as social releasers for this behavior.

Twenty-eight squirrel monkeys, equally divided between sex and rearing condition, were placed in individual cages (Roy and Grannis, 1978). Following an acclimation period of a few days, an observer (without glasses and who neither reared nor cared for any of the animals) rapidly entered the room, approached a cage, and positioned himself directly in front of and at eye level to the subject. Ten such trials occurred over a two-week period. The presence and latency of any genital displays which occurred within 30 seconds were scored (see MacLean, 1972). Clitoric enlargement in females was not recorded. Feral-reared *Saimiri* of either sex never performed a thigh spread, and males never had an erect penis when approached. Nursery-reared subjects, in contrast, regularly showed these behaviors. The monkeys reared by humans showed significantly more urination and more vocalizations than those reared by their mothers in the wild. Both the frequency and the latency of erections in our males were comparable to those reported for males who display to their reflections.

Marmosets

Even less is known about marmosets than about squirrel monkeys. This includes behavior both within the laboratory and in the field. Though marmosets are now being bred in captivity, few of the offspring have been systematically monitored behaviorally. Information which is available from those infants reared apart from conspecifics suggests, however, that atypical behaviors and social preferences can occur as they do in *Saimiri*.

The social behavior of marmosets can be described as that of a tightly knit group with much reciprocity among its members (Stevenson and Poole, 1976). A species-typical behavior which is emitted in nonhostile situations (e.g., grooming and breeding) is "tongue flicking." This involves the rapid movement of the tongue in and out, sometimes in the presence of head shaking. Tongue flicking is interpreted as a greeting display and is directed toward conspecifics or mirror reflections (Hampton et al., 1966).

Infant marmosets receive care from both parents (Fess, 1975a; Hampton et al., 1966) and older siblings (Box, 1977; Kleiman, 1977) beginning at birth. The infants (twin births are the norm) spend much of their time with the father, moving to the mother to nurse. Infants move between caretakers unaided or with assistance. Prior to weaning at about two to three months, the infants are regularly played with, groomed, and retrieved by the parents. After weaning, feral infants retain contact with the parents and any other siblings for many months, at least until after younger siblings are born.

Marmosets can be maintained for long periods in the laboratory, but the production of viable offspring who are reared successfully is less than that found in *Saimiri* (Deinhardt, Wolfe, and Ogden, 1976). This lowered rate of success probably reflects both the excitability of this primate species and the need for compatible, experienced parents. Hand rearing of infants as young as one day has been successful when the neonate is housed inside or outside an incubator (Berkson, Goodrich, and Kraft, 1966; Fess, 1975a,b; Hampton and Hampton, 1967; Pook, 1974; Stevenson, 1976).

Atypically Reared Marmosets

There are few marmosets that have been reared past a few weeks of age either apart from or with reduced conspecific contacts for which we have behavioral information. The number is less than 25. The following reports, as a result, should be viewed as suggestive and no more. Adding to the problem of a small sample is the fact that few of the observations appear to have been conducted systematically.

Hampton and Hampton (1967) report the development of seven cotton-top marmosets (*Oedipomidas oedipus*) who were reared in an incubator-type system. The infants were removed from their parents when they were from one to 18 days old and placed into individual cages. All were hand-fed, and they treated cloth diapers as surrogates. Parent-reared juveniles were added to their cages when the isolated infants were as young as 45 days, with the intent to promote normal socialization. "In animals reared singly, their own face and head skin would occasionally be grasped in periods of apparent frustration" (p 8), but this behavior was only seen in juveniles. This general type of head grasping has also been reported for *Saimiri* (Kaplan, 1977c; Roy et al., 1978). Unlike parent-reared marmosets, these seven human-reared subjects had no fear of humans. No abnormalities were noted in social interactions with other marmosets and nonnutritive orality was absent. In contrast to this report of the absence of nonnutritive orality, both Le Clere (1975) and Pook (1976) report its occurrence in cotton-top marmosets reared by hand.

Berkson et al. (1966) discuss two common marmosets (*Callithrix jacchus*) which were raised in a germ-free incubator with scheduled feedings by humans. One infant was removed from its parents at day 14. The other was removed at birth, but lived with a sibling for 10 days before the latter's death. Each infant was in its own cage, but there was visual, auditory, olfactory, and possibly physical contact between infants. Each infant used a surrogate. The only abnormal, self-directed behavior seen was when one subject would "bite his knee in a repetitive manner during a sequence of activity in which he also bit at his water bottle" (p 493). Various cage stereopathies present in these two marmosets included repetitive backward somersaults, pacing, and jumping repetitively on all fours. A third subject, a male golden marmoset (*Leontocebus rosalia*), was delivered to the laboratory very young and caged alone. Later he also was seen to perform a cage stereopathy. This animal could not be placed with conspecifics, although he could be handled with ease and he presented to humans when grooming was desired.

No abnormal behaviors were seen in four nursery-reared common marmoset infants (*C. jacchus*) reported by Stevenson (1976). All infants were separated from their parents soon after birth, but, either through group caging or through daily play periods with each other, all received social contacts with conspecifics on a regular basis. All subjects were hand-fed and given surrogates.

There are two reports of marmosets being raised as house pets without access to conspecifics. Fess (1975a) raised a male cotton-top (*Oedipomidas oedipus*) from day two to 14 months of age. This marmoset was believed to be imprinted on people in that it would always greet humans with the characteristic tongue flicking. It did not interact normally either socially or sexually when presented with a mate. A subsequent report of a male common marmoset (*C. jacchus*) showed the same type of preference for humans (Fess, 1975b). This male was over two years old when it was caged with a captive-born, parent-reared female. Repeated attempts at pairing were unsuccessful, and much fighting occurred. During these periods the male "directed lip-smacking and tongue-flicking displays at the observer and groomed women visitors" (p 19).

Conclusions

It would be misleading at this time to offer any general statements about characteristics of New World primate species. The diversity in social and individual behaviors just between the two species discussed in this chapter means that, as with many Old World types, generalizations across species must be made with caution.

A discussion of each New World species is thus more appropriate until systematic studies are completed.

The observations that species-typical innate behaviors were directed toward humans by both squirrel monkeys (with the genital display) and marmosets (with tongue flicking) when reared by humans suggest that the social releaser for such actions is influenced by postnatal social experiences. This conclusion has support in that some of the marmosets even displayed for grooming toward humans. The fact that nursery-reared *Saimiri* eventually emit a genital display when interacting with conspecifics (Kaplan, 1977a; Roy et al., 1978) does not negate the position of an acquired social releaser in this species because these monkeys as infants had various conspecific contacts either before or during placement in the nursery. Thus there was ample opportunity for two groups of organisms (i.e., humans and squirrel monkeys) to acquire social-releaser status. It would be of interest to see if a squirrel monkey reared without *any* contact with conspecifics would still display toward *Saimiri* when given the opportunity as an adult.

It is an open question as to whether the fact that marmoset and squirrel-monkey innate behaviors can be "triggered" by a nonspecies member reflects a distorted species identity. More research involving isolation rearing and objective social preference testing is needed, and a wider range of behaviors will have to be monitored. Nevertheless, the behavior of the two home-reared marmosets reported by Fess (1975a,b) strongly suggests that a normal species identity was *not* present in these pets.

Nursery rearing can contribute to the development of various abnormal behaviors in both squirrel monkeys and marmosets, but perhaps not as readily as these behaviors are induced in some Old World species. The mere presence of aberrant behaviors in *Saimiri* lasting past 12 months of age is important in itself. Previous reports that squirrel monkeys and marmosets were relatively free of abnormal behaviors may have been inadvertently based on subjects which were not raised atypically enough or which were not observed under similar situations.

The essentially normal social and maternal behaviors of adult *Saimiri* reported by Roy et al. (1978) suggest that this species shows few long-term effects of rearing in the nursery. This may be the case, at least for the type of nursery rearing and post-nursery rearing experiences (which included minimal contact with humans) these particular monkeys received. A direct comparison with rhesus monkeys or chimpanzees unfortunately is not possible, since there have been no reports of an ongoing social group composed solely of these nursery-reared subjects. In addition, maternal behavior in female *Saimiri* is not very complex or involved; thus it may not readily reflect any residual effects of nursery rearing.

In summary, the presence of abnormal behaviors and social displays in squirrel monkeys and marmosets that have been reared with little or no experience with conspecifics does not directly influence our understanding of a species identity in these species. The presence, however, of such behaviors indirectly suggests that, when different enough, atypical social experiences early in life can distort the development of normal social preferences in these primate types. This would imply a distorted species identity. Postnatal social experiences apparently have some as yet undetermined influence on species identification in both marmosets and squirrel monkeys.

Acknowledgment

Work on this chapter and research mentioned within were conducted while the author was a postdoctoral research fellow in the Neurology Department of the Delta Regional Primate Research Center, Tulane University, Covington, Louisiana. Drs. Peter Gerone, Ken Brizzee, and Bernice Kaack greatly supported this endeavor.

Notes

1. Many Old World primates have been used as subjects in studies which have manipulated early rearing experiences (see Roy, 1976, 1977a, b). Since the majority of investigations have used either chimpanzees (*Pan*) or rhesus monkeys (*Macaca mulatta*), these two types will be used synonymously with "Old World species." This is done in full awareness of subspecies differences in susceptibility to nursery rearing (see Sackett et al., 1976).
2. Zoological classifications for many New World primates differ among individuals (see Hershkovitz, 1974; Simpson, 1945; Thorington, 1976). As a result, the names "squirrel monkeys" or "*Saimiri*" will be used to refer to all subspecies of squirrel monkeys, while "marmosets" will collectively refer to marmosets, tamarins, and pinches. Family and subfamily names will be mentioned when they are particularly relevant.

3. The incidence of behavior abnormalities in these subjects was surprising given the literature which reported that they did not occur. This inconsistency may be attributed to our more atypical rearing conditions. Compared to Kaplan's (1974, 1977b,c) subjects who were not reported as abnormal, our *Saimiri* were removed from their mothers earlier, a nursing nipple was not available as often or as long, they were observed at an older age, and they had less exposure to a surrogate object. Furthermore, the surrogate used by Kaplan (1974) was more like the real mother both in its texture and in its position. It was affixed to the cage at an angle which permitted the infant to "ride it."

References

Baldwin, J. F. and S. I. Baldwin. 1971. Squirrel monkeys (*Saimiri*) in natural habitats in Panama, Columbia, Brazil, and Peru. *Primates 12*, 45–61.

Berkson, G., J. Goodrich, and I. Kraft. 1966. Abnormal stereotyped movements of marmosets. *Percep. Mot. Skills 23*, 491–498.

Box, H. O. 1977. Quantitative data on the carrying of young captive monkeys (*Callithrix jacchus*) by other members of their family groups. *Primates 18* (2), 475–484.

Deinhardt, F., L. Wolfe, and J. Ogden. 1976. The importance of rearing marmoset monkeys in captivity for conservation of the species and for medical research. In *First inter-American conference on conservation and utilization of American nonhuman primates in biomedical research*. Washington, D.C.: W. H. O. Publ. 317.

Dumond, F. 1968. The squirrel monkey in a seminatural environment. In L. A. Rosenblum and R. W. Cooper (Eds.), *The squirrel monkey*. New York: Academic Pr.

Fess, K. J. 1975. Observations on a breeding pair of cotton-top pinches (*Oedipomidas oedipus*) and nine twin births and three triplet births. *J. Marmoset Breeding Farm 1*, 4–12. (a)

Fess, K. J. 1975. Observations of feral and captive *Cebuella pygmaea* with comparisons to *Calithrix geoffroyi*, *Callithrix jacchus*, and *Oedipomidas oedipus*. *J. Marmoset Breeding Farm 1*, 13–22. (b)

Hampton, S. H. and J. K. Hampton, Jr. 1967. Rearing marmosets from birth by artificial laboratory techniques. *Lab. Anim. Care 17*(1), 1–10.

Hampton, J. K., Jr., S. H. Hampton, and B. T. Landwehr. 1966. Observations on a successful breeding colony of the marmoset, *Oedipomidas oedipus*. *Folia Primatol. 4*, 265–287.

Hershkovitz, P. 1974. A new genus of late Oligocene monkey (Cebidae, Platyrrhini) with notes on postorbital closure and platyrrhine evolution. *Folia Primatol. 21*, 1–35.

Hinkle, K. and L. Session. 1972. A method for hand rearing of *Saimiri sciureus*. *Lab. Anim. Sci. 22*(2), 207–209.

Kaplan, J. N. 1974. Growth and behavior of surrogate-reared squirrel monkeys. *Dev. Psychobiol. 7*(1), 7–13.

Kaplan, J. N. 1977. Breeding and rearing squirrel monkeys (*Saimiri sciureus*) in captivity. *Lab. Anim. Sci. 27*(4), 557–567. (a)

Kaplan, J. N. 1977. Perceptual properties of attachment in surrogate-reared and mother-reared squirrel monkeys. In S. Chevalier-Skolnikoff and F. E. Poirier (Eds.), *Primate bio-social development*. New York: Garland. (b)

Kaplan, J. N. 1977. Some behavioral observations of surrogate- and mother-reared squirrel monkeys. In S. Chevalier-Skolnikoff and F. E. Poirier (Eds.), *Primate bio-social development*. New York: Garland. (c)

Kaplan, J. N., D. Cubiciotti, and W. K. Redican. 1977. Olfactory discrimination of squirrel monkey mothers by their infants. *Dev. Psychobiol. 10*, 447–453.

Kaplan, J. and M. Russell. 1974. Olfactory recognition in the infant squirrel monkey. *Dev. Psychobiol. 7*, 15–19.

Kaplan, J. and R. J. Schusterman. 1972. Social preferences of mother and infant squirrel monkeys following different rearing experiences. *Dev. Psychobiol. 5*, 53–59.

King, J. D., J. T. Forbes, and J. L. Forbes. 1974. Development of early behaviors in neonatal squirrel monkeys and cotton-top tamarins. *Dev. Psychobiol. 7*, 97–109.

King, J. E. and P. A. King. 1970. Early behaviors in hand-reared squirrel monkeys (*Saimiri sciureus*). *Dev. Psychobiol. 2*, 251–256.

Kleiman, D. G. 1977. Progress and problems in lion tamarin reproduction. *Int. Zoo Yearb. 17*, 92–97.

Latta, J., S. Hopf, and D. Ploog. 1967. Observation on mating behavior and sexual play in the squirrel monkey (*Saimiri sciureus*). *Primates 8*, 229–246.

LeClere, K. 1975. A successful reintroduction of a hand-reared cotton-top marmoset, *Saguinus oedipus*, to its family. *The Keeper 1*, 3–5.

MacLean, P. D. 1964. Mirror display in the squirrel monkey, *Saimiri sciureus*. *Science 146*, 950–952.

MacLean, P. D. 1972. Cerebral evolution and emotional processes: New findings on the striatal complex. *Ann. N.Y. Acad. Sci. 193*, 137–149.

MacLean, P. D. 1975. Role of pallidal projections in species-typical display behavior of squirrel monkey. *Trans. Am. Neurol. Assoc. 100*, 25–28.

Ploog, D. W. and P. D. MacLean. 1963. Display of penile erection in squirrel monkey (*Saimiri sciureus*). *Anim. Behav. 11*, 32–39.

Pook, A. G. 1974. The hand-rearing and reintroduction to its parents of a saddle-back tamarin (*Saguinus fuscicollis*). *Eleventh annual report of the Jersey Wildlife Preservation Trust*, 35–39.

Pook, A. G. 1976. Some notes on the development of hand-reared infants of four species of marmoset (*Callitrichidae*). *Thirteenth Annual Report of the Jersey Wildlife Preservation Trust,* 35–39.

Rosenblum, L. A. 1968. Mother-infant relations and early behavioral development in the squirrel monkey. In L. A. Rosenblum and R. W. Cooper (Eds.), *The squirrel monkey*. New York: Academic Pr.

Rosenblum, L. A. and S. Alpert. 1977. Response to mother and stranger: A first step in socialization. In S. Chevalier-Skolnikoff and F. E. Poirier (Eds.), *Primate bio-social development.* New York: Garland.

Roy, M. A. 1976. Early rearing of infrahuman primates with reduced conspecific contacts: A selected bibliography. Part I. *J. Biol. Psychol. 18*(2), 36–42.

Roy, M. A. 1977. Early rearing of infrahuman primates with reduced conspecific contacts: A selected bibliography. Part II. *J. Biolog. Psychol. 19*(1), 21–27. (a)

Roy, M. A. 1977. Early rearing of infrahuman primates with reduced conspecific contacts: A selected bibliography. Part III. *J. Biol. Psychol. 19*(2), 29–30. (b)

Roy, M. A. and S. Grannis. 1978. Atypical social preferences in nursery-reared squirrel monkeys (*Saimiri sciureus*). Paper presented at the annual meeting of the Midwestern Psychological Association, Chicago, May.

Roy, M. A., R. H. Wolf, L. N. Martin, S. R. S. Rangan, and W. P. Allen. 1978. Social and reproductive behaviors in surrogate-reared squirrel monkeys (*Saimiri sciureus*). *Lab. Anim. Sci. 28*(4), 417–421.

Rumbaugh, D. M. 1965. Maternal care in relation to infant behavior in squirrel monkey. *Psychol. Rep. 16,* 171–176.

Sackett, G. P., R. A. Holm, and G. L. Ruppenthal. 1976. Social isolation rearing: Species differences in behavior of macaque monkeys. *Dev. Psychol. 12*(4), 283–288.

Simpson, G. G. 1945. The principles of classification and classification of mammals. *Bull. Am. Mus. Nat. Hist. 85,* 1–350.

Stevenson, M. F. 1976. Maintenance and breeding of the common marmoset with notes on hand-rearing. *Int. Zoo Yearb. 16,* 110–116.

Stevenson, M. F. and T. B. Poole. 1976. An ethogram of the common marmoset (*Calithrix jacchus*): General behavioral repertoire. *Anim. Behav. 24,* 428–451.

Thorington, R. W., Jr. 1976. The systematics of New World monkeys. In *First Inter-American Conference on Conservation and Utilization of American Nonhuman Primates in Biomedical Research.* Washington, D.C.: W. H. O. Publ. 317.

Vandenbergh, J. G. 1966. Behavioral observations of an infant squirrel monkey. *Psychol. Rep. 18*(3), 683–688.

Chapter 9

Old World Monkeys: Consequences of Atypical Rearing Experiences

Gilbert W. Meier
Victoria Dudley-Meier

In a communal setting macaque monkeys develop an early recognition of mother, siblings, other close relatives, nonrelatives, and nonmonkeys. In this context the infant monkey returns promptly to its mother upon any social agitation, even at 3 to 4 weeks of age when making its first uncertain ventures away from its usual ventrum–ventrum relation with the parent. It subsequently shows specific familiarity with its siblings, less with aunts and age peers, and even less with unrelated adults. During this early period others of the social group identify the infant as the offspring of the particular mother and respond differentially to the infant according to its age and gender and to the mother's parity and social rank. These specific orientations among offspring, parent, siblings, and unrelated individuals are maintained for the entire preadult period and, in the case of the female offspring at least, well into the reproductive life of the offspring herself (see Kaufmann, 1966; Koford, 1965; Lindburg, 1971). During the preadult period, close affiliations with mother and siblings dominate the daily life of the individual. Furthermore, they direct the acquisition of food habits and feeding behaviors (Meier, 1971; Meier and Devanney, 1972), as well as the qualitative and distributional patterns of play behaviors (Gard and Meier, 1977; Meier and Devanney, 1974). During the adult period, such relations are reflected in mating choices (Koford, 1965; Missakian, 1972; Sade, 1972) and possibly in patterns of cohesion and cleavage of the group structure (Missakian, 1972, 1973).

In the laboratory, isolate-reared monkeys show somewhat similar specific orientations to unique features of their environments, as determined by maturity, previous experience, and the complexity and variety of the immediate environment. An isolate-reared infant will show behaviors well directed toward a piece of toweling, a wire cylinder, a cloth-covered cylinder, cloth-covered heated cylinder (Mitchell, 1970), a cloth-covered moving cylinder (Mason and Berkson, 1975), a dog (Mason, 1972), or another infant monkey (Harlow and Harlow, 1965). Each choice depends upon the unique familiarity generated by previous rearing experiences. Given a choice, the infant will readily orient to and approach the familiar object—its mother surrogate—rather than nonsurrogates and even the human caretaker rather than strange conspecifics (Sackett, 1970a; Sackett, Porter, and Holmes, 1965). In later years the isolate-reared monkey will show difficulty in sustaining species-typical relations with conspecifics, in effecting successful sexual relations, and in giving care to offspring (Meier, 1965b; Mitchell, 1970; Sackett, 1970a,b; Sackett and Ruppenthal, 1974).

Do these orientations to features of the social environment or to features of the isolated laboratory environment represent the establishment of a species identification on the part of the socially reared animal or the absence of such an identification in its isolate-reared counterpart? We do not think either case is true, at least not in the strictest, most cognitive sense of the concept of species identification. Rather, the field and laboratory observations together reveal much about the species-typical stimulation and the species-typical responses characteristic of early postnatal macaque development. These revelations inform us of the developmental biology of these monkeys in that they suggest a limitation of concepts regarding the implications of the usual environment in which these life processes occur (see Bekoff, 1976). Ultimately they constrain us in our application of such abstractions as species and species identity when attached to the behaviors of organisms other than human or of human beings of a different culture. These concepts are of recent origin, having evolved for classificatory purposes in the Western scientific world. If, on the other hand, we place our definition of species identity in strictly observable behavioral terms of social orientation and adaptation, we may recognize the fundamentals of our concept in some of these early behaviors of infant macaques. We may grasp, further, through the analysis of available data on the elaboration of social behavior something of the reality of the nonhuman primate world, the contributions that each member offers at the time of its entry into that world, and the usual outcomes represented by subsequent changes in both the target animal and its environment which follow upon their continuing interactions. Just possibly this analysis will provide some insights into the understanding of ourselves as a biologic grouping, representatives of a species defined according to the rules of current Western science.

Consequences of Varied Rearing Experiences

Before examining some of the data on the consequences of typical and atypical rearing in macaques, let us first offer this caveat: none of these studies, at least none known to these authors, was undertaken to investigate the ontogeny of a concept of species identity or anything like it. Insofar as is known, no author has made a claim of such a purpose, implicitly or explicitly, in the written report of his investigations. Each author instead undertook his investigation with other goals in mind, such as the understanding of the processes of affiliation, attachment, communication, emotion, learning, memory, and reinforcement or such as the evaluation of the contributions of early experiences to these processes and the establishment of critical periods germane to each or to all. Only recently have these researchers acknowledged the relevance of the concepts of ontogeny and phylogeny and of species and taxa for their investigations.

This chapter will, further, be limited largely to those descriptions of isolate-reared animals which have been exposed during the infantile period to minimal human intervention and frank physical restriction (Sackett, 1968). These are the studies in which the animals were reared under conditions of partial social isolation or, as we prefer calling them, under conditions of essential social (or haptic) isolation. In them the infant was placed in a relatively small cage, with towel or surrogate and a bottle of liquid diet (proprietary milk formula) with appropriate nipple, in a smallish room (the nursery) with other monkeys similarly confined (van Wagenen, 1950). These are the conditions which have been most widely used and for which the greatest amount of long-term behavioral data are available. (A meaningful definition of species identity must include some consideration of the implications of the particular behaviors for reproduction, i.e., for species propagation and continuance, thus the concern for long-term study. The studies on the sequelae of essential social isolation do indicate that reproduction is possible with these rearing experiences, although the detailed conditions for success are far from clear [cf. Harlow and Harlow, 1962; Meier, 1965a,b; Sackett and Ruppenthal, 1974].) The socially reared animals, the controls, in the typical experiment were feral-reared (i.e., wild-born) until approximately the age that the isolate-reared animals were removed from the most severe impoverishment and placed in more conventional laboratory conditions. Although the conditions of rearing varied from experiment to experiment and from laboratory to laboratory, the experiences of these socially reared animals after about 18–24 months of age were usually the same as for the isolate-reared animals. Consequently any appraisal of the adult behaviors of these animals must reflect that controls and experimentals alike had had some lengthy period in juvenile isolation and social impoverishment and therefore may not have behaved in ways wholly like feral-maintained age peers, merely adapted to the laboratory regimen (cf. Meyer-Holzapfel, 1968).

Laboratory Rearing

Now consider the actual reports: The most prominent body of research comes from the Wisconsin laboratories, primarily those under the direction of Harlow (e.g., Harlow and Harlow, 1965; Harlow, Harlow, Schiltz, and Mohr, 1971). He and his collaborators have advanced the procedure of isolate-rearing (e.g., Blomquist and Harlow, 1961) initially for the purposes of experimental manipulation and control (e.g., Harlow, 1958). Others have followed suit often for identical reasons (e.g., Fleischman, 1963). The procedure usually entails the removal of the infant from the mother at the time of birth or shortly thereafter and maintaining the infant in a laboratory confinement patterned, apparently, after a modern American nursery for human newborns. After a period of three to six months the infant is removed from the more restrictive environments of incubator and small cage to larger quarters—both cage and cage enclosure—but is still kept in haptic isolation from other members of its species. The sights, sounds, and odors of conspecifics, especially of age peers, may be present, however. This animal observed at two and a half years of age and thereafter, shows a number of behaviors only rarely seen among feral-reared animals of these ages: stereotypies oriented to the self and to details of the environment, exaggerated responses to fixed and novel aspects of the environment, and extreme withdrawal from significant environmental change. The behaviors shown idiosyncratic for each animal result in bizarre approaches to conspecifics and often in confounded or delayed reproduction. The qualitative and quantitative natures of these behaviors vary with the gender of the animal, the age when observed, the conditions and duration of observation, the duration of adaptation to the observational setting, and to the as yet unspecified conditions of animal maintenance and operations which typify and distinguish individual laboratories (Meier, 1970). Recent data reveal the subtlety of the differences in adult behavior of the isolate-reared animals in social situations.

Communication Characteristics

In the context of an examination of the assimilation and transmission of significant information in an aversive conditioning paradigm, Miller and his associates at the University of Pittsburgh (1971) noted that the isolated-reared adult rhesus monkey can transmit but not receive information effectively. That is, in this affect communication situation, those animals that had been isolate-reared did respond appropriately to the onset of the conditioned stimulus (a light signaling the onset of electric shock after a fixed interval) with species-typical facial–postural changes which could be discriminated by the socially reared conspecific. When placed, however, in the operant or receiver role in which differential responding to such facial–postural displays was necessary for successful shock avoidance (a bar press to terminate the signal and to preclude the onset of shock for that trial), the isolate-reared animals were no match for their socially reared controls. Briefly, the isolates could not read the displays from the sending animal with the same effectiveness as shown by the socially experienced animal, nor with the same facility which they themselves had shown in sending the requisite message to the socially reared receivers (Miller et al, 1967).

A feature of the affect communication studies that deserves emphasis, certainly much more than has been given previously, is that the isolate-reared monkey as receiver did respond to its yoked control; it oriented and it performed. The differences between the performances of the two groups of animals appear to be largely quantitative, i.e., differences in the probability of responding when the probabilities of both groups were greater than chance. In this regard these studies are consistent with others in which the behaviors of the isolate-reared animals as adults and sometimes as juveniles were contrasted with those of socially reared age peers. For example, Sackett examined the preferential behaviors of rhesus monkeys which had been isolate-reared, using a circus apparatus in which the tested animal could orient or approach (depending upon the situation and the experiment) one of three (or six, again depending upon the situation and the experiment) stimulus compartments. In this apparatus Sackett presented to the test animal, in the center of this six-sided configuration having the stimulus compartments at the periphery, other relevant animate stimuli: cagemates, peermates, conspecifics, macaques of other species, adults, infants, etc. Sackett generally found that infant animals which had been isolate-reared oriented more commonly toward and approached stimulus animals which had been reared like the target animal itself. More specifically, the infant isolate oriented toward the infant isolate, and the juvenile isolate oriented toward and approached the isolate-reared stimulus animal of the same gender as well as of the same rearing experience. At about four to five years of age, when the animals reached sexual maturity, the target animal more often oriented to the like-reared stimulus animal of the opposite gender, as did the socially reared young adults (Sackett, 1970b).

To illustrate still other features of the adult behaviors of isolate-reared macaques and the adaptability–plasticity that they can show, we will describe some recent data from our own laboratory. The first experiment, conducted as a study for a master's thesis by Strond at the University of Nebraska at Omaha, represents in part an update of the emotionality studies of Hebb (1946) and in part an analysis of the influence of social context on the display of emotional behaviors in a laboratory setting. That isolate-reared monkeys were involved makes the investigation especially pertinent here. Strond used four animals, all rhesus monkeys of about six years of age. Two (one male and one female) had been reared in isolation under the conditions considered typical in this review until about one and a half years of age, then in pairs in standard adult cages until about two and a half years of age, and thereafter in individuated cages as subadults and adults in a moderate-sized colony room. Two (one male and one female) had been reared in a communal setting (one feral and one simulated) for at least three years and in individuated cages for at least two years thereafter. After training the animals to respond by approach and retrieval of reinforcement (raisins) upon presentation of the test apparatus, Strond then paired the presentation of the food with the simultaneous presentation, to the rear of the food well, of a ceramic life-sized model of a rhesus head giving an emotional display (open-mouth threat, threat, neutral, or sleeping). He next contrasted the monkeys' behaviors with those shown during the simultaneous presentation of a ceramic blank (a sphere of roughly the same size as the ceramic models; see Figure 1). For this testing, the target animal was maintained in its individuated cage, now isolated from the colony in a nearby room.

A testing session began with the wheeling up of the apparatus to the front of the cage. Each trial began with the opening of the doors which separated the animal from the food well and the ceramic stimulus. Each trial ended with a successful retrieval of the reinforcement or the lapse of five minutes in the case of no retrieval and then the closing of the doors. The intertrial interval was also five minutes.

Figure 1. Ceramic models of rhesus-monkey facial displays, from left to right: sleeping, neutral, threat, open-mouth threat.

Each ceramic head was shown to the test animal once daily in a balanced order, with presentations of the blank between successive presentations of the heads and at the beginning and at the end of each session, on each of four days.

During two additional blocks of four days, each of the animals was paired in the same test cage with each of the others of opposite gender.

For both the individual and the paired conditions Strond recorded the incidence of responding (orienting, approaching, and retrieving) successfully to the test situation, the latency of response, and the occurrence of other behaviors (e.g., threats, withdrawals, shifts of location). The data on latency of successful response in the individual condition indicated clear differences related to early rearing experiences and to the nature of the ceramic stimuli. All animals responded differentially to the blank as contrasted with the models of the facial displays: each retrieved the raisins more slowly in the presence of the facial displays than in the presence of the control stimulus. The isolate-reared animals progressively responded more rapidly to all the facial displays over successive presentations of the emotional, inhibiting displays. In contrast, the socially reared animals showed this disinhibition most clearly to the sleeping face and only slightly less so to the neutral face; they evidenced little significant change in the speed of their responding to the most intense displays (the threat and open-mouth-threat models). No differences in latencies which could be attributed to the gender of the animal tested were recorded.

Similar data from the paired condition revealed complex interactions in the behaviors as related to the gender (see Mitchell and Caine, Chapter 10, this volume) and the experience factors. Although all animals demonstrated inhibition of response despite their recoveries (disinhibition) under the unpaired condition, the presence of a male generally inhibited further the response of the females. The males, on the other hand, were more likely to indulge in other behaviors than they were in the unpaired condition. These trends were least evident in the pairing of the socially reared female with the isolate-reared male and most evident in the pairing of the socially reared female with the socially reared male. In the former pairing the female was the more likely member of the pair to approach the apparatus and retrieve the raisins; in the latter pairing, neither approached the food well: the female huddled in the corner and the male threatened the ceramic model. For all animals in the paired condition, the latencies were considerably longer than for those in the unpaired condition. Some of this increase can be attributed to the high incidence of stereotypic behaviors among the isolate-reared animals and to the parallel increase in the incidence of threat and submissive displays among the socially reared animals which were directed to the stimuli and to each other apparently in response to the stimulus presentation. The ceramic heads clearly were effective in eliciting emotionally toned behaviors (including the inhibition of successful response), the intensity and persistence of which were related to particular features of the three-dimensional ceramic stimuli. The precision of such behaviors, moreover, was related to the developmental experiences of the particular animal. Equally important, these data indicate the strong influence of social context; the responses to particular stimuli were exaggerated, if anything, by the presence and the gender of the conspecific. To some extent, at least, even the isolate-reared animals recognized the presence of a species-significant stimulus (the macaque facial display) and the developmental and gender characteristics of the conspecific with whom they were paired. The differential behaviors shown under the single and the paired conditions provide strong support for the notion that emotional behaviors are largely social in nature and that they are a form of communication, typically with a conspecific. Espousing such a notion, the authors were not surprised by the rapid diminution of the effect with repeated presentation in the unpaired condition nor by the relatively sustained effect in the paired condition. We do recognize, nevertheless, that all animals—isolate-reared and socially reared alike—had several years of social experience of some sort. For those that were socially reared this experience included a period of intense social interaction typical of the feral state plus the more recent experience restricted to visual, auditory, and possibly olfactory interactions. For those that were isolate-reared, this experience included the year (between one and a half and two and a half years) paired with an animal of like age, gender, and early rearing experiences plus the early and recent experience restricted to visual, auditory, and olfactory interactions. (The more recent experiences in individuated cages were interrupted only infrequently by the social interactions requisite for the time breeding program in which both groups were involved.) The failure to find even more striking differences between the two groups as related to their early postnatal experiences may be attributed by some to the brevity and the timing of that early experience and to the lengthy period of some form of social experience that followed. Although both suppositions are plausible, they are contrary to the existing data and to the speculations regarding the social sensitivity of a selected early postnatal period and the role of haptic versus visual–auditory stimulation in the experience during that period (Harlow and Harlow, 1965; Ruppenthal et al., 1976). That both groups recognized the fundamentals of the simulated primate displays may attest to the significance of the absolute dimensions of those simulated displays relative to the dimensions of the viewers, as well as to the qualitative characteristics of the models. Perhaps these perceptual absolutes diminished the significance of the differences in early social experiences.

The qualitative nature of Strond's data is consistent with the observations on social strategies in socially reared and isolate-reared juvenile rhesus monkeys reported by Anderson and Mason (1978) and Anderson, Kenney, and Mason (1977). Although incomplete for the purposes of this review—the animals were not adult at the time of data collection—they do indicate that the isolate-reared animals approached the competitive test situations with unique and generally less complex strategies than did their socially reared counterparts. The behaviors of the isolate-reared animals may change with further laboratory experience in a direction suggesting the use of more complex strategies. Whether, however, the group differences will disappear may depend upon the sensitivity and limitations of the measures used. Compare, for example, the observations of Mason (1960) on breeding behaviors of juvenile males with those of Meier (1965a) and of others alluded to later in this review from the authors' laboratory on adult monkeys. Inadequacies in reproductive behavior may be reduced or removed entirely by the longer laboratory existence, as detected by measures of reasonable but narrow scope. In any case we must acknowledge that laboratory existence need not entail unique or readily quantified additions in social experience, relevant or otherwise, for demonstrable change to be effected.

Reproductive and Caretaking Behaviors

Data on the reproductive behaviors of isolate-reared animals lend further emphasis to the need to consider the social environment in the analysis of the behavioral sequelae of early rearing experience. Furthermore, they may provide additional information on the recognition of species identity in isolate-reared animals. Among socially reared animals the quality of mother–infant interactions depends, in large part, upon the nature of the environment in which the pair is placed. Most influential in this regard are the variety and the complexity of the social environment. Infants, for example, maintained with mothers under conditions of social isolation and comparative physical austerity (i.e., in a barren cage) show limited exploratory behavior and considerable physical contact with the socially reared mother. Infants maintained with socially reared mothers under similar conditions of social isolation but with modest physical affluence (i.e., in a cage with objects for more complex manipulation and locomotion) show much exploratory behavior and limited maternal contact, but receive considerable maternal rejection (Jensen et al., 1968). In contrast, infants reared in a larger social context with other adults, juveniles, and infants are much more limited in their interactions with the environment but have much contact with the mother which restricts the infant to her immediate proximity (Wolfheim et al., 1970). The group seemingly exerts a pressure on the mother–infant pair which tends to keep them together physically and to delay the infant's interactions with the total group. Apparently this pressure is by way of social control of the mother's behaviors (e.g., increasing the tendency to confine the infant, to restrict the infant's locomotor activities, and to retrieve the infant promptly should it chance to wander off) and is exerted through the usual channels of communication, specifically, the vocal and facial-postural displays. The sectioning of the facial nerve (to paralyze the face), for example, in adults of a mixed social group significantly reduces the interactions between the infants and their mothers in the group and permits unusually large distances of separation to be tolerated by both members of the pair (Meier, Izard, and Cobb, 1971).

Variants in the species-typical patterns of reproduction, as seen in the laboratory that is, should tell us much regarding the processes of communication in isolate-reared monkeys and the nature of acceptable communicators and conspecifics (defined behaviorally, as discussed earlier). Female monkeys, however, that have been reared in essential social isolation for the first 12–18 months of postnatal life show relatively poor reproductive performance (from conception through pregnancy, delivery, and care-giving) upon reaching sexual maturity (Harlow and Harlow, 1962; Mitchell, 1970; see also Sackett, Holm, and Ruppenthal, 1976, for information on striking species differences). Although conception and satisfactory delivery are possible (Meier, 1965a), the female rhesus macaques are more readily disturbed under other-than-normal delivery conditions (e.g., following cesarean delivery, Meier, 1965b). In such pairs, moreover, as in pairs in which the mother had been socially reared, both the mother and the infant modify and are modified by the behavioral interactions with each other. The presence of the infant, for example, may reduce sharply the frequency and the severity of stereotypic and other bizarre behaviors typically displayed by the isolate-reared adult female (Chappell and Meier, 1974).

In devising a follow-up experiment on observations of reproductive behaviors of isolate-reared animals, we speculated that under conditions of parturition and caregiving, the isolate-reared animals would respond in idiosyncratic ways to the signals generated by other adults in the immediate vicinity, as well as by the infant at breast or otherwise in close proximity. We maintained that such adult–adult and adult–infant interactions, however bizarre, would be reliable and would ensure the adequate development of the offspring of the isolate-reared female, at least under the restrictions and simplicity of the usual laboratory environment. The unique sequences of behaviors and the contingencies of events would yield a seemingly normal outcome in the infant's developmental status. We presumed, further, that the conditions which lead to effective care-giving behaviors exist in the immediate physical and social environment and exert their control on the adult female (the mother), as determined by her immediate behavioral repertoire and her history of environmental reinforcements of those behaviors.

In the experiment (unpublished) a within-subjects design was incorporated—each subject being a dyad of adult female monkeys—with four replications, one each for four different historical conditions. Focus was placed on the individual behaviors of the females, one of whom was pregnant in each dyad, during the prepartal period and, subsequently, on behaviors of the triad now consisting of these females and the infant born to one of them. The analysis was of the contingent relations among these individual behaviors and of the controls exerted by the behaviors of one upon those of the other(s). Each pregnant female rhesus monkey (the mother) was placed in one compartment of a two-compartment wire cage; the nonpregnant female (the aunt) was placed in the other. Each dyad represented one of two rearing conditions (isolate or social) and one of two pregnancy conditions (primigravidous or multigravidous) relative to the adult females involved. The dyads, or eventual triads, were observed daily for one or two ten-minute sessions and scored simultaneously for individual behaviors at 15-second intervals on a standardized checklist (Meier and Devanney, 1975).

This experiment did not determine whether infant care-giving does or does not occur or whether certain care-giving behaviors are good or bad. Rather, it revealed something of the variety of infantile and maternal behaviors that can yield a sustaining care-giving relationship. Five of the eight pregnancies (six to isolate and two to socially reared females) observed in this study were followed by the successful sustenance of the infant until two months of age, when the infant first ventured into the aunt's cage and when our period of most intense observation ended. The three failures, all to isolate-reared females, were rejections of the infants immediately after delivery. The females involved, all primiparae, behaved toward their infants in ways that we have seen in primiparous socially reared females. (One of these three females sustained the infant from a subsequent pregnancy.) Neither these females nor the other isolate-reared females showed any greater difficulty in becoming pregnant, which would suggest reproductive inadequacy or complication, than did the socially reared females. (Note: the sires in these pregnancies were also isolate-reared; cf. Mason, 1960; Meier, 1965a.)

Among the isolate-reared pairs the frequency of stereotypic behaviors decreased with the approach of parturition and very sharply after the actual delivery of the infant. It increased after the infant became mobile, especially after the infant wandered into the aunt's cage and was duly entertained there.

All aunts, isolate-reared and socially reared alike, demonstrated keen interest in the infants as indicated by the increased grooming of the mother after the infant's delivery and especially by the grooming of the mother's ventrum and at times by the seemingly surreptitious grooming of the infant. Immediately after the birth of the infant the effect of the aunt in controlling the mother's behaviors (e.g., retrieving, vigorous cradling, hugging, and patting of the infant) was evident in all triads. The effect was least prominent in the triad with the isolate-reared primipara. Here the aunt did not show keen interest in the mother–infant pair until the infant became active on the mother's ventrum at about ten days of age.

Most of these aunt–mother–infant behaviors revealed clear factorial interactions between the age-parity and rearing-experience conditions. The most evident control of the mother's behaviors, for example, was in the triad with the socially reared primipara; the least evident, but in a context of frequent sisterly interactions, was in the triad with the socially reared multipara. The two isolate-reared triads were aligned as other data would suggest: the less evident control, in a context of very little interaction of any sort, was in the triad with the isolate-reared primipara; more evident control, second to the triad with the socially reared primipara, was displayed by the triad with the isolate-reared multipara. More striking, however, were the qualitative differences in emotional tone of the behaviors of the isolate-reared versus the socially reared mothers. In most respects, these differences seemed to be exaggerations of those seen in Strond's study: the occurrence of stereotypic behaviors associated with the presence of another adult animal; the vigorous behaviors which promptly increased, as far as physically possible, the separation between the two adults; and the occurrence of restrictive behaviors directed toward the infant or, with equal likelihood, of punitive behaviors and withdrawal from the infant. In contrast with the pairs of socially reared adults, the behaviors of the pairs of isolate-reared adults seemed to lack coordination and focus. Often the females in each pair oriented and reacted toward diverse features of their immediate environment in ways not seen in the socially reared pairs.

Summary and Conclusions

The studies reviewed here collectively suggest that the quantitative differences between isolate-reared and socially reared monkeys have a developmental nature indicative of a retarded behavioral maturation of the isolate-reared animal (e.g., Mason, 1968) which may be compensated, in part, by postisolation rearing in conventional maintenance conditions (e.g., individual caging in colony rooms with others similarly caged). The qualitative differences suggest different learning experiences, such that unusual behaviors are established and appear at adulthood under conditions which elicit approach or withdrawal behaviors in all macaques. Such conditions include the presence of a conspecific, adult or infant, or of a simulated conspecific, such as a ceramic head. The thresholds of orientational behaviors appear to be similar regardless of the rearing experiences; the nature of the behaviors in these circumstances is distinctive, however. The mechanisms for these instances of behavioral variability, represented by both the quantitative and the qualitative differences between the two rearing groups, are not yet clear, although two immediately suggest themselves—differential reinforcement and stimulus differentiation.

The particular behaviors observed in these isolate-reared animals reveal inexperience with significant, species-typical reinforcers and with the wide range of animate stimulation and social possibilities typical of the preadult life of a feral-reared macaque. In the laboratory-rearing conditions some of these reinforcers and a limited portion of the range of significant stimulation do exist for the partially isolate-reared animals discussed in this review. They exist in the toweling and the surrogate provided for the infant, in the character of the early bottle-feedings, and in the early and continued interactions with the few human caretakers. They exist, in fact, in all of those features of the artificial rearing procedures required (i.e., they work) for the sustenance of the infant. That they are sufficient is indicated by the survival of the infant and the occasional reproductive successes of the adult. The limits of those successes, however, indicate that the maintenance conditions may not be truly necessary or sufficient.

As an older infant, juvenile, and subadult, the laboratory-reared monkey has other experiences, all relatively stable and unchanging. These occur with still other monkeys in similar confinements; with a larger group of human caretakers each with his own pattern of caretaking, interaction, and manipulation; with a prepared (dry) monotonous diet; with very predictable sources of food and water in unvarying amounts; and with a larger cage in a more spacious outside-of-the-cage environment. We must recognize in our analysis of these reports that after the infantile period the isolate-reared monkey is presented with experiences which tend to be much more variable across time and laboratories, and which are dictated by the convenience and the foibles of the experimenters. Therefore they are only rarely recorded and never reported. Given the vicissitudes of experimenters and laboratories and the time necessary to reach adulthood, the laboratory monkey may have a variety of experiences which could, possibly, affect the experimental outcomes and their interpretations. This variety, however, is a mere shadow of the reality which is the feral state, in which each possible dimension has many, many hues and brightnesses.

We should be impressed that in our artlessness we have chanced upon those conditions sufficient for the sustenance of these infant animals and which, through apparently minor variations, permit or endanger the continuity of the race through adequate or inadequate reproductive behaviors. We do know the general nature of those conditions sufficient for the very young infant monkey: the temperature and light levels in the cage environment; the size of that environment; the availability of a surrogate of specifiable dimensions and surface character with peripheral supports; and the presence, location, and nature of a source of liquid diet (Meier and Berger, 1965; Meier, Berger, and Garcia-Rodriguez, 1967). These conditions require a minimum of human contact or manipulation (two to three times during the first 12 postnatal hours). They change, of course, with the maturation of the infant, generally in the direction of even less human involvement. Because success in rearing socially isolated infant macaques has been so adventitious, the authors cannot conclude that the successful maintenance conditions are equally so—that they are an empirically random but a theoretically rational selection from among the many possibilities seemingly available to mammalian life. These conditions, rather, are fixed and reflect the most fundamental features of a macaque mother (and later of a macaque "other") in a controlled environment. They are represented as a constellation of specifiable physical qualities meted out under stated conditions at equally predictable times.

The examination of these studies has revealed, first, the inherent response of the developing monkey to selected environmental features and, second, the flexibility in possible outcomes. This examination, in effect, has permitted us to draw the prototypic pattern of the species which is identified by the isolate-reared macaque and is revealed through its behaviors. We should not think ourselves maudlin to acknowledge the expertness of the primate mother; she is the most efficient and the most available packaging of those required conditions and contingencies that we have readily available for the ontogeny of primate offspring adequate for laboratory purposes. Our manipulations, except under the most radical and irrational circumstances, invariably have failed to remove those vestiges of primate behavior necessary for reproduction. These failures, however, have given us some appreciation of the complex species-oriented behavior involved in reproduction, the bases of which are built both into the organism, to be revealed at birth and thereafter, and into the environment, to be selected and thereupon reinforcing.

In summary, we have found in a comparison of the communal–social and the laboratory-isolate situations and of their outcomes indications that the factors which lead to the survival of the newborn macaque and to the reproductively successful rearing of this primate are both species-specific and individually specific. During the infantile period these factors include the characteristics of light, odor, sound, and temperature in the distal environment and the tactile, vestibular, and spatial dimensions of the mother and of the mother-surrogate in the proximate environment. Restriction of these characteristics may be evident in the development of the infant which, nevertheless, may show some species-specific orientation and some social success when it is later introduced to a conspecific which, of course, looks and acts somewhat like itself. We conclude that the general characteristics of the conspecific in the species orientation of macaques are built in, i.e., they are intrinsic to the organism and its degree of maturity. Species, culture, and unique postnatal experiences will determine the individually unique features of the conspecific to which the particular monkey will respond. The interactions with these experiences and its maturity will determine the nature of those directed responses.

References

Anderson, C. O., A. M., Kenney, and W. A. Mason. 1977. Effects of maternal mobility, partner, and endocrine state on social responsiveness of adolescent rhesus monkeys. *Dev. Psychobiol. 10*:421–434.

Anderson, C. O. and W. A. Mason. 1978. Competitive social strategies in groups of deprived and experienced rhesus monkeys. *Dev. Psychobiol. 11*:289–299.

Bekoff, M. 1976. The social deprivation paradigm: Who's being deprived of what? *Dev. Psychobiol. 2*:499–500.

Blomquist, A. J. and H. F. Harlow. 1961. The rhesus monkey program at the University of Wisconsin Primate Laboratory. *Proc. Anim. Care Panel 11*:57–64.

Chappell, P. F. and G. W. Meier. 1974. Behavior modification in a mother–infant dyad. *Dev. Psychobiol. 7*:296.

Fleischman, R. W. 1963. The care of infant rhesus monkeys (*Macaca mulatta*). *Lab. Anim. Care 13*:703–710.

Gard, G. C. and C. W. Meier. 1977. Social and contextual factors of play behavior in juvenile rhesus monkeys. *Primates 18*:367–377.

Harlow, H. F. 1958. Behavior of the infant monkey. In W. F. Windle (Ed.), *Neurological and psychological deficits of asphyxia neonatorum*. Springfield, Ill.: Thomas.

Harlow, H. F. and M. K. Harlow. 1962. The effect of rearing conditions on behaviour. *Bull. Menninger Clin. 26*:213–224.

Harlow, H. F. and M. K. Harlow. 1965. The affectional systems. In A. M. Schrier, H. F. Harlow, and F. Stollnitz (Eds.), *Behavior of nonhuman primates*. Volume 2. New York: Academic Pr.

Harlow, H. F., M. K. Harlow, K. A. Schiltz, and D. J. Mohr. 1971. The effect of early adverse and enriched environments on the learning ability of rhesus monkeys. In L. E. Jarrard (Ed.), *Cognitive processes of nonhuman primates*. New York: Academic Pr.

Hebb, D. O. 1946. Emotion in man and animal: Analysis of the intuitive processes of recognition. *Psychol. Rev. 53*:86–106.

Jensen, G. D., R. A. Bobbitt, and B. N. Gordon. 1968. Effects of environment on the relationship between mother and infant pig-tailed monkeys (*Macaca nemestrina*). *J. Comp. Physiol. Psychol. 66*:259–263.

Kaufmann, J. H. 1966. Behavior of infant rhesus monkeys and their mothers in free-ranging band. *Zoologica (N. Y.) 51*:17–28.

Koford, C. B. 1965. Population dynamics of rhesus monkeys on Cayo Santiago. In I. Devore (Ed.), *Primate behavior: Field studies of monkeys and apes*. New York: Holt, Rinehart and Winston.

Lindburg, D. G. 1971. The rhesus monkey in North India: An ecological and behavioral study. In L. A. Rosenblum (Ed.), *Primate behavior: Developments in field and laboratory research*. Vol. 2. New York: Academic Pr.

Mason, W. A. 1960. The effects of social restriction on the behavior of rhesus monkeys. I. Free social behavior. *J. Comp. Physiol. Psychol. 53*:582–589.

Mason, W. A. 1968. Early social deprivation in the nonhuman primates: Implications for human behavior. In D. C. Glass (Ed.), *Environmental influences*. New York: Rockefeller Univ. Pr./Russell Sage Found.

Mason, W. A. 1972. *How baby monkeys construct their mothers and vice versa*. Paper presented to the International Society for Developmental Psychobiology, Houston, Tex.

Mason, W. A. and G. Berkson. 1975. Effects of maternal mobility on the development of rocking and other behaviors in rhesus monkeys: A study with artificial mothers. *Dev. Psychobiol. 8*:197–211.

Meier, G. W. 1965. Other data on the effects of social isolation during rearing upon adult reproductive behaviour in the rhesus monkey (*Macaca mulatta*). *Anim. Behav. 13*:228–231. (a)

Meier, G. W. 1965. Maternal behaviour of feral- and laboratory-reared monkeys following the surgical delivery of their infants. *Nature 206*:492–493. (b)

Meier, G. W. 1970. Commentary. In M. R. Jones (Ed.), *Miami symposium on the prediction of behavior, 1968: Effects of early experience*. Coral Gables, Fla. Univ. Miami Pr.

Meier, G. W. 1971. Operant cycles and imitational learning in a social setting. *Primates 12*:221–227.

Meier, G. W. and R. J. Berger. 1965. Development of sleep and wakefulness patterns in the infant rhesus monkey. *Exp. Neurol. 12*:257–277.

Meier, G. W., R. J. Berger, and C. Garcia-Rodriguez. 1967. *The maintenance of the rhesus neonate*. Unpublished manuscript (available from senior author).

Meier, G. W. and V. D. Devanney. 1972. *The ontogeny of food-reinforced operant behavior: The definition of a culture*. Paper presented at the Fourth International Congress of Primatology, Portland, Ore.

Meier, G. W. and V. D. Devanney. 1974. The ontogeny of play within a society: Preliminary results. *Am. Zool. 14*:289–294.

Meier, G. W. and V. D. Devanney. 1975. *Social and experiential factors in the control of mother-infant behaviors*. Paper presented to the International Society for Developmental Psychobiology, New York.

Meier, G. W., C. E. Izard, and C. Cobb. 1971. *Facial displays in primate communication*. Paper presented to the Southeastern Psychological Association, Miami.

Meyer-Holzapfel, M. 1968. Abnormal behavior in zoo animals. In M. W. Fox (Ed.), *Abnormal behavior in animals*. Philadelphia: Saunders.

Miller, R. E. 1971. Experimental studies of communication in the monkey. In L. A. Rosenblum (Ed.), *Primate behavior: Developments in field and laboratory research*. Vol. 2. New York: Academic Pr.

Miller, R. E., W. F. Caul, and I. A. Mirsky. 1967. The communication of affects between feral and socially isolated monkeys. *J. Pers. Soc. Psychol. 7*:231–239.

Missakian, E. A. 1972. Genealogical and cross-genealogical dominance relations in a group of free-ranging rhesus monkeys (*Macaca mulatta*) on Cayo Santiago. *Primates 13*:169–180.

Missakian, E. A. 1973. The timing of fission among free-ranging rhesus monkeys. *Am. J. Phys. Anthropol. 38*:624–627.

Mitchell, G. 1970. Abnormal behavior in primates. In L. A. Rosenblum (Ed.), *Primate behavior: Developments in field and laboratory research*. Vol. 1. New York: Academic Pr.

Ruppenthal, G. C., G. L. Arling, H. F. Harlow, G. P. Sackett, and S. J. Suomi. 1976. A 10 year perspective of motherless-mother monkey behavior. *J. Abnorm. Psychol. 85*:341–349.

Sackett, G. P. 1968. Abnormal behavior in laboratory-reared rhesus monkeys. In M. W. Fox (Ed.), *Abnormal behavior in animals*. Philadelphia: Saunders.

Sackett, G. P. 1970. Unlearned responses, differential rearing experiences, and the development of social attachments by rhesus monkeys. In L. A. Rosenblum (Ed.), *Primate behavior: Developments in field and laboratory research*. Vol. 1. New York: Academic Pr. (a)

Sackett, G. P. 1970. Innate mechanisms, rearing conditions, and a theory of early experience effects in primates. In M. R. Jones (Ed.), *Miami symposium on the prediction of behavior, 1968: Effects of early experience*. Coral Gables, Fla.: Univ. Miami Pr. (b)

Sackett, G. P., R. A. Holm, and G. C. Ruppenthal. 1976. Social isolation rearing: Species differences in behavior of macaque monkeys. *Dev. Psychol. 12*:283–288.

Sackett, G. P., M. Porter, and H. Holmes. 1965. Choice behavior in rhesus monkeys: Effect of stimulation during the first month of life. *Science* *147*:304–306.

Sackett, G. P. and G. C. Ruppenthal. 1974. Some factors influencing the attraction of adult female macaque monkeys to neonates. In M. Lewis and L. A. Rosenblum (Eds.), *The effect of the infant on its caregiver*. New York: Wiley.

Sade, D. S. 1972. A longitudinal study of social behavior of rhesus monkeys. In R. Tuttle (Ed.), *Functional and evolutionary biology of primates*. Chicago: Aldine.

van Wagenen, G. 1950. The monkey. In E. J. Farris (Ed.), *The care and breeding of laboratory animals*. New York: Wiley.

Wolfheim, J. H., G. D. Jensen, and R. A. Bobbitt. 1970. Effects of group environment on the mother-infant relationships in pig-tailed monkeys (*Macaca nemestrina*). *Primates* 1970. *11*:119–124.

Chapter 10

Macaques and Other Old World Primates

G. Mitchell
Nancy G. Caine

Old World monkeys and particularly rhesus monkeys (*Macaca mulatta*) have been studied more extensively than any other nonhuman primates. Their behavior and other biological characteristics place them taxonomically somewhere between the New World monkeys and the great apes.

The rhesus-monkey experiments in early social deprivation initiated by Harlow and Mason at the University of Wisconsin Primate Laboratory became catalysts for most of the primate behavioral research of the last two decades. This very volume indeed owes much to the groundbreaking work of Harlow and Mason (cf. Mason, 1960).

The rhesus monkey, however, is only one of a dozen macaque species and only one of 15 different genera of Old World monkeys. One Old World monkey genus alone (*Cercopithecus*) consists of over 20 different species of monkeys. Some of these genera (in the subfamily Colobinae) are primarily arboreal, leaf-eating monkeys. Rhesus monkeys are terrestrial omnivores. The rhesus monkey clearly cannot represent all Old World monkeys. Since, however, most research on atypical rearing and social preference has been done on the rhesus, we shall begin with that species.

Portions of this chapter are based upon material to be published in Mitchell, G. 1979. *Behavioral sex differences in nonhuman primates.* New York: Van Nostrand-Reinhold.

Rhesus Monkeys

Rhesus in the Wild

Rhesus monkeys (*Macaca mulatta*) are diurnal and primarily terrestrial, living in large multimale groups in northern India. Social groups include many adult males, adult females, subadults, juveniles, and infants. Adult females and their young are found in the center of a troop along with the most dominant adult males. Young adult and subadult males frequent the periphery of the troop. Subgroups based upon kinship are often found. Outside the main troops, all-male groups and solitary males are sometimes seen. Young adult males (but not females) sometimes change troops during the breeding season (Lindburg, 1971).

Rhesus monkeys, unlike some other nonhuman primates, show extreme physical sexual dimorphism; the adult male is much larger than the adult female and has much larger canine teeth. Prenatal androgen from the fetal testes apparently causes this sexual differentiation and in addition predisposes rhesus males toward more frequent and rougher play, more aggression, less infant care, and more foot-clasp mounting (cf. Mitchell, 1979).

The following are other gender differences in rhesus monkeys: males display more vigilance, troop protection, intertroop behavior, distant visual communication (e.g., branch shaking), and aggression; females evince more infant care, social alliances, grooming, soft social vocalizations, and in many cases greater learning ability (cf. Mitchell, 1979).

Infant rhesus monkeys receive more maternal care than New World monkeys but less than chimpanzees. Mothers frequently groom, cradle, retrieve, and sometimes punish infants. Weaning is completed by one year of age (usually when another infant is born); but infants will stay near their mothers until the subadult period (age three) for males and into adulthood for females (Lindburg, 1971).

Rhesus in Captivity

Infant rhesus monkeys can distinguish their mothers from other females at an early age and can form strong emotional attachments to familiar cloth surrogates if reared apart from their mothers (Harlow and Zimmerman, 1959). When hand-raised by humans rhesus monkeys survive well, and even when reared individually in wire cages they can grow to maturity if given a small bottle from which to nurse and if given ad-lib food (Blomquist and Harlow, 1961). Thus their physical development is normal.

When reared in the nursery, however, with reduced or no contact with conspecifics, rhesus monkeys develop many *behavioral* abnormalities which change with age and vary with the sex of the animal. These behavioral abnormalities may provide some insight into species identification processes in these subjects.

Social Deprivation and Separation

Social deprivation is a catch-all term which refers to such phenomena as rearing in total or partial social isolation, separation from mother or from other attached objects, and the effects of abnormal maternal behavior on male and female infants. Roy (1976, 1977a, 1977b) has published an extensive bibliography on the early rearing of infrahuman primates with reduced conspecific contacts.

Rearing a rhesus monkey infant from birth to 6 or 12 months of age in social isolation produces marked behavioral abnormalities in the animal. During the first year, the isolate develops self-clinging, repetitive stereotyped rocking, self-mouthing and digit-sucking, and crouching. When given the opportunity to interact with others prior to puberty, the isolate-reared rhesus hides its face and crouches and rocks. It withdraws from others in fear (Mitchell, 1970).

At puberty the behavior of the isolate-reared rhesus changes. Continued pacing replaces repetitive rocking, self-biting replaces digit sucking, and abnormal social aggression gradually replaces social fear (Mitchell, 1970).

During adulthood the isolate-reared rhesus becomes even more aggressive and self-mutilative. It is, in addition, unable to display normal sexual posturing (Mitchell, 1970).

In Chapter 1 of this volume, it is pointed out that early experiences with conspecifics appear to be crucial for determining a species identity for at least some species. We think that the rhesus monkey is one of these species. It was also noted in Chapter 1, however, that males and females of a given species may not be equally influenced by early experiences and that each sex of a given species may develop its own species identity at a different rate or even in a different way. The next section of this chapter considers the question of sex differences in response to early deprivation.

Sex Differences in Response to Deprivation

With regard to sex differences in response to social deprivation, male rhesus monkeys exhibit more disturbances than females; self-mouthing and self-clasping are more frequent in young isolated rhesus males than in young isolated rhesus females. More rocking occurs in isolate-reared males than in their female counterparts (Suomi, Harlow, and Kimball, 1971). Older male isolates show more abnormalities in sexual behavior. Female isolates can eventually learn appropriate sexual behavior, whereas males usually do not. Male isolates also do more self-biting. According to Cross and Harlow (1965), male isolates show higher levels of self-aggression than female isolates. Gluck and Sackett (1974) and Chamove and Harlow (1970) report the same finding. In social behavior male isolates are more disturbed than female isolates (Sackett, 1974).

In group formation studies done by Bernstein, Gordon, and Rose (1974), 18 socially deprived rhesus monkeys were used as a core group. The isolates were initially immobilized. They displayed awkwardness, bizarre movements, stereotyped movements, and low social interaction rates. The isolates seemed to have no coherent group structure and did not act as a unit against an intruder. When aggression occurred it persisted. When 16 isolates were placed into a free-ranging environment with a wild troop, only 2 of 6 isolate males survived, whereas 7 of 9 females survived (Sackett, 1974). The more enriched the early environment, the more easily an animal operates in a team or coalition (Suomi, 1974).

In terms of sexual behavior, rhesus isolates display strange behavior indeed (but see Maple, 1977, for descriptions of the range of unusual sexual behaviors in primates). According to Mason, Davenport, and Menzel (1968), "the female appears to be less seriously handicapped than the male by social deprivation" (p 25). Male isolate-reared rhesus monkeys only 2 years old particularly show markedly deficient sexual behavior (Mason, 1960). In a group of isolates studied by Missakian (1969), "no socially deprived male, experienced or naive, executed an appropriately oriented mount'' (p 403).

Female isolates vocalize more than male isolates (Mitchell, Raymond, Ruppenthal, and Harlow, 1966), particularly when stimulated by a human observer (Cross and Harlow, 1965); they also make more fear grimaces, but threaten less as adults than isolate males do (Mitchell, 1968).

As noted earlier, aggression increases with age in both isolate-reared and socially reared rhesus monkeys (Mitchell, 1975). There is a sex difference in aggression in both rearing groups, males showing more aggression than females. Among the isolates, this means abnormal aggression. Thus adult male isolates show more abnormal aggression than adult female isolates (Mitchell, 1974; Mitchell, 1976). This is evident in visual threat behavior as well as in direct attacks (Mitchell and Redican, 1972; Brandt, Stevens, and Mitchell, 1971).

In longitudinal studies of the development of isolate-reared infants from birth to adulthood (see Baysinger, Brandt, and Mitchell, 1972), sex differences in behavioral abnormalities are studied as they develop. Self-clasping and rocking, which develop early, occur in males more than in females, whereas fear grimaces occur in females more than in males. When brought out of isolation, female isolates emit more distress calls than males (Brandt, Baysinger, and Mitchell, 1972). Isolates of both sexes, however, emit fewer distress calls than normals.

When isolate-reared infants are paired with older, normal preadolescent animals, male isolates elicit more aggression from the older socially reared animals than female isolates do (Brandt and Mitchell, 1973). Female control animals are better able to establish social contact and an emotional attachment to deprived animals than males are; but there is also some evidence for superiority here of heterosexual control-isolate dyads over isosexual control-isolate dyads (Maple, Brandt, and Mitchell, 1975). Normal preadolescents paired with isolated infants of the opposite sex seem to be more upset (as evidenced by increases in coo vocalizations at separation) when separated from their isolates than normal preadolescents paired with isolated infants of the same sex do. At 3 years of age, isolate-reared males still exhibit more self-clasping behavior than isolate-reared females, but isolate-reared females exhibit more miscellaneous bizarre behaviors than the males. Three-year-old isolation-reared males spend more time looking at themselves than 3-year-old isolation-reared females do.

The average total duration of looking at other animals is also particularly high in 3-year-old male isolates (Erwin, Maple, Mitchell, and Willott, 1974).

Isolate-reared mothers ("motherless mothers") are brutal or indifferent to their first offspring (Arling and Harlow, 1967) but not to their second and third infants. Apparently the female isolate becomes somewhat less aggressive toward infants with age and/or maternal experience with infants (Arling, Ruppenthal, and Mitchell, 1969; Mitchell 1971). Abnormal maternal punishment is interesting because "motherless mothers" are more brutal to male than to female infants (Sackett, 1974). Because of differential excessive punishment of males, young males themselves become hyperaggressive as they mature (Mitchell, Arling, and Møller, 1967).

If the "parent" is an isolate-reared male, there is also a danger of excessive punishment of the infant, but not in isolate-reared males who are at least 10 years old (Gomber and Mitchell, 1974). Infants reared by older isolate-reared males apparently develop much as normal infants do; however, like the adopted infants of normally reared males, they are played with very frequently and very intensely (Gomber, 1975). Since all infants in the isolate-male "parent" study have been female, there is no information on infant sex differences in such a situation. (For published accounts of these experiments, consult Mitchell, 1974; Mitchell, Redican, and Gomber, 1974; Mitchell, 1977.)

To discover the specific kinds of missing input related to various behavioral abnormalities appearing in deprived infants, Mason and Berkson (1975) raised rhesus infants on either mobile or stationary surrogate mothers. In both groups of animals with surrogate mothers, infant males were more disturbed than infant females (evidenced in more frequent distress calls and barking). At 1½ years of age the contrast between the two sexes in disturbance and in deviant social contact (males *more*) was greatest in the infants raised on stationary surrogates.

In tests where looking behavior is a measure of preference, monkeys reared by real mothers discriminated more sharply between the sexes than did monkeys reared by surrogate mothers; however, those reared on moving surrogate mothers discriminated better than those reared on immobile surrogates. The first two groups showed a visual preference for a strange adult male to a strange adult female, whereas the immobile group showed no preference at all (Eastman and Mason, 1975). Is this an example of faulty gender identity in the rhesus?

It is interesting that surrogate-reared males will sexually mount surrogate models as they mature, including thrusting to ejaculation. As is known from the earlier isolate studies, however, they will not mount real females. Surrogate-reared females similarly will present only to *surrogate* models (Deutsch and Larsson, 1974). Do these animals feel they belong to a species of surrogates?

In summary of rhesus sex differences in response to deprivation, male rhesus monkeys appear to be more adversely affected by early deprivation than female rhesus monkeys are. Even in adulthood it appears that social separation is more serious for the male. This result is surprising in the light of the female's greater sociability and the male's tendency to be peripheralized, to change troops, and even to become solitary.

Social Deprivation in Other Monkeys (with Sex Differences)

Very few other monkeys have been studied at all with regard to the effects early social deprivation may have on them. Testa and Mack (1977), however, have studied the effects of early social isolation on the sexual behavior of the crab-eating macaque (*Macaca fascicularis*). They found that male isolates showed an abnormal form of mounting, and females displayed incomplete sexual presentations. In pigtail macaques (*M. nemestrina*) raised in captivity male infants show more thumbsucking than females (Jensen, 1966). Jensen (1969) claims that behavioral sex differentiation is severely disturbed by early deprivation and that males are more vulnerable to deprivation than females. Pigtail males need a richer environment for development than pigtail females (Jensen, Bobbitt, and Gordon, 1966). In the stumptail macaque (*M. arctoides*), also, deprivation does not seem to affect males and females equally. In fact, Riesen, Perkins, and Struble (1977) found that isolated males and females tended to be farther apart in behavior than control males and females. Males apparently were the most severely affected.

According to Rosenblum (1974), who has studied squirrel monkeys (*Saimiri sciureus*), bonnets (*M. radiata*), and pigtail macaques (*M. nemestrina*) extensively, deprivation devastates male more than female exploration. He believes it is during the period after the eleventh week of life that the major effects of different rearing conditions and the sex of the infant can be discerned in the squirrel monkey.

Sackett, Holm, and Ruppenthal (1976) have compared pigtail macaques reared in total social isolation to rhesus isolates. Pigtail monkeys were more social, more passive, and less exploratory than the rhesus; the rhesus isolates displayed abnormal behavior to a much greater extent than the pigtail isolates. Such results point out the need for more comparative studies prior to one making generalities for all primates. Sex differences in these pigtail isolates were also not as clear-cut as in the rhesus.

In summary, the few studies available on deprivation in monkeys other than rhesus macaques indicate that the male seems more seriously affected by early deprivation than the female, although some studies of the pigtail macaque may provide exceptions to this general rule.

Social Deprivation in the Apes: Sex Differences

In a summary of the effects of early deprivation on the sexual behavior of chimpanzees, Mason et al. (1968) concluded that sexual roles were not as sharply differentiated in chimpanzees as they were in rhesus monkeys, consequently, "early social deprivation may have more severe and lasting consequences for masculine sexual development in the monkey than in the chimpanzee" (p. 26). In the same article, Mason et al. presented some evidence that, for the chimpanzee, puberty may be a critical "period for the acquisition or refinement of sex skills" (p. 26). However, even for chimpanzees who are more able to recover from isolation than rhesus monkeys, males seem more adversely affected by restricted rearing than females (Rogers and Davenport, 1969). Among chimpanzees in captivity, "The female's behaviour was closer to that of their wild-living counterparts" (Tutin and McGrew, 1973, p 255).

In the captive gorilla (*Gorilla gorilla*), as in the macaque, there is some evidence for improved mothering of the second infant (Nadler, 1975b) because of either learning or age. There is in addition some evidence that early rearing with peers can compensate somewhat for the lack of a mother (Nadler, 1975a; see also

Chapter 13, this volume). A mother gorilla who is hand-raised may show restlessness and carelessness with her infant (Faust, 1977; Scollay, Joines, Baldridge, and Cuzzone, 1975).

Deprived orangutan mothers show inappropriate maternal responses. But an infant orangutan separated from its mother cannot learn appropriate sexual behavior and, in turn, becomes a poor parent itself (Zucker, Wilson, Wilson, and Maple, 1977). Whether such early deprivation is more serious for male than for female gorillas and orangutans is not yet known. In Chapter 13, however, it is reported that Carmen, a nursery-reared orangutan with one previous offspring which died at birth, has conceived (by a feral-born male) and recently delivered a viable infant which she is caring for normally, at least for the first month.

Among great apes, at least among chimpanzees, individuals gradually come to recognize not only their own species, but their own sex and their own selves. Apes raised by humans can identify, classify, and discriminate behaviorally in response to human males and females. They can therefore conceptualize the idea of gender (cf. Chapter 12, this volume). While socially reared monkeys can also classify according to gender, they cannot classify themselves as being of one gender or the other. Using these differences in abilities between apes and monkeys, a clever scientist should be able to learn much about gender identification.

Interspecies Attachment and Species Identity

Maple (1974) has made a convincing plea for the use of interspecies interactions in comparisons of species-specific behavioral tendencies. Maple's studies actually involved intergeneric rather than interspecific interactions. He paired juvenile baboons (*Papio anubis*) with juvenile macaques (*M. mulatta*).

Choice tests before and after intergeneric exposure proved that the rhesus monkeys (even after the first year of life) came to prefer their baboon friends to conspecific strangers. The baboons also reversed their conspecific attachments and acquired a strong social attachment to an alien (in this case a rhesus). When these juvenile intergeneric pairs were separated, a distress call was emitted by both members of each pair. Likewise, a 12-year-old adult male rhesus eventually (after being paired 12 months) developed a strong bond with a 12-year-old adult female baboon (*P.*

anubis). The attachment was evident in increased proximity, in reciprocal grooming, and in copulation.

That interspecific attachments need not occur very early in life is also shown in research reported by Mason and Kenney (1974). These researchers demonstrated a redirection of filial attachment in rhesus monkeys to dogs which were used as mother surrogates for the monkeys.

Old World monkeys certainly have the potential to become attached to an alien species (even to inanimate objects). To what extent, however, does this occur in the wild? According to some researchers hybrid baboons occur naturally, and in central Ethiopia a fairly wide natural hybrid zone is expanding (Shotake, Nozawa, and Tanabe, 1977). The same authors have asserted that the seven different "species" of Asian macaques they studied are actually all one species (Nozawa, Shotake, Ohkura, and Tanabe, 1977). Thus not only do Old World monkeys apparently have the potential for *transpecific identity,* they may also use that potential in the wild.

In the chapter by Cooke, the importance of intraspecific and interspecific studies to the topic of species identity was shown in studies of the familial plumage of the snow goose. While snow geese were capable of using familial plumage as a means of recognizing their own species, they could also generalize from the individual appearance of family members to more general species-specific characters. Studies similar to those on coloration in snow geese could be done on various species of the lesser ape (genus *Hylobates*).

In the genus *Hylobates* different gibbon species have different pelage coloration and different vocalizations. Despite these differences interspecies hybridization occurs in nature. Mixed-species groups and hybrids are found. These hybrids often have different pelage colors and unique vocalizations (cf. Brockelman, 1976). This may be an interesting species group in which to study mechanisms of species identity.

On the other hand there are also some species of Old World primates e.g., bonnet monkeys, which tend to display very little outbreeding and may therefore be considered to show a relatively strong species identity (DeVor, 1977). It may be interesting to compare the development of species identity in a species such as this (*M. radiata*) with a species where species identity was more fragile (e.g., the rhesus). We already know, for example, that bonnet monkeys are less distressed when separated from their mothers than other macaques are (cf. Rosenblum and Kaufman, 1968).

How Are Old World Monkeys and Apes Different?

One of the most fascinating aspects of the species-identity question is that of self-identity. Apes know who they are, monkeys apparently do not (Gallup, 1970; see also Chapter 11, this volume). But how extensive is this self-knowledge possessed by the great apes? Certainly they know their own species and they appear to know to which gender they belong (Shapiro, personal communication, 1978). It seems that we must ask more questions about their identities: their species identities, their gender identities, and their selves. We may most profitably do this by utilizing a sign language of some sort. Or perhaps we could investigate the development of their concepts of self by using Piagetian techniques (cf. Parker, 1977; Chevalier-Skolnikoff, 1977).

Animal behaviorists using the term "behavior" ordinarily do not assume a "behaving" nonhuman organism is conscious of its behavior. They are mostly assuming that the typical primate does not intend to perform an act before the act is performed. They believe the average primate has no conceptual representation of its own "actions" (cf. Reynolds, 1976). From research over many decades, however, it has become obvious that at least some nonhuman primates are conscious of who they are, do intend to perform given actions, do have a concept of their actions, and in fact *are* aware of their own existence. In the following pages we will show the importance of this ability to an animal's species identity.

As has been seen, there is substantial individual variation in nonhuman primate behavior. There is so much variability from genus to genus, from species to species, from troop to troop, from individual to individual, and from situation to situation within an individual that it has become wise for primatologists to be very careful about using the words "always" and "never." In addition, classifying certain behaviors according to age or sex may also be risky (cf. Burton, 1977) because of individual variability.

When Roy (1977c) first organized this symposium at the American Psychological Association meetings in San Francisco,

California, the topic was "Early Experience and the Development of an Adequate Species Identity." This symposium grew out of the growing body of evidence that atypical or deprived rearing conditions in early life can produce a failure to develop a normal orientation toward one's own species. Complex organisms like the nonhuman primates evidently require physical social contact with conspecifics in order to develop adequate species affiliation. Species-typical behavioral patterns depend upon an awareness by the organism that it belongs to the same species as does its partner. (See description of Toto in Chapter 1, this volume.)

Primates are born, however, with *some* knowledge about their own species; they are not completely dependent upon early experience. Infant rhesus monkeys show selective attention to monkey in preference to nonmonkey pictures. Even human infants prefer schematic human faces (Stone, Smith, and Murphy, 1974). This partial innate recognition, however, cannot account for complete species identity, since as we already know, rearing in social isolation interferes with it.

We have seen that species identity is not always completely exclusive or specific. Cross-species affinities are common in the laboratory and even occur in the wild (cf. Maple, 1974). There is therefore some generalizability of attachment processes within the primate order (Mitchell, 1976).

Abnormal behavior invariably results from early social isolation, i.e., social behaviors directed toward conspecifics become abnormal. In the case of isolated rhesus monkeys, the social abnormality gradually changes from one in which the socially isolated infant withdraws from others in fear to one in which the adult aggresses against his or her conspecific (Mitchell, 1975).

Thus the monkey reared in social isolation changes as it matures from an animal showing abnormal social fear to one evincing abnormal social aggression. As these changes occur, so do changes in self-directed behaviors. The rhesus monkey reared in social isolation shows exaggerated self-clinging, self-sucking, self-rocking, and other idiosyncratic *self*-directed movements and postures. As the isolate-reared monkey gets older, its self-rocking changes to more whole-bodied swaying, somersaulting, repetitive pacing, or circling, and its self-sucking behavior gradually changes to self-chewing, self-hitting, and self-biting (Mitchell, 1975).

It is interesting that the changes in the isolates' self-directed behaviors seem to parallel those seen in the development of their abnormal social behaviors. Early in development, the isolate-reared monkey hides from its own arms and legs as well as from other monkeys. When it looks at its bodily parts, it often makes fear grimaces as though afraid of them. As the animal grows older, however, it begins to respond to its own limbs with threats and bites. But perhaps more important for this particular topic, the isolate-reared rhesus appears to respond to its own bodily parts as if they were not part of himself or herself. It seems as though the animal cannot discriminate self from nonself (Mitchell, 1975). But does a monkey have a self it can recognize? None of the following Old World monkey and/or lesser ape species recognize themselves in mirrors: rhesus monkeys, crab-eater monkeys, stumptail monkeys, spider monkeys, capuchins, mandrills, hamadryas baboons, and gibbons (cf. Gallup, 1977b). In all of these species, the tested individuals respond to the mirror image as though they were in the presence of another conspecific individual. Does this mean that there is a prewired "innate" difference between the great apes and other nonhuman primates in self-recognition ability? The available evidence suggests that the difference is not completely dependent upon innate factors. Chimpanzees are not born with this ability, as shown by work with isolate-reared subjects (see Chapter 11, this volume). Remedial social experience (social "therapy"), however, increases chimpanzee self-recognition to some extent.

Thus the development of an individual identity in a chimpanzee requires social interaction with others. This interaction surprisingly need not be with a conspecific. Home-reared (human-reared) chimpanzees recognize themselves in mirrors, even though they prefer human company and show disdain for conspecifics (see Chapter 12, this volume). Thus an accurate species identity is not necessary for accurate self-recognition (or for self-awareness). Vicki, a chimpanzee who was hand-raised by Hayes, recognized herself and declared herself human. Vicki learned in her human home to sort snapshots into human versus animal piles. She placed her own picture on the human pile (Gallup, Boren, Gagliardi, and Wallnall, 1977).

With the exception of the other great apes and man, attempts to demonstrate self-recognition in primate species have failed. It is also interesting that only the great apes among nonhuman primates are capable of learning sign language. Some researchers (e.g., Terrace and Bever, 1976) believe that self-recognition may be necessary for the development of sign language: "All the ingredients for human language are present in other species—they do not become language until an animal learns that it can refer to itself symbolically" (Terrace and Bever, 1976, p 580).

Because there are sex differences in human verbal ability (Maccoby and Jacklin, 1974), with females showing superiority over males, it may be interesting to learn whether there are sex differences also in sign-language ability and self-recognition in chimpanzees. Terrace, Pettito, and Bever (1977) have taught signs to a male chimpanzee (Nim) and found that the rates at which he acquired signs (and the contents of his vocabulary) did not differ from those of a female chimpanzee (Washoe). Both Nim and Washoe made their first signs at the beginning of their fourth month.

Another interesting question concerning self-recognition and signing is the extent to which the pygmy chimpanzee is capable of performing these two complex skills. Since *Pan paniscus* (the pygmy chimpanzee) in some respects appears to be more human-like than *Pan troglodytes* (the common chimpanzee) (Savage and Bakeman, 1976), it might not be surprising that this species could develop more complex language skills and more integrated and complex self-concepts than *Pan troglodytes*. Finally, no one knows yet what might occur if an Old World monkey were raised with a mirror from birth. Would it then come to recognize itself?

What about Humans?

Amsterdam (1972) has traced the development of self-recognition in human infants, employing methods similar to those which Gallup (1970) used on monkeys and apes (see also Papoušek and Papoušek, 1974; Schulman and Kaplowitz, 1977).

Using videotape, Amsterdam and Greenberg (1977) demonstrated that 20-month-old human infants were different from 10- and 15-month old infants in their display of self-consciousness (when seeing their own image). Self-consciousness appears some time after 10 months of age and occurs with greatest frequency in response to a simultaneous self-image. On the subject of sex differences, self-consciousness in a young child is less likely to occur in the presence of an adult female than in the presence of a male (Amsterdam and Greenberg, 1977).

Self-observation in a mirror during the first two years of life has been studied by Schulman and Kaplowitz (1977). Self-observation does not vary between one and 24 months of age on a videotape. Somewhere between 19 and 24 months of age, however, self-recognition is evident. Before this age (at 13–18 months) an avoidance reaction (perhaps due to self-consciousness) is apparent. No sex differences are reported.

In recent years self-recognition, self-awareness, and consciousness have become respectable subjects in comparative psychology. Powers (1973), for example, has developed a model for consciousness based upon feedback principles. Shafton (1976) has written an extensive monograph on subjective factors in the social adaptations of man and other primates, in which he outlines the "conditions of awareness." The recent emphasis on awareness is affecting behavioral primatology and running headlong into another major theoretical framework for our discipline, that of sociobiology.

As Tobach (1976) has pointed out, the comparative method is useful in studies of behavior only when "the questions to be answered are based on stated assumptions which are testable and when the levels of the phenomena being compared are equivalent" (p 185). Like Tobach (1976), Porges (1976) has warned that in ontogenetic as well as in phylogenetic comparisons there are problems associated with response equivalence. Often the way in which a 2-year-old responds to a given situation cannot be directly compared to how an 8-year-old responds. (One suggests here that the same problems can occur in species and gender comparisons.) In phylogenetic comparisons this principle applies to both the "objective" concepts of sociobiology and to the more "subjective" concepts of the species-identity and self-awareness advocates of behavioral primatology. Those behavioral primatologists most interested in self-recognition or self-awareness are quick to point out the weaknesses of the sociobiological approach to comparative behavior in this regard. Self-awareness is not in the realm of sociobiological theorizing at this time, i.e., sociobiology cannot really explain awareness. It is up to those interested in proximate (as opposed to ultimate) mechanisms to account for the role of awareness in evolution. In any case, the proof of the existence of self-awareness in chimpanzees demands that we also develop sophisticated complements of sociobiology. Questions evolving from the constructs of identity (species, gender, and self) may provide such alternatives.

What does all of the above have to do with species identity? Awareness of self suggests that behavior is conscious, intentional, and purposive. Awareness of self also suggests that an individual is aware of his species and can intentionally, purposefully, and consciously act as he feels he should or wants to. While the existence of self-awareness does *not* mean that phylogeny, genetics, physiology, and, indeed, biology have nothing to do with behavioral sex differences, it does presuppose that the individual is capable of a different level, if not a wider range, of behaviors than would be the case without self-awareness. No wonder that great apes and humans often have so many unpredictable characteristics. The possible range of behaviors within each sex is very large, so large that some researchers, for example, have even stated that natural-science approaches to human sexuality and sex differences are inappropriate (Whitsett, 1977).

It is not our intention to make the present chapter a disclaimer for the concept of species identification. Species identity in primates (without intent or purpose) certainly does exist. So does a species identity in humans (e.g., awareness of human-ness seen in infancy occurs before self-recognition has developed). In fact, it is probable that there is even a species identity in humans in which purpose and intent play an obvious role. An area of research of most interest, however, would be one involving research on sex differences in self-awareness. Since there appears to be a relationship between verbal skills and self-awareness in nonhumans and retarded humans, it may be that female humans develop gender and self-recognition earlier than males. Who knows? They might also develop an earlier and more robust *species* identity.

References

Amsterdam, B. 1972. Mirror self-image reactions before age two. *Dev. Psychobiol. 5*, 297–305.

Amsterdam, B. and L. M. Greenberg. 1977. Self-conscious behavior of infants: A videotape study. *Dev. Psychobiol. 19*(1), 1–6.

Arling, G. L. and H. F. Harlow. 1967. Effects of social deprivation on maternal behavior of rhesus monkeys. *J. Comp. Physiol. Psychol. 64*, 371–378.

Arling, G. L., G. C. Ruppenthal, and G. Mitchell. 1969. Aggressive behavior of the 8 year old nulliparous isolate female monkey. *Anim. Behav. 17*, 190–213.

Baysinger, C. M., E. M. Brandt, and G. Mitchell. Development of infant social isolate monkeys (*Macaca mulatta*) in their isolation environments. *Primates 13*(3), 257–270.

Bernstein, I. S., T. P. Gordon, and R. M. Rose. 1974. Aggression and social controls in rhesus monkey (*Macaca mulatta*) groups revealed in group formation studies. *Folia Primatol. 21*, 81–107.

Blomquist, A. J. and H. F. Harlow. 1961. The infant rhesus monkey program at the University of Wisconsin Regional Primate Research Center and Laboratory. *Proc. Anim. Care Panel II 2*, 57–64.

Brandt, E. M., C. Baysinger, and G. Mitchell. 1972. Separation from rearing environment in mother-reared and isolation-reared rhesus monkeys (*Macaca mulatta*). *Int. J. Psychobiol. 2*(3), 193–204.

Brandt, E. M. and G. Mitchell. 1973. Pairing preadolescents with infants (*Macaca mulatta*). *Dev. Psychol. 8*, 222–228.

Brandt, E. M., C. W. Stevens, and G. Mitchell. 1971. Visual social communication in adult male isolate-reared monkeys (*Macaca mulatta*). *Primates 12*(2), 105–112.

Brockelman, W. Y. 1976. *Preliminary report on relations between Hylobates lar and Hylobates pileatus*. Paper presented at the International Primatological Society meeting, Cambridge, England, August.

Burton, F. D. 1977. Ethology and the development of sex and gender identity in non-human primates. *Acta Biotheor. 26*(1), 1–18.

Chamove, A. S. and H. F. Harlow. 1970. Exaggeration of self-aggression following alcohol ingestion in rhesus monkeys. *J. Abnorm. Psychol. 75*(2), 207–209.

Chevalier-Skolnikoff, S. 1977. A Piagetian model for describing and comparing socialization in monkey, ape, and human infants. In S. Chevalier-Skolnikoff and F. E. Poirier (Eds.), *Primate biosocial development*. New York: Garland, 159–188.

Cross, H. A. and H. F. Harlow. 1965. Prolonged and progressive effects of partial isolation on the behavior of macaque monkeys. *J Exp. Res. Pers. 1*, 39–49.

Deutsch, J. and K. Larsson. 1974. Model-oriented sexual behavior in surrogate-reared rhesus monkeys. *Brain Behav. Evol. 9*, 157–164.

DeVor, E. J. R. 1977. *Genetic differentiation in the genus Macaca*. Paper presented at the American Association of Physical Anthropologists meeting, Seattle, Wash. April.

Eastman, R. F. and W. A. Mason. 1975. Looking behavior in monkeys raised with mobile and stationary artificial mothers. *Dev. Psychobiol. 8*(3), 213–221.

Erwin, J., T. Maple, G. Mitchell, and J. Willott. 1974. A follow-up study of isolation and mother-reared rhesus monkeys which were paired with preadolescent conspecifics in late infancy: Cross-sexed pairings. *Dev. Psychol. 10*, 423–428.

Faust, R. 1977. Eighth gorilla born at Frankfurt. *AAZPA Newsl. 18*, 14.

Gallup, G. G. 1970. Chimpanzees: Self-recognition. *Science 167*, 86–87.

Gallup, G. G. 1977. Absence of self-recognition in a monkey (*Macaca fascicularis*) following prolonged exposure to a mirror. *Dev. Psychobiol. 10*, 281–284. (a)

Gallup, G. G. 1977. Self-recognition in primates: A comparative approach to the bidirectional properties of consciousness. *Am. Psychol. 32,* 329–338. (b)

Gallup, G. G., J. L. Boren, G. J. Gagliardi, and L. B. Wallnau. 1977. A mirror for the mind of man, or will the chimpanzee create an identity crisis for *Homo sapiens? J. Hum. Evol. 6,* 303–313.

Gallup, G. G., M. K. McClure, S. D. Hill, and R. A. Bundy. Capacity for self recognition in differentially reared chimpanzees. *Psychol. Rec. 21,* 69–74.

Gluck, J. D. and G. P. Sackett. 1974. Frustration and self-aggression in social isolate rhesus monkeys. *J. Abnorm. Psychol. 83*(3), 331–334.

Gomber, J. 1975. *Caging adult male isolation-reared rhesus monkeys (Macaca mulatta) with infant conspecifics.* Ph.D. thesis. Davis, Univ. California.

Gomber, J. and G. Mitchell. 1974. Preliminary report on adult male isolation-reared rhesus monkeys caged with infants. *Dev. Psychol.* 9, 419.

Harlow, H. F. and R. R. Zimmerman. 1959. Affectional responses in the infant monkey. *Science 130,* 421–432.

Hoyt, A. M. 1941. *Toto and I.* New York: Lippincott.

Jensen, G. D. 1966. Sex differences in developmental trends of mother-infant monkey behavior (*M. nemestrina*). *Primates 7*(3), 403.

Jensen, G. D. 1969. *Environmental influences on sexual differentiation: Primate studies.* Paper presented at Symposium on Environmental Influences on Genetic Expression, Bethesda, Md., April.

Jensen, G. D., R. A. Bobbitt, and B. N. Gordon. Sex differences in social interaction between infant monkeys and their mothers. *Rec. Adv. Biol. Psychiatry,* 9, 283–293.

Lethmate, J. and G. Ducker. 1973. Untersuchungen zum Selbsterkennen im Soiegel bei Orang-utans und linigen anderen Affenarten. *Z. Tierpsychol. 33,* 248–269.

Lindburg, D. G. 1971. The rhesus monkey in North India: An ecological and behavioral study. In L. A. Rosenblum (Ed.), *Primate behavior: Developments in field and laboratory research.* Vol. 2. New York: Academic Pr., 1–106.

Maccoby, E. E. and C. N. Jacklin. 1974. *The psychology of sex differences.* Stanford, Cal.: Stanford Univ. Pr.

Maple, T. 1974. Basic studies of interspecies attachment behavior. Ph.D. thesis. Davis, Univ. California.

Maple, T. 1977. Unusual sexual behavior of nonhuman primates. In J. Money and H. Musaph (Eds.), *Handbook of sexology.* New York: Elsevier, 1167–1186.

Maple, T., E. M. Brandt, and G. Mitchell. 1975. Separation of preadolescents from infants (*Macaca mulatta*). *Primates 16*(2), 141–153.

Mason, W. A. 1960. The effects of social restriction on the behavior of rhesus monkeys. I: Free social behavior. *J. Comp. Physiol. Psychol. 53*(6), 582–589.

Mason, W. A. and G. Berkson. 1975. Effects of maternal mobility on the development of rocking and other behaviors in rhesus monkeys: A study with artificial monkeys. *Dev. Psychobiol.* 8(3), 197–211.

Mason, W. A., R. K. Davenport, Jr., and E. W. Menzel, Jr. 1968. Early experience and the social development of rhesus monkeys and chimpanzees. In G. Newton and S. Levine (Eds.), *Early experience and behavior.* Springfield, Ill.: Thomas, 277–305.

Mason, W. A. and M. D. Kenney. 1974. Redirection of filial attachments in rhesus monkeys: Dogs as mother surrogates. *Science 183,* 1209–1211.

Missakian, E. A. 1969. Reproductive behavior of socially deprived male rhesus monkeys (*Macaca mulatta*). *J. Comp. Physiol. Psychol. 69,* 403–407.

Mitchell, G. 1968. Persistent behavior pathology in rhesus monkeys following early social isolation. *Folia Primatol. 8,* 132–147.

Mitchell, G. 1970. Abnormal behavior in primates. In L. A. Rosenblum (Ed.), *Primate behavior: Developments in field and laboratory research.* Vol. 1. New York: Academic Pr., 195–249.

Mitchell, G. 1971. Parental and infant behavior. In E. S. E. Hafez (Ed.), *Comparative Reproduction of laboratory primates.* Springfield, Ill.: Thomas, 382–402.

Mitchell, G. 1974. Syndromes resulting from social isolation of primates. In J. H. Cullen (Ed.), *Experimental behavior: A basis for the study of mental disturbance.* Dublin: Irish Univ. Pr., 216–223.

Mitchell, G. 1975. What monkeys tell us about human violence. *The Futurist* 9(2), 75–80.

Mitchell, G. 1976. Attachment potential in rhesus macaque dyads (*Macaca mulatta*): A sabbatical report. *JSAS: Catalogue of selected documents in psychology 6, 7,* MS1177.

Mitchell, G. 1977. Parental behavior in nonhuman primates. In J. Money and H. Musaph (Eds.), *Handbook of sexology.* New York: Elsevier/North Holland Biomedical Pr., 749–759.

Mitchell, G. 1979. *Behavioral sex differences in nonhuman primates.* New York: Van Nostrand–Reinhold.

Mitchell, G., G. L. Arling, and G. W. Møller. 1967. Long-term effects of maternal punishment on the behavior of monkeys. *Psychonomic Sci. 8,* 209–210.

Mitchell, G., E. J. Raymond, G. C. Ruppenthal, and H. F. Harlow. 1966. Long-term effects of total social isolation upon behavior of rhesus monkeys. *Psychol. Rep. 18,* 567–580.

Mitchell, G. and W. K. Redican. 1972. Communication in normal and abnormal rhesus monkeys. In *Proceedings of the XXth International Congress of Psychology.* Tokyo: Sci. Coun. Japan, 171–172.

Mitchell, G., W. K. Redican, and J. Gomber. 1974. Lesson from a primate: Males can raise babies. *Psychol. Today* 7(11), 63–68.

Nadler, R. D. 1975. Determinants of variability in maternal behavior of captive female gorillas. In S. Kondo, M. Kawai, A. Ehara, and S. Kawamura (Eds.), *Proceedings of the Symposium of the 5th Congress of the International Primatological Society*. Tokyo: Japan Science Pr., 207–216. (a)

Nadler, R. D. 1975. Second gorilla birth at the Yerkes Regional Primate Research Center. *Int. Zoo Yearb. 15,* 134–137. (b)

Nozawa, K., T. Shotake, Y. Ohkura, and Y. Tanabe. 1977. Genetic variations within and between species of Asian macaques. *Jap. J. Genet. 52,* 15–30.

Papoušek, H. and M. Papoušek. 1974. Mirror image and self-recognition in young human infants: I. A new method of experimental analysis. *Dev. Psychobiol. 7,* 149–157.

Parker, S. T. 1977. Piaget's sensorimotor series in an infant macaque: A model for comparing unstereotyped behavior and intelligence in human and nonhuman primates. In S. Chevalier-Skolnikoff and F. E. Poirier (Eds.), *Primate biosocial development.* New York: Garland, 43–112.

Porges, S. W. 1976. Ontogenetic comparison. *Int. J. Psychol. 11,* 203–214.

Powers, W. 1973. *Behavior: The control of perception.* Chicago: Aldine.

Reynolds, V. 1976. *The biology of human action.* San Francisco: Freeman.

Riesen, A. H., M. Perkins, and R. G. Struble. 1977. *Open-field behavior in socially deprived stumptail monkeys.* Paper presented at the American Society of Primatologists meeting, Seattle, Wash., April.

Rogers, C. M. and R. K. Davenport. 1969. Effects of restricted rearing on sexual behavior of chimpanzees. *Dev. Psychol. 1,* 200–204.

Rosenblum, L. A. 1974. Sex differences in mother-infant attachment in monkeys. In R. C. Friedman, R. M. Richart, and R. L. Vande Wiele (Eds.), *Sex differences in behavior.* New York: Wiley, 123–145.

Rosenblum, L. A. and I. C. Kaufman. 1968. Variations in infant development and response to maternal loss in monkeys. *Am. J. Orthopsychiatry 38*(3), 418–426.

Roy, M. A. 1976–1977. Early rearing of infrahuman primates with reduced conspecific contacts: A selected bibliography. Part I. *J. Biol. Psychol.,* 1976, *18*(2), 36–42. Part II. *Op cit,* 1977, *19*(1), 21–27.(a) Part III. *Op cit,* 1977, *19*(2), 29–30. (b)

Roy, M. A. 1977. *Early experience and the development of an adequate species identity.* Symposium presented at the American Psychological Association meeting, San Francisco, Cal., August. (c)

Sackett, G. P. 1974. Sex differences in rhesus monkeys following varied rearing experiences. In R. C. Friedman, R. M. Richert, and R. L. Van de Wiele (Eds.), *Sex differences in behavior.* New York: Wiley, 99–122.

Sackett, G. P., R. A. Holm, and G. C. Ruppenthal. 1976. Social isolation rearing: Species differences in behavior of macaque monkeys, *Dev. Psychol. 12*(4), 283–288.

Savage, E. S. and R. Bakeman. 1976. *Comparative observations on sexual behaviour in Pan paniscus and Pan troglodytes.* Paper presented at the International Primatological Society meeting, Cambridge, England, August.

Schulman, A. H. and C. Kaplowitz. 1977. Mirror-image responses during the first two years of life. *Dev. Psychobiol. 10,* 133–142.

Scollay, P. A., S. Joines, C. Baldridge, and A. Cuzzone. 1975. Learning to be a mother. *Zoonooz, 48*(4), 4–9.

Shafton, A. 1976. *Conditions of awareness.* Portland, Ore.: Riverstone Pr.

Shotake, T., K. Nozawa, and Y. Tanabe. 1977. Blood protein variations in baboons. I. Gene exchange and genetic distance between *Papio anubis, Papio hamadryas,* and their hybrid. *Japan J. Genet. 52,* 223–237.

Stone, L. J., H. T. Smith, and L. B. Murphy (Eds.). 1973. *The complete infant.* New York: Basic Books.

Suomi, S. J. 1974. Social interactions of monkeys reared in a nuclear family environment versus monkeys reared with mothers and peers. *Primates* 15(4), 311–320.

Suomi, S. J., H. F. Harlow, and S. D. Kimball. 1971. Behavioral effects of prolonged partial social isolation in the rhesus monkey. *Psychol. Rep. 29,* 1171–1177.

Terrace, H. S. and T. G. Bever. 1976. What might be learned from studying language in the chimpanzee? The importance of symbolizing oneself. *Ann. N.Y. Acad. Sci. 280,* 579–588.

Terrace, H. S., L. Pettito, and T. G. Bever. 1977. Project Nim: Progress report. I. Unpublished manuscript.

Testa, T. J. and D. Mack. 1977. The effects of social isolation on sexual behavior in *Macaca fascicularis.* In S. Chevalier-Skolnikoff and F. E. Poirier (Eds.), *Primate biosocial development.* New York: Garland, 407–438.

Tobach, E. 1976. Evolution of behavior and the comparative method. *International J. Psychol.* 11(3), 185–201.

Tutin, C. E. G. and W. C. McGrew. 1973. Chimpanzee copulatory behavior. *Folia Primatol. 19,* 237–256.

Whitsett, G. 1977. *A critique of the natural science approach to human sexuality.* Paper presented at the Animal Behavior Society meeting, University Park, Pa., June.

Zucker, E. L., M. E. Wilson, S. F. Wilson, and T. Maple. 1977. *The development of sexual behavior in infant and juvenile male orang-utans (Pongo pygmaeus).* Paper presented at the American Society of Primatologists meeting, Seattle, Wash., April.

Chapter 11

Chimpanzees and Self-awareness

Gordon G. Gallup, Jr.

Self-awareness has always been an elusive topic. Yet the concept of self occupies a central position in many areas of psychology. The basic problem even in man continues to be one of definition and measurement. In this chapter a technique for operationalizing the self-concept in animals will be described, and then the implications of the concept of self for a species identity will be briefly discussed.

In order to conceive of ourselves, we have to have a sense of identity, otherwise there would be nothing to conceive. By identity is meant a sense of continuity over time and space (cf. Descartes, 1662). "I" persist over time and space, therefore "I" exist. Memory represents one form of continuity over time and space. But by itself, memory neither presupposes nor necessarily provides for such a sense. In this regard human consciousness is typically bidirectional. Not only can "I" be aware of objects and events in the world around me, "I" can become the object of my own attention. My brain can speculate about the mechanisms of its own functioning. Based on my sense of self, "I" can even contemplate my own individual demise. In other words, the bidirectional properties of consciousness translate into consciousness and self-consciousness.

For some time man has been held unique in his capacity to form a self-concept (Ardrey, 1961; Dobzhansky, 1972; Lorenz, 1971). The history of science, however, can be viewed in part as having brought about gradual changes in man's conception of man, and with such changes man may eventually have to relinquish or at least temper his claim to special status.

Chimpanzees and Men

Primate research has increasingly threatened many traditional and cherished notions about man. In spite of the widespread acceptance of Darwinian theory, man has always conceived of himself as being not only different from other animals, but endowed with special qualities. At one time or another man has been held unique because of the following traits: tool use, tool fabrication, culture, cross-modal perception, self-awareness, and language.

We now know, however, that under natural conditions chimpanzees use and fabricate tools in ingenious ways and, consistent with the concept of culture, such information is often passed on from one generation to the next (Beck, 1975; Goodall, 1968). In recent times it has also become fashionable to view man's evolutionary history as being distinctive among primates because of his presumed hunting and predatory behaviors. But it has been shown that chimpanzees not only eat meat, they are also more than merely opportunistic predators. Chimpanzees engage in systematic and cooperative hunting (Teleki, 1973). It is curious to note that hunting among chimpanzees is usually confined to males, and meat is one of the few foods chimpanzees are willing to share with each other. Male-oriented, cooperative hunting patterns which terminate in sharing have been held to be a hominoid hallmark and an evolutionary precursor of the development of modern man. It is also known that chimpanzees are capable of cross-modal perception (Davenport and Rogers, 1970), and it appears that with proper training they can even master many of the rudiments of syntax and symbolic communication (Gardner and Gardner, 1969; Premack, 1971).

Mirrors

Since the differences between chimpanzees and men seem to be a matter of degree rather than of kind, one of the last substantive hold-outs for human uniqueness has been self-awareness. Even many psychologists have a very deep-seated feeling that consciousness and self-awareness are simply not amenable to objective analysis in animals (e.g., Gardiner, 1974). According to Klüver (1933), a noted primatologist, the content of experience or of consciousness itself is not a reasonable object of scientific study: "scientifically they do not and cannot exist."

Mirrors, however, provide a means of assessing the presence of self-awareness in animals. *Mirror-image stimulation* (Gallup, 1968) refers to a situation in which an organism is confronted with its own reflection in a mirror. Many organisms react to themselves in mirrors as if they were seeing other animals and engage in a variety of species-typical social responses directed toward the reflection (e.g., aggressive behaviors, sexual postures, greeting gestures). In other words, mirror-image stimulation seems to function, at least initially, as a social stimulus. Moreover, when an animal first sees itself in a mirror, it finds an animal with particular facial features it has never seen before. And, indeed, the initial reaction often consists of responses that would typically be made to strangers. Mirrors also have incentive properties for many species, and animals can be taught to engage in a variety of instrumental responses reinforced by nothing more than brief visual access to their own reflections (see Gallup, 1968, 1975).

Some animals even appear to show a peculiar preference for viewing mirrors instead of other animals. For example, when goldfish are individually placed in an underwater alley and allowed to view a mirror at one end or another goldfish at the other, they show a decisive three-to-one preference for the mirror (Gallup and Hess, 1971). In an attempt to determine the generality of these findings European weaver finches were individually tested in a square enclosure with a perch mounted on each wall (Gallup and Capper, 1970). Each perch was associated with a different visual incentive and was wired to a timer which measured the amount of time a bird spent on any particular perch. By standing on one perch the bird could see another finch behind glass, on another it could see itself in a mirror, and on still another it could eat. Although the initial tendency was to view the other bird, after nine days of testing all the finches showed a reliable preference for mirror-image stimulation. The effect in parakeets is even more striking. As shown in Figure 1, not only do the birds show a beginning preference for the mirror, but this preference becomes even more exaggerated over days. The picture for monkeys, however, is quite different. When placed in a cage containing doors which could be opened to view mirrors or other monkeys, feral

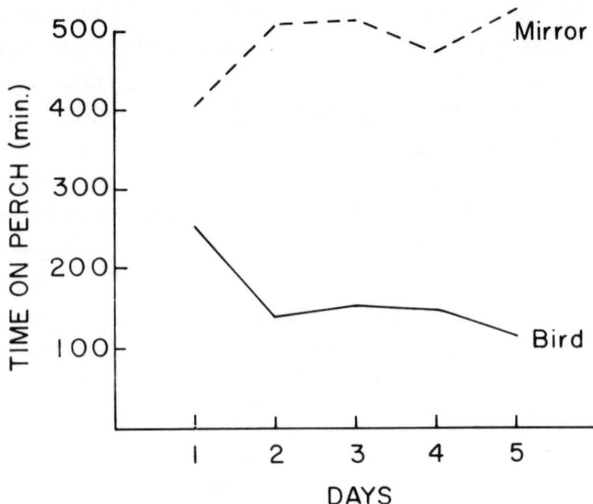

Figure 1. Distribution of viewing time in parakeets on a perch providing access to a mirror and one in front of another bird (from Gallup and Capper, 1971).

rhesus monkeys preferred viewing other monkeys (Gallup and McClure, 1971). But laboratory-born, isolation-reared rhesus monkeys resembled fish and birds in showing a significant preference for viewing mirrors.

Self-recognition

The most interesting point about mirrors and probably the most obvious for humans is that they provide information about the self. Mirrors enable visually capable organisms to see themselves as they are seen by others. In front of a mirror an animal is an audience of its own behavior. Yet, for some curious reason, most animals seem incapable of recognizing the dualism inherent in such stimulation and even after prolonged exposure fail to discover the relationship between their behavior and the reflection of that behavior in a mirror. For most animals mirrors represent an exclusive instance of self-sensation. Although ostensibly stimulated by themselves, they fail to recognize that their behavior is the source of the behavior depicted in the mirror and persist in showing other-directed rather than self-directed behavior in response to the reflection. Animals, in other words, tend to respond to mirrors as though they were seeing other animals. Humans, however, are capable of self-perception as evidenced by our use of mirrors for grooming and purposes of self-inspection. Whether self-sensation translates into self-perception depends on self-recognition.

Aside from the fact that most of us undoubtedly are given explicit verbal instructions by our parents about the identity of the reflection, the ability to recognize one's own image would seem to be in part a function of prolonged exposure to mirrors. After all, mirror surfaces, at least in this culture, are an everyday part of the human experience. Infants and adult humans who have never seen mirrors before react just like animals and initially respond to the reflection as if confronted with another person (Amsterdam, 1972; von Senden, 1960). Maybe if animals were given the same opportunity for extended self-inspection, they might also come to recognize themselves in mirrors.

Self-recognition in Chimpanzees

To test this conjecture, in the original study (Gallup, 1970) a number of preadolescent, wild-born chimpanzees were individually exposed to mirrors for eight hours a day over a period of 10 consecutive days. In all instances their initial reaction to the mirror was as though they were seeing another chimpanzee, and for the first few days they engaged in a variety of social gestures while watching the reflection. After about 3 days, however, the tendency to treat the reflection as a companion disappeared. This was replaced by the emergence of what appeared to be a self-directed orientation. That is, rather than respond to the mirror as such, after about the third day the chimpanzees began to use the mirror to respond to themselves. Under conditions of self-directed responding they used the reflection to gain visual access to and experiment with otherwise inaccessible information about themselves, such as grooming parts of the body that had not been seen before, making faces at the reflection, inspecting the insides of their mouths, manipulating anal–genital areas, etc. Figure 2 portrays an instance of self-directed responding.

Prior to the development of self-directed behavior all the chimpanzees showed an avid interest in the mirror (see Figure 3); but with the emergence of a self-directed orientation, this interest diminished. Why did viewing time decrease? It may be that with the advent of self-recognition the mirror produces a state of objec-

Figure 2. A chimpanzee viewing his mirror-image and using the reflection to remove extraneous material from his nose.

tive self-awareness, and studies with humans suggest that prolonged periods of objective self-awareness may be aversive (Duval and Wicklund, 1972).

It was possible, however, that other behavioral scientists might not be persuaded by these subjective interpretations of what had transpired. So in an attempt to clarify and objectify these impressions, an unobtrusive and more rigorous test of self-recognition was instituted. Following the last day of mirror exposure, each chimpanzee was anesthetized and the mirror was removed. After the chimpanzee was completely unconscious, the experimenter painted the uppermost portion of an eyebrow ridge and the top half of the opposite ear with a bright red, odorless, nonirritating alcohol-soluble dye (see Figure 4). The animal was then placed back into its cage and allowed to recover in the absence of the mirror.

The significance of this technique is threefold. First, the chimpanzees had no way of knowing about the application of the marks, since the procedure was accomplished under deep anesthesia. Second, the dye was selected for its complete lack of tactile and olfactory properties, as determined by applying it to the experimenter's own skin several days prior to testing. Finally, the marks were placed at predetermined points on the chimpanzee's face so that they could only be seen in a mirror.

Following recovery, all subjects were observed to determine the number of times any marked portion of the skin was touched in the absence of the mirror. The mirror was then reintroduced as an explicit test of self-recognition. Upon seeing themselves in the mirror, all the chimpanzees showed mark-directed responses or attempted to touch the marked areas on themselves while watching their reflections. As shown in Figure 3, there was also a dramatic increase in viewing time in the test of self-recognition. Something about the presence of the red marks greatly enhanced their visual attention to the reflection. In addition to mark-directed responses, there were a number of noteworthy attempts

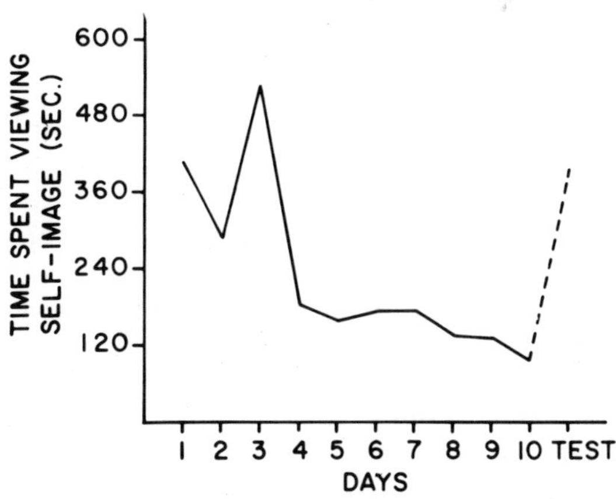

Figure 3. Average amount of time, during two 15-minute sessions, that chimpanzees spent viewing themselves in the mirror as a function of days (from Gallup, 1970).

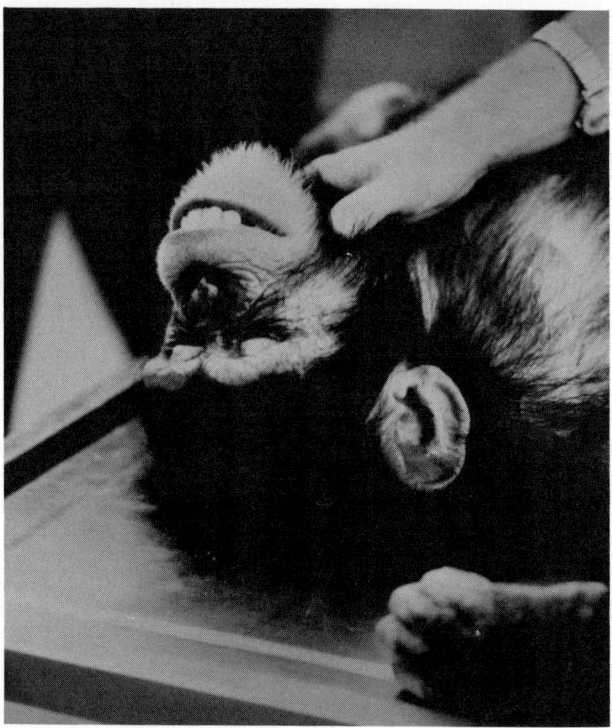

Figure 4. An anesthetized chimpanzee after being marked in preparation for the test of self-recognition.

to examine visually and to smell the fingers which had touched the marked parts of the face, even though the dye was indelible. I suspect that you would probably show a similar pattern of behavior if, upon awaking one morning, you saw yourself in a mirror with strange red marks on your face.

An occasional objection to this procedure as a test of self-recognition is the hypothesis that the chimpanzees had simply learned that they controlled the behavior of the "other" animal in the mirror. Therefore, when they saw the other animal with red marks on its face on the test of self-recognition, they touched the appropriate parts of their own bodies in an attempt to direct its attention to these areas. If this were the case, however, there would be no reason for the chimpanzees to visually inspect and to smell their own fingers, since ostensibly these would not have been the fingers to have made contact with the red marks seen in the mirror.

In a further attempt to eliminate any doubt about the source of these reactions, several comparable chimpanzees which had never seen themselves in mirrors were also anesthetized and marked. When given access to the mirror for the first time, there were no mark-directed responses whatsoever, patterns of self-directed behavior were completely absent, and the dye was ostensibly ignored. Throughout the test their orientation to the mirror was unmistakably social rather than self-directed. They all responded to the reflection as if confronted with another chimpanzee. These data strongly support the proposition that self-recognition was learned by the first group of chimpanzees sometime during the initial ten days of mirror exposure.

Self-recognition in Other Primates

These findings have now been replicated several times by a number of investigators (Gallup, McClure, Hill, and Bundy, 1971; Hill, Bundy, Gallup, and McClure, 1970; Lethmate and Dücker, 1973), and extended by Lethmate and Dücker to include orangutans as well as chimpanzees. With the exception of man and the great apes, however, attempts to demonstrate self-recognition in all other primates have uniformly failed even after their extended exposure to mirrors.

To date, spider monkeys, capuchins, squirrel monkeys, stumptail macaques, rhesus monkeys, crab-eating macaques, Java monkeys, pigtail macaques, mandrill, olive, and hamadryas baboons, and two species of gibbons have all been systematically tested for self-recognition (Benhar, Carlton, and Samuel, 1975; Gallup, 1970, 1977a; Lethmate and Dücker, 1973),[1–5] but none has shown any sign of realizing that its own behavior is the source of the behavior seen in the mirror. Among nonhuman primates, the great apes seem to be the only ones who derive this abstract information from mirrors.

In a recent attempt to salvage the conceptual integrity of monkeys, the author gave a female crab-eating macaque five months, or over 2400 hours of exposure, to her image in a mirror (Gallup, 1977a), but to no avail. At the end of that period she failed to show any signs of self-directed behavior or mark-directed response following the application of red dye. Throughout the five months she continued to respond to her image as if she were seeing another monkey. More recently, Thompson and Radano[4] conducted a study in which a pigtail macaque was kept in a cage containing a mirror for 11 months, but the monkey still gave no evidence of self-recognition. In sharp contrast to these findings, chimpanzees require only about three days of mirror exposure before they begin showing signs of self-directed behavior.

Although monkeys can learn to use mirrors to manipulate objects (Brown, McDowell, and Robinson, 1965), they appear totally incapable of learning to integrate features of their own reflection sufficiently in order to use mirrors to respond to themselves. Again, it is not that they cannot learn to respond to mirrored cues. When looking at the reflection of a human or a bit of food, they can detect the inherent dualism as it pertains to objects other than themselves and, after adequate experience, do respond appropriately by turning away from the mirror to gain more direct access to the object of the reflection (Tinklepaugh, 1928). Yet, for some curious reason, they completely fail to interpret mirrored information about themselves correctly. On other more traditional tasks and tests of cognitive processes in primates (e.g., learning set, problem solving, match-to-sample), one typically finds a continuum or gradation of ability as one moves up the phyletic scale from one primate species to another. Yet for self-recognition there is this abrupt, almost unprecedented stepwise change at the level of the great apes.

The Significance of Self-recognition

It would appear, therefore, that monkeys and other animals lack an essential cognitive category for processing mirrored information about themselves. What is the nature of this cognitive deficit? Self-recognition is predicated on a sense of identity, otherwise there would be no basis for recognition. A unique feature of mirrors is that the identity of the observer and his reflection in a mirror are necessarily one and the same. Therefore the capacity to infer the identity of the image correctly must presuppose an already existent identity on the part of the organism making that inference (Gallup, 1977b). Without an identity of one's own, the source and significance of the reflection would forever remain unknown. In other words, the monkey's inability to recognize himself in a mirror could be due to the absence of a sufficiently well-integrated self-concept. With the exception of man and the great apes, we eventually may have to entertain the possibility that primate consciousness may be unidirectional. On the other hand, to the extent that self-recognition implies a rudimentary concept of self, the data on chimpanzees and orangutans seems to show that man may not have a monopoly on the concept of self.

Another question often raised is that a visual sense of self may simply be different from an olfactory or auditory self. According to this view, animals other than the great apes, while lacking a visual self, may have a sense of self via some other sensory modality. This may be a mistaken distinction; one's conception of self is not modality-specific. The sense of self, for example, does not disappear when you close your eyes or cover your ears. The self is a multimodal phenomenon. Self-directed and mark-directed behaviors presuppose considerable intermodal equivalence (e.g., visual, proprioceptive, tactile, and—in the case of smelling fingers which had been used to touch marked portions of the skin—olfactory equivalence).

Naturally we are a long way away from being able to specify a neurological basis for the sense of self. Self-recognition, however, may represent an emergent phenomenon that only occurs once a species acquires a certain number of cortical neurons with sufficiently complex interconnections. One could alternatively view these data from the standpoint of a threshold model. Different organisms may very well have differing degrees of self-awareness; but only with an explicit sense of identity does self-recognition become possible. The threshold for self-recognition may be quite high compared to other forms of self-conception. It is also possible that self-recognition may be an indirect consequence of selection for other cognitive skills and as such would represent a correlated trait (e.g., self-awareness may be a precondition for language).

As far as chimpanzees and orangutans are concerned, this author does not think their sense of self in any way emerges out of experience with a mirror. Rather, a mirror simply represents a means of mapping what the chimpanzee already knows and provides him with a new and more explicit dimension of knowing about himself, in the sense that he now has an opportunity to see himself as he is seen by other chimpanzees.

Ways to Facilitate Self-recognition

Related to the issue of trying to demonstrate an awareness of self in other primates is the question of how to accelerate the development of self-recognition. Most attempts to show the phenomenon in monkeys have employed animals which were adolescent or older. However, a number of studies (e.g., Itani, 1958; Welker, 1956) suggest that curiosity and innovativeness among primates may be, within limits, inversely proportional to age. As yet there have been no systematic attempts to investigate self-recognition in primates (other than man) as a function of age. Raising an infant monkey in the presence of its own reflection might prove worthwhile. If the absence of self-recognition in monkeys proves genuine, the question of whether access to a mirror might partially offset the otherwise debilitating effects of social isolation becomes relevant.

Several other possibilities should also be mentioned. In most studies the mirror is kept out of the animal's reach simply to prevent it from being broken. If, however, the animal were allowed to contact a metal mirror and discover its inanimate nature, this might provide more information as to the source of the reflection. When given physical access to mirrors many great apes initially attempt to look behind the mirror as if searching for what they perceive as being another animal (Gallup, 1968).

Another means of providing more immediate and explicit information about the identity of the reflection would be to give a pair of familiar animals simultaneous exposure to the same mirror. Given the well-documented capacity for individual recognition (Hediger, 1976), each member of the pair should be able to identify or at least recognize the reflection of its companion correctly. Therefore, by more direct implication, the question is raised as to the origin of the other unfamiliar animal each one sees in the mirror.

Finally, the possibility of using some training regime based on operant conditioning should be mentioned. An organism could gradually be shaped to display patterns of self-directed responses in the presence of its own reflection, and perhaps in the process it may come to interpret correctly what it sees in the mirror. It should be acknowledged, however, that at least one such attempt has been made with baboons (Benhar, Carlton, and Samuel, 1975) and proved ineffective. Alternatively, a monkey placed in a restraining chair and fitted with a wide opaque collar which prevents it from seeing its body could be placed in front of a mirror and taught to touch different marked portions of its own body for reinforcement. Once this was learned one could unobtrusively mark the forehead and remove the collar in the presence of the mirror to see if the monkey would touch the dye. The fact, however, that social responsiveness to mirrors continues even after months of exposure (e.g., Gallup, 1977a) poses another problem. If this technique were successful, the continued presence of social behavior directed toward the reflection would confound any simple interpretation of the results. In the absence of other instances of self-directed behavior, one could argue that the monkey had merely learned a complex operant response to mirror-image reversal, without any explicit realization as to the significance of what it was seeing in the mirror. After having touched the mark on its forehead, for example, would the monkey look at or smell its fingers? Would it use the mirror to inspect other portions of its face? Although mark-directed responses constitute an objective test, other self-directed behaviors in a less formal setting are needed to validate an interpretation based on self-recognition.

Possibilities for Self-recognition in Other Species

Given the apparent difference in the capacity for self-recognition between great apes and all other primates, it is unfortunate that there have been no systematic attempts to document the phenomenon in gorillas. This is a particularly important bit of missing information since in addition to chimpanzees and orangutans (and maybe man, depending on one's taxonomic bias), gorillas are the only other species of great apes. Gibbons are classified as apes rather than great apes, and attempts to show self-recognition in gibbons have failed (e.g., Lethmate and Dücker, 1973).[3]

Because of their exceptionally large, complex brains and intricate social behavior, the cetaceans (porpoises and whales) represent an intriguing group of mammals other than the great apes which might be capable of self-conception. Porpoises in particular appear to have the capacity to process and store information in ways that are strikingly human (e.g., Thompson and Herman, 1977). Another class of big-brained mammals that might be interesting to test for self-recognition are elephants. Even though they lack arms and fingers, their unique prehensile trunks would represent an ideal vehicle for exhibiting mark-directed and self-directed behaviors in the presence of a mirror.

Social Influences on Self-conception

A prevailing view of self-concept formation in humans is that it only develops out of a social milieu and as such is dependent upon interaction with others. According to Mead (1934), for instance, in order for the self to emerge as an object of conscious inspection it requires the opportunity to examine itself from another's point of view. Figuratively speaking, in order to conceive of yourself you may need to be able to see yourself as you are seen by others. In keeping with this general idea, it was found in another study (Gallup, McClure, Hill, and Bundy, 1971) that chimpanzees reared in social isolation seemed incapable of recognizing themselves in mirrors.

As shown in Figure 5, wild-born chimpanzees reared with cagemates evidenced a high degree of initial curiosity in the mirror; but, as before (see Figure 3), this interest waned over days. The isolates, however, paid unabated attention to the reflection for the entire period. When tested for self-recognition following the anesthetization and marking procedure, the socially reared chimpanzees all showed mark-directed responses, while isolates did not. Furthermore, as shown in Figure 5, the normal chimpanzees showed a dramatic increase in viewing time when they first saw themselves in the marked condition, and this abrupt reinstatement of visual attention replicated my earlier findings. But isolate-reared chimpanzees, as further evidence that they remain completely oblivious to the source and significance of the

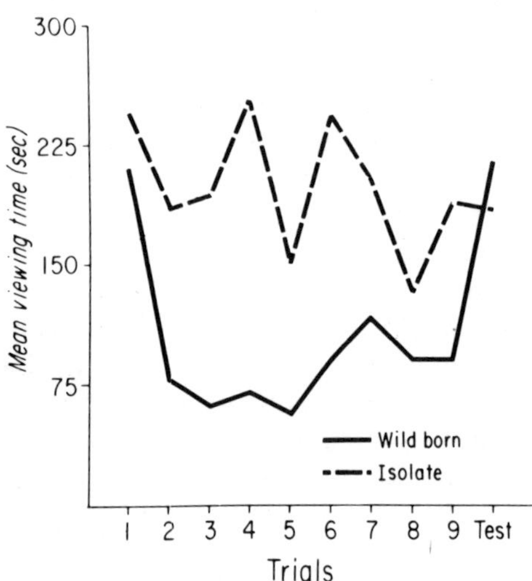

Figure 5. Average time during ten 30-minute sessions that wild-born and isolated chimpanzees spent viewing themselves in a mirror (from Gallup, McClure, Hill, and Bundy, 1971).

reflection, showed virtually no change in viewing time in the test of self-recognition. There is also preliminary evidence that remedial social experience, accomplished by housing isolates together in the same cage, is sufficient to provide at least suggestive signs of self-recognition (Hill, Bundy, Gallup, and McClure, 1970).

If explicit self-awareness depends on the opportunity to examine oneself from another's point of view, then some intriguing possibilities arise. Jung (1958), for example, argued that a truly objective view of man would only be possible if we could see ourselves from another species' perspective. This is a very compelling idea. It follows that the prospect of achieving an objective view of ourselves is rather bleak, at least for the foreseeable future. If, however, the present analysis of self-conception is correct and the opportunity to examine oneself were restricted to another species' point of view, this ought to distort one's concept of self profoundly. It is reported, for example, that when Washoe, who was reared with humans, was first confronted with other chimpanzees, she referred to them in sign language as "black bugs" (Linden, 1974). As another illustration, it is well known that being reared in social isolation may have devastating effects on primate sexual behavior (e.g., Mason, 1960). It is less well known, however, that human-reared chimpanzees are actually more impaired sexually than those reared in complete and abject social isolation (Rogers and Davenport, 1969). Why? Maybe it is because they think they are human, as if to say, "Who wants to mate with one of those nasty beasts?"

Then there is the informal experiment conducted with another home-reared chimpanzee named Vicki (Hayes and Nissen, 1971). Among other things, Vicki was taught to sort stacks of snapshots into a human and an animal pile. One day, unbeknown to Vicki, her own photograph was placed in the stack, and when she came to her picture, without hesitation she picked it up and placed it on the human pile. It seems to me that there are at least two ways to interpret this behavior. First, man may not be the only one to appreciate the similarities between chimpanzees and men. The other is that maybe Vicki thought she was human.

Temerlin (1975) recently reported that as a result of rearing a chimpanzee in his home (see Temerlin, Chapter 12, this volume) his son began to develop doubts about his own species identity. To the extent that this is a reliable effect, it would clearly follow from the notion of social influences on self-conception. The imprinting literature makes it clear that affiliation tendences are often tied to early social experiences. For man and the great apes, the content of individual identity may be similarly subject to social influences. In other words, the presence or absence of social interaction may determine whether or not one has a self-concept, while the quality and nature of those interactions determine the content of the concept.

Self-perception and Species Identity

While the behavioral or functional consequences of imprinting may be pretty much the same for most animals, the data on self-recognition can be used to argue that the cognitive consequences of imprinting may be quite restricted, and could create an identity crisis of sorts for those with an intellectual investment in the concept of a species identity. It seems, however, that the formation of a species identity is logically predicated on a sense of individual identity.

Contrary to the stance which many people have taken in the literature (e.g., Gottlieb, 1971), affiliation tendencies are far from being isomorphic with species identity. Species identity implies a sense of belonging or membership, while species-specific affiliation patterns, following responses, and attachment behaviors may merely reflect the operation of primitive familiarity effects and genetic predispositions. Just because "birds of a feather flock together," there is hardly a strong inferential basis for postulating the existence of such a formative concept as a species identity.

An explicit sense of species identity would have to involve the cognitive capacity to equate oneself with other organisms or the intellectual equivalent of thinking "I am one of them." But to be able to think "I am one of them" presupposes a sense of "I." If a visually capable animal cannot identify itself in a mirror, how could it possibly be expected to identify itself as being a member of some particular class of organisms? It seems like a logical impossibility. Therefore, species identity and individual identity may not be mutually exclusive; and without a concept of self, the designation "species identity" is a misnomer. As traditionally conceived, it carries too much excess conceptual baggage. In instances of organisms other than the great apes, the concept of species identity should be replaced by some less misleading and more descriptive term such as "affiliation tendencies."

Table 1 depicts the progression from genetic predispositions to species identity. Inborn tendencies to interact with members of the same species are widespread and well documented. Take, for example, the parasitic egg layer called the cowbird. The female cowbird lays a single egg in the nest of a host species different from her own species. The host treats the egg as its own, incubates it, and later fosters the baby cowbird. Since it is raised by a species other than its own, in order to pick an appropriate mate the cowbird has to have information about what another cowbird looks like, or at least sounds like (King and West, 1977). Species-specific predispositions in turn eventuate in social interaction. But if and only if, as a byproduct of that interaction, the organism acquires an individual identity, does the development of a species identity become possible. A predisposition to engage in species-specific affiliation patterns precludes the need for a species identity. It may be, therefore, that the loss of genetic constraints on affiliation patterns in great apes prompted the development of an individual identity which in turn provides for a species identity.

It is also intriguing to note that while the development of an individual identity in great apes probably requires social interaction with others, the object of that interaction need not involve others of the same species. Many home-reared chimpanzees, for example, show obvious signs of self-recognition in response to mirror images, but an active avoidance or even disdain for members of their own species (e.g., Temerlin, 1975). Thus the existence of self-awareness in chimpanzees does not appear to depend on a particular or even an accurate species identity. The development of an individual identity, in fact, may constitute a better model of early-experience effects than affiliation tendencies;

Table 1. Hypothetical Progression from Genetic Predispositions to Species Identity

Species-Specific Predispositions

↓

Social Interaction

↓

Individual Identity

↓

Species Identity

since, as already mentioned, the latter are often encumbered for good evolutionary reasons by genetic constraints. According to some of the early Lorenzian views of imprinting, if animals became indelibly imprinted to the first moving object they encountered during the critical period, one would expect, among other things, to see a lot more ducks imprinted on leaves which had been blown by the wind at the magic moment of primary socialization. Even in many primates, as Sackett's (1970) work has clearly shown, affiliation patterns are partially predetermined. The content of an individual identity, however, may be relatively unconstrained. For this reason it is possible that self-awareness has the capacity to emancipate organisms, intellectually at least, from some of the otherwise deterministic and unrelenting forces of evolution (Slobodkin, 1977). I suspect that it is this emancipation which makes the kind of analysis in this chapter possible.

Acknowledgment

The author would like to thank L. M. Tornatore for comments on an earlier draft of this paper.

Notes

1. Bertrand, M., personal communication, 1972.
2. Geberer, N. N., personal communication, 1978.
3. Pribram, K., personal communication, 1972.
4. Thompson, R. L., personal communication, 1977.
5. Wallnau, L. B., personal communication, 1977.

References

Amsterdam, B. 1972. Mirror self-image reactions before age two. *Dev. Psychobiol.* 5, 297–305.

Ardrey, R. 1961. *African genesis.* New York: Dell.

Beck, B. B. 1975. Primate tool behavior. In R. H. Tuttle (Ed.), *Socioecology and psychology of primates.* The Hague, Netherlands: Mouton.

Benhar, E. E., P. L. Carlton, and D. Samuel. 1975. A search for mirror-image reinforcement and self-recognition in the baboon. In S. Kondo,

M. Kawai, and A. Ehara (Eds.), *Contemporary primatology: Proceedings of the Fifth International Congress of Primatology*. Basel, Switzerland: Karger.

Brown, W. L., A. A. McDowell, and E. M. Robinson. 1965. Discrimination learning of mirrored cues by rhesus monkeys. *J. Genet. Psychol. 106*, 123–128.

Davenport, R. K. and C. M. Rogers. 1970. Intermodal equivalence of stimuli in apes. *Science 168*, 279–280.

Descartes, R. 1662. *De homine*. Leiden, The Netherlands.

Dobzhansky, T. H. 1972. On the evolutionary uniqueness of man. In T. H. Dobzhansky, M. K. Hecht, and W. C. Steere (Eds.), *Evolutionary biology*. Vol. 6. New York: Appleton-Century-Crofts.

Duval, S., and R. A. Wicklund. 1972. *A theory of objective self awareness*. New York: Academic Pr.

Gallup, G. G., Jr. 1968. Mirror-image stimulation. *Psycho. Bull. 70*, 782–793.

Gallup, G. G., Jr. 1970. Chimpanzees: Self-recognition. *Science 167*, 86–87.

Gallup, G. G., Jr. 1975. Towards an operational definition of self-awareness. In R. H. Tuttle (Ed.), *Socio-ecology and psychology of primates*. The Hague, Netherlands: Mouton.

Gallup, G. G., Jr. 1977. Absence of self-recognition in a monkey (*Macaca fascicularis*) following prolonged exposure to a mirror. *Dev. Psychobiol. 10*, 281–284. (a)

Gallup, G. G., Jr. 1977. Self-recognition in primates: A comparative approach to the bidirectional properties of consciousness. *Am. Psychol. 32*, 329–338. (b)

Gallup, G. G., Jr. and S. A. Capper. 1970. Preference for mirror-image stimulation in finches (*Passer domesticus domesticus*), and parakeets (*Melopsittacus undulatus*). *Anim. Behav. 18*, 621–624.

Gallup, G. G., Jr. and J. Y. Hess. 1971. Preference for mirror-image stimulation in goldfish (*Carassius auratus*). *Psychonomic Sci. 23*, 63–64.

Gallup, G. G., Jr., and M. K. McClure. 1971. Preference for mirror-image stimulation in differentially reared rhesus monkeys. *J. Comp. Physiol. Psychol. 75*, 403–407.

Gallup, G. G., Jr., M. K. McClure, S. D. Hill, and R. A. Bundy. 1971. Capacity for self-recognition in differentially reared chimpanzees. *Psychol. Rec. 21*, 69–74.

Gardiner, W. L. 1974. *Psychology: A story of a search*. Belmont, Cal.: Wadsworth.

Gardner, R. A. and B. T. Gardner. 1969. Teaching sign language to a chimpanzee. *Science 165*, 664–672.

Goodall, J. 1968. The behavior of free-living chimpanzees in the Gombe Stream Reserve. *Anim. Behav. Monogr. 1*, 161–311.

Gottlieb, G. 1971. *Development of species identification in birds*. Chicago: Univ. Chicago Pr.

Hayes, K. J. and C. H. Nissen. 1971. Higher mental functions in a home-raised chimpanzee. In A. M. Schrier and F. Stollnitz (Eds.), *Behavior of nonhuman primates.* Vol. 3. New York: Academic Pr.

Hediger, H. 1976. Proper names in the animal kingdom. *Experientia 32,* 1357–1364.

Hill, S. D., R. A. Bundy, G. G. Gallup, Jr., and M. K. McClure. 1970. Responsiveness of young nursery-reared chimpanzees to mirrors. *Proc. La. Acad. Sci. 33,* 77–82.

Itani, J. 1958. On the acquisition and propagation of a new food habit in the troop of Japanese monkeys at Takasakiyama. *Primates 1,* 131–148.

Jung, C. G. 1958. *The undiscovered self.* Boston: Little, Brown.

King, A. P. and M. J. West. 1977. Species identification in the North American cowbird: Appropriate responses to abnormal song. *Science 195,* 1002–1004.

Klüver, H. 1933. *Behavior mechanisms in monkeys.* Chicago: Univ. Chicago Pr.

Lethmate, J. and G. Dücker. 1973. Untersuchungen zum Selbsterkennen im Spiegel bei Orang-utans und einigen anderen Affenarten. *Z. Tierpsychol. 33,* 248–269.

Linden, E. 1974. *Apes, men, and language.* New York: Penguin.

Lorenz, K. 1971. *Studies in animal behavior.* Vol. 2. Cambridge, Mass.: Harvard Univ. Pr.

Mason, W. A. 1960. The effects of social restriction on the behavior of rhesus monkeys: I. Free social behavior. *J. Comp. Physiol. Psychol. 53,* 582–589.

Mead, G. H. 1934. *Mind, self and society: From the standpoint of a social behaviorist.* Chicago: Univ. Chicago Pr.

Premack, D. 1971. Language in chimpanzee? *Science 172,* 808–822.

Rogers, C. M. and R. K. Davenport. 1969. Effects of restricted rearing on sexual behavior of chimpanzees. *Dev. Psychol. 1,* 200–204.

Sackett, G. P. 1970. Unlearned responses, differential rearing experiences, and the development of social attachments by rhesus monkeys. In L. A. Rosenblum (Ed.), *Primate behavior.* Vol. 1. New York: Academic Pr.

Slobodkin, L. B. 1977. Evolution is no help. *World Archaeol. 8,* 332–343.

Teleki, G. 1973. *The predatory behavior of wild chimpanzees.* Lewisburg, Pa.: Bucknell Univ. Pr.

Temerlin, M. K. 1975. *Lucy: Growing up human.* Palo Alto, Cal.: Science and Behavior Books.

Thompson, R. K. R. and L. M. Herman. 1977. Memory for lists of sounds by the bottle-nosed dolphin: Convergence of memory processes with humans? *Science 195,* 501–503.

Tinklepaugh, O. L. 1928. An experimental study of representative factors in monkeys. *J. Comp. Psychol. 8,* 197–236.

Von Senden, M. 1960. *Space and sight: The perception of space and shape in the congenitally blind before and after operation.* Glencoe, Ill.: Free Press.

Welker, W. I. 1956. Some determinants of play and exploration in chimpanzees. *J. Comp. Physiol. Psychol.* 49, 223–226.

Chapter 12

The Self-concept of a Home-reared Chimpanzee

Jane W. Temerlin

This chapter describes the development of self–other awareness in a species-isolated, home-reared chimpanzee, and the changes that occurred in her self-concept and her identifications with others when she was introduced to other chimpanzees, both in the human home where she grew up and subsequently in Africa. The chimpanzee, Lucy, lived in our household from her third day of life until she was 12 years old. During these 12 years I was concurrently involved in the development and management of a primate colony at the University of Oklahoma, where I observed infant chimpanzees with their chimpanzee mothers and with human surrogate mothers. These experiences form the basis for this chapter. Lucy was an integral part of my life for a long time, and I cared for her as I cared for my human son, with whom Lucy was raised as a sibling. I was and remain deeply attached to her and so cannot claim a high level of objectivity in the sense of being emotionally far removed from the subject matter. However, my observations of captive chimps and the reports of other laboratory and field workers have provided a basis of comparison and some measure of balance in these conclusions about Lucy's inner experience and the meaning of her actions.

Born in Florida, Lucy was taken from her chimpanzee mother at the age of two days, presumably before she was capable of learning anything about chimpanzee behavior from her mother. She then lived as a member of the Temerlin family until she was 11 years, 9 months old. (Further details of Lucy's development as interpreted and shared by my husband are available in M. K. Temerlin, 1975). For her first 8½ years, Lucy saw no other chimpanzees. At that time an attempt was made to introduce her to a male chimpanzee, somewhat older than herself, who was gentle with both humans and other chimps. This interaction was so frightening for her that it was discontinued after 24 hours. No further attempts were made to expose Lucy to her own species for another two years; then, when she was 10½, she was introduced to a 3-year-old female, Marianne, who had been born and raised in the Yerkes Laboratory.[1] This attempt was successful, and the two chimps lived together with Lucy's human family for a year while their attachment to each other and the transmission of culture between them was studied. After a year both chimps were taken to Africa as part of a chimpanzee rehabilitation project developed by Brewer and her father. The aim of this project is to teach once captive chimpanzees to live free in areas protected by the governments of Gambia and Senegal. This project is described by Brewer (1978) and in a special report of the International Primate Protection League (Brewer, 1976).

The "rehabilitation" of Lucy is being conducted by Janis Carter, who accompanied Lucy, Marianne, and the Temerlin family to Africa. Carter remains in Africa, observing Lucy and Marianne and facilitating the formation of a stable social group presently consisting of the two female chimps and two male chimps. The current plan is that these chimpanzees will be released in a protected area and given whatever human support is necessary until they are able to survive in the wild on their own. The plan is to follow Lucy's adaptation indefinitely and the transmission of her "human culture" to the other chimps and/or Lucy's offspring. My original purpose in raising Lucy was to study the effects of early rearing apart from her own species on her adult social and sexual behavior. Of particular interest was the effect of human family life and culture on Lucy's higher mental processes. Within the limits that she have no contact with other chimpanzees, every effort was made to meet her emotional and social needs in whatever way possible and allow the fullest development of her biological endowment in a human context.

Species Identity and Individual Identity

In Chapter 1, species identity was defined as the consistent preference by an organism "for a group of other organisms which have similar physical and behavioral characteristics between themselves, but who may or may not be similar in characteristics to the target organism." Species identity is inferred from "the target organism's behavior in social situations" and is measured by "the courting of and/or attempted breeding with, the imitation or copying of, the ability to communicate with, the degree of mutual grooming or reciprocal interactions with, and/or the presence of allelomimetic behavior with the species with which an identification has occurred." A species identity appears in different species with differing capacities for "cognitive processing" and varying degrees of dependence upon environmental factors for the development of social and communicative behavior.

The definition of species identity in Chapter 1 is useful for understanding much of Lucy's social behavior because, if we consider her species identification to be with human beings, it is consistent with her observable preference for human beings as partners in grooming, eating, communication, and sex and as models to imitate. Lucy clearly has preferred humans to all other species for most of her life and I therefore think she has a species identity as a human being; this is what would be expected from her life experience and lack of familiarity with chimpanzees. Humans cared for her, comforted her, protected her, and were her first playmates. They were the first interpersonal mirrors of her self.

Since Lucy could not observe her own species reacting to her, she had to think of herself as human. (It is asking too much of her cognitive abilities, however much we respect them, to conclude that she thought of herself as a chimpanzee surrounded by humans; in other words, there is less cognitive dissonance for her if she simply ignores the physical differences and concludes that she is one of us.) While Lucy could perceive that her body shape and hair distribution differed from ours, it is unlikely she concluded that she was fundamentally different on such a superficial basis since she was treated as if she were one of us, anymore than she would conclude that my husband and I were of different species because he is taller, darker, and has more facial hair than I do. In short, it appears that Lucy thought of herself as human and this identification of herself (her species identity) was formed before her identification of herself as Lucy, a particular human (her individual identity). This identification seems reasonable because Lucy, like human beings, did not have an identification prior to birth. In species with the cognitive capacities of chimps and humans, the first identifications arise early in life out of the primary attachment to the caretaker or mothering individual. During the long period of dependency of either the chimpanzee or the human infant upon the caretaker, the interactions between infant and caretaker and the emotional climate in which they take place have far greater importance in determining the first cognitions (identifications) and experience of the self than any visual perception of physical differences between mothering individual and the infant.

The First Observations

While Lucy's species identity as human appeared to precede her identification as a *particular* human, both identifications were inferred from Lucy's overtly observable behavior. Lucy was first observed on her second day of life. She was clinging tightly to her chimpanzee mother, Joanne, grasping a wad of her mother's hair in each little fist. Lucy would suck at her mother's breast every few minutes. Each nursing bout was initiated by Lucy with short, jerky rooting movements which resulted in her discovery of the nipple and vigorous sucking. Lucy's rooting or searching behavior reminded me of the rooting movements made by my son when he was a nursing infant. However, such behavior could have been accomplished both by my human son and my adopted chimpanzee daughter with no sense of identity, either individual or species.

Joanne, on the other hand, seemed to know what she was doing. Lucy was not her first infant, and her actions seemed to have purpose. She was relaxed, if somewhat excited by the attention she was receiving. She always supported Lucy with at least one arm, adjusting her position from time to time while simultaneously regarding with interest the admiring humans standing by her cage. If Lucy vocalized, Joanne would adjust herself and Lucy until Lucy was quiet again. She would look down at Lucy and groom her whenever one of the humans approached, occasionally looking up at us and making a soft, panting sound.

Joanne was given a carbonated drink which was dosed with a hypnotic. As the drug took effect, Joanne carefully positioned herself and the infant so that, as she went to sleep, Lucy was safely cradled in her arms and protected from the weight of her body. When Joanne was asleep, Lucy was taken from her arms and given to me. I immediately placed her against my chest and covered her with a receiving blanket. This was Lucy's introduction to the human world where she would spend her next 12 years.

From an external point of view, neither an individual identity nor a species identity had emerged at this time. If any awareness existed, it seemed to be only of the difference between comfort and discomfort. Lucy's behavior consisted simply of grasping and nursing, digesting and eliminating, which could have been accomplished on an instinctive basis. She showed no distress as long as she was held securely, fed when hungry, and kept dry and at a fairly even temperature. She gave no sign that she perceived any difference between her biological and adoptive mother. Human caretaking seemed as satisfying as that given by her chimpanzee mother. From the fourth day of her life, Lucy impressed me as an active, responsive organism *with a vague awareness of some sort*. Perhaps this was the origin of her first identification, for from then on she responded, usually with smiling and soft panting, to whatever social stimulation I provided. In those first few days she did not seem to notice any differences among the humans who held her, but gradually she developed a pattern of preferences for family members. This process seemed to parallel the development of primary attachments in human infants and was the first overt behavior from which her species identification could be inferred.

These first responses were not entirely automatic. If I held her away from my body, for example, even though she was held securely, she would squirm and flail about, making a whiny, high-pitched cry of unmistakable distress. These same movements and sounds would be made when Lucy was placed in her crib on her back or on her stomach without being tucked in securely. Ventral contact and the opportunity to cling were particularly important in maintaining a feeling of security. In nature, the infant chimp clings to its mother's ventral surface almost continuously for the first six months of life (van Lawick-Goodall, 1968). The ventral contact and the manner in which it is gradually attenutated by the infant's increasing independent motility have important effects on the development of the infant's experience as an individual separate from the mother. Body contanct also can be an important organizer of experience for the mother—at least it was for me.

Before Lucy was born, a 3-year-old male chimp, Charlie Brown, had lived in our home for a year before his accidental death. Charlie was an affectionate chimp and often would leap into my arms and hug me for comfort, reassurance, or sheer joy. It felt good to both of us. After he died, I experienced a variety of the "phantom limb" phenomenon, a kind of aching tingling on the skin of my arms and ventral surface. I could sometimes still feel him hugging me. To comfort myself, I would sometimes hold a month-old infant chimp but she, being much smaller, did not fill Charlie's space in my arms and I continued to feel the ghostly outlines of his body against mine for some weeks. About five years later a co-worker (Savage, 1975) and I watched a chimpanzee mother have what appeared to be a similar experience. The chimpanzee mother had been drugged with a hypnotic and, while she was unconscious, her infant had been taken from her. As she gradually regained consciousness, she groggily looked at her chest and pelvic areas and with her hand touched the areas of her body where her infant had previously clung. She continued to search for her lost infant in this way for some hours.

The point is that between infants and their mothers—whether human, chimpanzee, or cross-species mothering is involved—the mutual holding and handling, and pulling away and returning, aid the infant's development of a sense of a self separate from the mother's body. Furthermore, this physical experience, with a necessary degree of cognitive processing, contributes to the formation of a concept of self. Such an experience is also used therapeutically with autistic children to help them form a more clear and stable concept of self by, for example, tracing the outline of the child's body on a large piece of paper. The child has the physical experience of his body outline being traced and may then observe the subsequent external representation of "me."[2]

The Development of Self

Lucy's use of vocalizations progressed from an early undifferentiated state to one of gradually increasing differentiation. Her earliest vocalizations, for example, were not discrete sounds such as pant hoots or barks, but a sequence of sounds which she seemed unable to stop until she had exhausted her entire repertoire. Later they became clearer and more discrete. So far as one could tell, she emitted most sounds made by the wild chimps described by van Lawick-Goodall (1968). But her first vocalizations sounded like an infantile version of all the sounds that adult chimps make, emitted one after the other and punctuated by startle barks or single explosive sounds. This may have been what Hayes (1951) referred to as Vicki's "babbling." In Lucy's case, these "chain reactions" were at times a response to humans talking to her or making their own version of chimp greeting sounds. At other times they were emitted in the absence of apparent social stimulus. Her vocalizations gradually became modified into identifiable units which she used consistently and appropriately (for a chimpanzee) in social situations. Although they seemed to be the same sounds made by other chimps, she made them far less frequently than chimps living in a captive colony. One speculation is that without the vocal and social stimulation of other chimps, she never reached the arousal level necessary for these sounds to occur with their usual frequency. Living with humans was a fairly calm affair, and many visitors familiar with both wild and captive chimps often commented on Lucy's calm personality. (Recent observations of Lucy and Marianne in Africa with the two male chimps support this: it was some weeks before Lucy, unlike Marianne, hooted in unison with the others.) Her calmness could have reflected a more than usual sense of her self.

Lucy's personality development might have been affected by several variables. The first was her biologically determined response patterns inherited from her chimpanzee parents. Her father was a trained circus performer and her mother was described as a very manageable, sweet chimp. Differences in responsiveness are apparent in very young chimpanzee infants. Charlie Brown at two and three years of age was a rowdy and rambunctious youngster, for example. Lucy at the same age was much less so, while her half-sister (also human-reared from birth in a human home until she died at the age of five) was more like Charlie Brown. Differences between Lucy and her half-sister were observable from a very early time.

The second factor in the development of a calm personality was the attention given to Lucy's contact and security needs. It was impossible to carry her as continuously as a chimpanzee mother would have done but, particularly during her first year, I always gave her a security blanket or a pillow to clutch (which she did with both hands and feet) when I could not hold her. When she would fuss while being diapered, for example, I would place a pillow on her stomach and let her clutch it until I could pick her up again. When she was alone in her crib, I would place her on her abdomen and tuck her in securely with a blanket or sheet until she could roll over on her own. When she began to crawl about, she always had her security blanket with her. The fact that Lucy could not be held and allowed to cling with the same continuity which rearing by her biological species would have provided meant that there were more and earlier physical separations from her primary caretaker. These separations took place in a complex human environment. Thus she had more opportunities for making self–not self distinctions in relation to that environment. Later, as she began to crawl, having a security blanket or "transitional object" (Giovacchini, 1979; Winnicott, 1953) allowed her not only to set her own pace in moving away from and returning to her mother, but also to do so with a sense of self-sufficiency.

Her early social contacts took place within a stable context. I took her to work each day until she was 3 years old, and at home her brother or father held her and played with her whenever I was not available. At the office there were students, secretaries, and animal caretakers who enjoyed helping with her. She had two identical cribs, one at home and one in the office. She was handled by a large number of people who were interested in her, enjoyed playing with her, and helped care for her. Thus she had a variety of social contacts and stimulation within the stable structure of her family and daily routine.

A third set of factors contributing to her tranquility was the attempts we made to facilitate the development of a sense of self. Often this was done with play. We showed a delight in, and rewarded her by our interest, any behaviors which seemed to define a developing sense of self. For example, beginning at about 15 to 18 months we played a game with her which was called "Lucy's nose." It was played by putting one's finger on Lucy's nose and then on one's own nose while saying, "Where is Lucy's nose? Where is Jane's nose?" In turn, the two eyes, two ears, and mouth of each participant were touched and identified. Then it was Lucy's "turn" and she would be asked to put her finger on each part of her own face and that of her human partner as she was asked, "Where is Lucy's nose?" or, "Put your finger on Jane's nose," and so on. Everyone who had regular contact with her was encouraged to play the same game, and eventually Lucy would initiate the game herself. This game became a regular part of her interactions with people; she initiated it more often than we did, and still does so. Her possessions also were carefully respected, and she eventually became possessive about many of the toys and articles of clothing we gave her. Conversely, she was expected to respect our possessions and she did.

Lucy was provided with verbal stimulation much as any middle-class human infant during feeding, diapering, or other caretaking activities. In her first few weeks, she began to identify family members by the sound of their voices. By three weeks of age she vocalized in response to family members speaking to her.

She would "greet" family members with soft panting while nonfamily members talking to her or speaking in her hearing might result in more vigorous and excited vocalizations on her part, often ending in barks or single explosive sounds. She also would smile when her bottle came into view and follow my movements visually from her crib. By three months she could identify each member of the family by both sight and voice, and she responded to them differently than to nonfamily members. She was comfortable when held by family members, but would "fuss" and squirm when held by others. It is interesting that she seldom responded in any way to my attempts to imitate chimpanzee greeting sounds, food barks, or hoots. This may have been because I was not good at reproducing them, or it could have reflected the beginnings of her cross-species identity. However, she was always responsive to my normal speaking voice.

My normal speaking voice would always calm rather than excite her. Similarly, my voice would soothe Lucy even when I was not holding her, whereas she would respond to the vocal stimulation of others (when I was not holding her) with an excitement which often appeared uncomfortable. Also when being held by "mother," she would be less stimulated by a sight or sound which would have excited or frightened her while being held by another person. This observation is consistent with Mason's (1965) finding that the threshold for painful stimuli is raised when chimps are held by familiar human handlers. Lucy's responsiveness to my voice indicates the emergence of a cross-species identity as "human" through her identification with me, in direct analogy with the emergence of self-awareness in human infants after they have first experienced a symbiotic union with the mother. Mahler (1975) describes the "symbiotic union" with the mother as a necessary experience if the human child is subsequently to differentiate a self at an emotional and perceptual level. Lucy seemed to experience herself and first Joanne and then me as a single psychological unit. Awareness gradually appeared and expanded to include self–other distinctions. Everyone else seemed to be in in the "other" category. Very quickly, however, her brother and father were included in the "me" category in the sense that she then responded to the three of us in the same way, and to all others differently. By 3½ months of age, Lucy had an identity as a member of the family; she readily identified and responded consistently to each family member. In short, by this time she had a species identity, even though it was an adopted species. There is some evidence that a family identity might have existed even earlier. At the age of 2 months she had her first temper tantrum when she was picked up and held by a human male whom she did not know—a behavior she had never exhibited in response to a member of her own family. In a human child such experience might be called "stranger anxiety," and its appearance in Lucy implies that she had made a previous distinction between us as family members (her species) and "them."

When she was 5 months old, she showed the first signs of aggression and it seemed to be toward conspecifics invading her territory. Two young brothers aged 5 and 3 came into my office where Lucy was playing in her crib. Her hair became pilo erect, she barked at the two boys, slapped the side of her crib, and hit out at them, even though they stayed across the room quite frightened. Her display seemed directed primarily at the younger child and did not subside until some time after the boys had left the room. Two months later, under the same circumstances, the two boys entered the room and her response was the same.

Individual Identity

In a series of experiments, Gallup (1970, 1977) has demonstrated that a chimpanzee looking into a mirror sees an image which it identifies as itself, not another chimpanzee. His experimental results were confirmed by observations of Lucy. We never saw her exhibit fear of her image or respond as if her image were that of another chimp. By the time Lucy was two years old, a mirror had become a highly prized and much used tool and toy. Although she may have used a mirror earlier (we cannot be sure when mirror gazing first appeared because there were mirrors all about the house), by two years of age she was using a mirror to adorn herself, explore her body, watch herself making play faces, and "practice" other expressions, as well as to aid in self-grooming. The time spent with a mirror gradually increased, and by adulthood she was using a mirror more than any other toy or tool. She used it to explore all body orifices and, in one of her more dramatic self-explorations, she placed it on the floor, positioned herself carefully, and urinated on the mirror while watching the entire process intently.

Lucy's use of mirrors to investigate and groom her body often suggested that she was aware of and curious about changes in her body. An example was her first menstrual period. On that day, as soon as she was given access to the bathroom, she jumped up on the counter, turned around, and crouched backwards, exposing her bottom to the mirror. Carefully watching her reflection in the mirror, she investigated her vulva and outer vagina with her fingers. She then sniffed and tasted the blood on her fingers and, still watching her image in the mirror, twice made the American Sign Language (ASL) sign for "hurt." (This signing to herself in the mirror, which she often did after her ASL sign training had begun, seemed to reflect a primitive kind of rhetorical activity, i.e., "talking" to herself.)

Lucy was so fascinated with the process of observing herself in the mirror that she often continued doing so even when watching her reflection made the activity more difficult. While striking a cardboard match, for example, she typically would watch her image in the mirror instead of her fingers as she tried to strike the head of the match against the sandpaper. Incidentally, she did learn to strike matches in this fashion, and she would hold them as long as possible, dropping them only when her fingers were almost singed. Then, after 10 years of human culture and with no urging from us, she spontaneously started a fire on the living-room floor.

Lucy's mirror behavior was fascinating as an index of her developing self-concept. While observing Lucy at home I also had the opportunity to observe the behavior of other infant chimpanzees removed from their mothers and given a human surrogate mother. The initial reaction of these chimpanzee infants might also, like mirror behavior, tell us something of their self-concept. For example, the response of a chimpanzee infant upon receiving a human surrogate mother for the first time might indicate whether the infant saw the human as "same" or "other."

The initial reaction of nine chimpanzee infants was observed when they were removed from their chimpanzee mothers and presented to a human surrogate mother. In some cases the infant immediately accepted the surrogate mother with no sign of distress once contact with her body was made, just as Lucy had done with the transfer from Joanne to her human mother. In other infants there was quite a different reaction. They screamed, struggled, bit, and tried with all their strength to escape from the human surrogate mother. For some of those who resisted human contact, the initial period of distress lasted only a few hours, and they then accepted comfort and nursing care from the human surrogate mother; for others, the distress of the transfer lasted

Table 1. Identity Inferred from Initial Response to Human Surrogate Mother[1]

Chimpanzee Infant[2]	*Sex*[3]	*Approximate Age at Separation in Days*	*Initial Response to Human Surrogate Mother*[4]	*Remarks*
A (Lucy)	F	2	Same as to chimp mother; accepted without distress	—
B	F	3	accepted without distress	—
C	M	3	accepted without distress	—
D	F	3–4	accepted without distress	—
E	F	5–6	accepted without distress	—
F	F	24	accepted without distress	—
G	F	21	accepted without distress	—
H	M	45	accepted without distress.	—
I	M	76	Hostile biting struggle to get away; threat and distress vocalizations; does not accept comfort.	Six hours of hostile biting struggle, threat, and distress vocalizations before infant accepted comfort from human surrogate mother.
J	F	125	Same as above	Two weeks before infant would accept comfort from surrogate mother.

[1]Some of these observations have been reported (Lemmon, 1971), though with a quite different interpretation.
[2]Eight infants were observed at the primate laboratory of the University of Oklahoma; two at other facilities.
[3]No differences in response by sex were noted.
[4]Human surrogate mother is defined as the first human being to lift and carry the infant after the chimp mother was rendered unconscious; in five cases, above, it was the author; in five others, workers whom the author observed. In eight cases the first human surrogate mother subsequently passed the infant to another foster mother, but the infant's reactions were the same as to the first human who had touched it.

much longer. These observations are presented in Table 1. Notice that for these 10 infants (including Lucy), the sensitive period occurred between 45 and 76 days. Before this sensitive period the infant responded to the human surrogate mother about the same as it did to its real mother; it allowed itself to be nursed and cared for as though no difference between mothers was perceived and all that mattered was that its biological needs were met. However, after the sensitive period for an uneventful separation had passed, the infant showed every sign of perceiving the foster mother as "different"—even threatening—implying the prior emergence of species identity in its experience.

Sexual Behavior

Since breeding across species lines is so rare, Lucy's sexual behavior provides an index of her species identification. Her first estrous was at the age of eight. For the next two years her cycling was regular and her sexual behavior during full engorgement was consistent: she approached human males who were unrelated to her (i.e., she avoided contact with her human "father" and "brother" during this period) and invited sexual contact by covering their mouth with hers, positioning the males so that she could mount them in a ventral–ventral position, and thrusting her pelvis against the waist, thigh, or chest of the human male (Temerlin, 1975a; Lemmon, Temerlin, and Savage, 1975). Mounting and thrusting in this fashion might last from 5 to 15 seconds, but be repeated again in 5 to 7 minutes, often to the consternation of our friends and guests. At times of maximum engorgement and when nonfamily were not available, Lucy would press her ventrum and pelvis against pillows, windows, walls, chairs, dolls, pictures, or almost anything except the body of a family member. Her positioning with both humans and inanimate objects was never observed in the manner known to be species-specific for the female *Pan troglodytes,* i.e., dorsal presentation allowing ventral intromission. She approached nonfamily females in this way less often. Lucy's failure to make a sexual approach in the dorsal, crouching position for *Pan troglodytes* possibly can be explained by the fact that, unlike her conspecifics in nature, she had never had to assume this position as either a gesture of submission or to enable her bottom to be explored and groomed.

By this time Lucy had experienced eight years of human culture and had had no contact with other chimpanzees. She was well adapted to life in a human home; she could use most household tools correctly and she seemed quite content and free of neurotic symptoms except during periods of full estrous when she seemed emotionally intense and driven to sexual discharge. Otherwise she had no stereotypies or other indices of neurotic or psychotic behavior, such as those sometimes seen in capitve animals.

Her human family had become very fond of Lucy and she indicated that she felt the same way about each of us. Nonetheless, she was totally dependent upon us and, while well adapted to human life in our household, her future development as a chimpanzee was uncertain. Chimpanzee social bonds are a prerequisite to the autonomous functioning of an adult chimp. In other words, Lucy needed to shift her species identity from humans to chimps if she were to realize her biological potential. After more than a year of often painful discussions and consideration of alternatives, it was decided to help her develop an identity as a chimpanzee. We began to plan experiences that we hoped would aid the shift in her identifications and concept of herself.

Becoming a Chimpanzee

When she was 8½, Lucy was introduced to a male chimpanzee who was about 10 years old and could easily be handled outside a cage by humans. He had to our knowledge never bitten anyone. The meeting was scheduled for neutral ground—an isolated grove of trees. The two were brought to the scene at the same time: Lucy was led by her human father and mother, and the male by a handler known to Lucy. She first saw the male from about 50 yards away, and she immediately became frightened. Her hair erected and she whimpered, bared her teeth, developed immediate diarrhea, and tightly clutched her father, who urged her to approach the male chimp. There was no evidence that she recognized a fellow member of her own species, or even a friend or a possibility of one; rather, it was as though she saw, projected ahead, the personification of her worst fears. We had never seen her so terrified or miserable. She was forced to approach within a few yards of the male, who then displayed. In the next moment they were holding one another, screaming and rolling about on the ground so rapidly that the details of their encounter could not be observed. They were pulled apart by a lead attached to a collar that each of them wore and, though their encounter had looked like a serious fight, no skin was broken on either one of them. The male was then caged for about eight hours outside Lucy's room in her home. She could see him if she chose, but no physical contact was possible. During this time Lucy whimpered, screamed, and stayed as far away from the male's area as possible. She was so utterly and constantly miserable that the attempt was abandoned.

In analyzing this failure later, it was decided that many mistakes had been made in our management of the introduction. It also seemed clear that Lucy had no species identity as a chimpanzee; she had been terrified by what was to her a strange animal, and by his greater size and strength. Since Lucy had always been sensitive to the feelings of humans it is likely that she also was reacting to the anxieties of the human beings present at the introduction.

Two years passed before an attempt to change Lucy's self-concept was made again. Dr. van Lawick-Goodall suggested that Lucy might be less frightened of a baby chimp, and if she saw me playing with a chimpanzee doll before the "baby" came, she might attempt to mother it. A young chimpanzee suitable to become Lucy's "baby" was therefore sought. Rather than exposing her to another chimp larger than herself who would want equal status at the least, we planned to show her a small, dependent, and playful version of herself. We hoped to tip the balance in favor of attraction over fear.

The next introduction was made possible by Dr. Bourne, director of the Yerkes Primate Center, and Dr. Savage-Rumbaugh. Bourne approved the use of a Yerkes chimpanzee for this phase of the research, and Savage-Rumbaugh, who had known Lucy well throughout her childhood and early puberty, picked the chimp most likely to get along with Lucy from the young ones available at Yerkes.

Marianne was 3½ years old at the time she arrived at our home for the introduction. Since Lucy was about 10, she was of course much larger than Marianne, "her baby." For three days Marianne was kept out of Lucy's sight so that she might get over the stresses of travel and become accustomed to her new human home before introduction to Lucy. Once Marianne seemed comfortable and the cameras were in place, they were introduced. Lucy did not react with the terror of several years before, when she had seen the male chimp. At first she "cringed" and whimpered, her gums maximally visible, and clutched me while alternately approaching and moving away from the much smaller, interested, but quite unafraid Marianne. Three or four times Lucy approached Marianne until only inches away, and then retreated in fear. On one of these occasions, with their faces inches apart and just as Lucy seemed about to conquer her fear, Marianne hit Lucy in the face. Lucy ran away, but within five minutes her curiosity overcame her fear and she returned to the encounter. This general sequence of Lucy's responses—fear, curiosity, approach, and fear-withdrawal—continued for about two hours, with a gradual diminution of Lucy's fear and an increase in her curiosity and attraction. Suddenly they were tickling one another and laughing, and their play continued until they both were fatigued. Lucy did not seem to fear Marianne again. They were watched closely for the next three days on the chance that aggression might suddenly erupt, resulting in injury to the much smaller chimp. However, this did not happen. While there were disagreements, quarrels were short-lived, and both Lucy and Marianne remained in our home for a year, during which time the following observations were made.

A Year of Shared Identifications

The interaction of Lucy and Marianne—a home-reared and a laboratory-reared chimpanzee—in a human home offered a possibility for observing cultural transmission, if any, of human and chimpanzee characteristics in an experimental situation. Van Lawick-Goodall (1973) previously discussed the cultural behaviors of the Gombe chimpanzees. Marianne's entry into a human home provided an opportunity to follow her adaptation to a new environment as well as the social influences on both chimps. Fouts (1973) had previously taught Lucy ASL, and she had acquired approximately 100 signs. The possibility that Lucy and Marianne would learn from one another was exciting. Lucy's ASL signs were a readily observable index of cultural transmission.

Marianne had no sign training, and we gave her none. Lucy's training had been discontinued before Marianne's arrival. Signs were still used with Lucy, and she used them to communicate with us. Within six months of joining the family, Marianne had acquired slightly more than three signs. They were "out," "hat," "in," and an almost complete rendition of the "baby" sign. She had received no tuition from humans, so her learning was the result of her own observations and imitative efforts.

None of Marianne's ASL signs were directed toward Lucy, as far as we observed, but only toward humans. Perhaps Marianne had observed Lucy signing to humans and she wished to receive the attention from them that Lucy did. Over time, our affection and attention became important to Marianne also, so that contact with us was rewarding and sought for more than food.

In summary, the following cultural transmissions occurred:

1. As stated, Marianne correctly learned and used four signs (out, hat, in, and baby) which she directed toward humans.
2. Lucy began to eat monkey chow, a food she had always refused, after observing Marianne do so. It is interesting that she tried to conceal the fact that she was eating monkey chow—a shyness which might have reflected a self-reflexive concern about how her "parents" would view her new behavior, to make an analogy with the behavior of human children.
3. Lucy displayed the same sexual behavior toward Marianne that she had toward nonfamily humans; she would hold her, usually against Marianne's wishes, and put her open mouth on Marianne's head and pant while thrusting her pelvis against Marianne's body. Marianne gradually became so accustomed to this behavior that she would continue eating throughout Lucy's advances, never missing a bite.
4. Marianne learned to use a few household tools, i.e., a cup, spoon, and bowl, but she was never able to acquire enough self-discipline and restraint to be more than a barbarian in a human home. She could not be restrained to a chair, taught to eat at a table and to avoid climbing on the walls, or kept from breaking things—reminiscent of feral human children whose socialization to humans was very slow if they were discovered past a sensitive period of development. All of these behaviors had been easily learned by Lucy at a younger age.
5. Lucy became protective or maternal toward Marianne. She would lead her about the house, and try to restrain or distract her. When we disciplined Marianne by loudly shouting "no," for example, Lucy would quickly lead her away, or at other times take our side and give Marianne a disciplinary bite or bark. As Marianne's incorrigibility became more obvious over time, Lucy less often spontaneously attempted to discipline Marianne and distracted her or modeled the appropriate behavior only after our insistence.

6. Marianne began to use a blanket and a bucket as toys, tools, and security objects.
7. There was a transmission of fears. Marianne was afraid of certain blankets and rugs which Lucy had used without fear in play, nest building, and sleep. After observing Marianne's fear reaction, Lucy became afraid and would not use that category of rug again. She would not carry or touch them and, if they were used as a door mat, she would leap over them to avoid stepping on them. It seemed that Marianne had transmitted her "superstitions" to Lucy.
8. Through her identifications with Marianne, Lucy showed a renewed interest in foods and toys that she had once enjoyed but had long since abandoned. She allowed Marianne to play with her possessions without protest.
9. Lucy became more playful. Before Marianne came, Lucy related to adult humans in an "adult" manner, primarily as sex objects or as persons to groom and by whom to be groomed. Marianne's approach to humans was playful, and as a consequence Lucy "regressed" and became playful again, much as she had been five or six years earlier.
10. Lucy maintained all her emotional bonds with her human family and friends and even developed new friendships with several humans during this period. In short, the emotional bond that developed between Lucy and Marianne did not exclude affective bonding with humans; nor did Lucy's previous attachment to humans interfere with her attachment to Marianne.

Africa and Rehabilitation

The second major step in Lucy's reeducation as a chimp was taken when she and Marianne were taken to Africa in September 1977. Lucy, Marianne, and two males, Lotus and Buddha, are now in a large enclosure at the Abuko Nature Reserve in Gambia. This is an intermediate step before their eventual release. Based on observations by Carter, Lucy was adapting well to the new climate and food after seven months, but she showed no sexual swelling until late March 1978. This was the first interruption in her estrous cycle since its occurrence and would seem to be a response to the stress of the transfer. Her introduction to the two males was obviously stressful, but she did not experience the terror of her first introduction to an adult male chimpanzee. She mostly ignored and avoided them while Marianne, in the same fearless if somewhat naive way with which she had first approached Lucy, asserted herself in interaction with them. No copulation has been observed between Lucy and either of the males but, during her recent estrous cycle, she paraded near Lotus and at times stopped to allow him to investigate her swelling. She never did this with human males. She has not approached him in the same way she did with human males as described earlier, but recently, with Carter observing, Lucy put her arm around Lotus's shoulder, covered his mouth with hers, and gave him a pant greeting. As she withdrew her arm, she patted him on the back. These are natural chimpanzee expressions of affection and reassurance. They can be interpreted as signs of Lucy's developing attachment to Lotus and an indication of her gradually shifting identifications of her self and her species.

Acknowledgments

I am indebted to many people who have been a part of Lucy's life over the years, and most recently to the dedication and skill of Stella Brewer and Eddie Brewer, who created the Africa rehabilitation projects. Without the active participation of Janis Carter, Lucy's move to Africa could not have been undertaken. She remains in Africa with Lucy, Marianne, and the other chimps, recording and facilitating their adaptation to the wild. Hers is the most difficult and demanding task of all. Her patience, scientific skills, and knowledge of chimpanzees have allowed her to develop a relationship with Lucy that is providing the bridge from a human to a chimpanzee culture. My husband and son have shared most intimately in the joys and frustrations of living with a chimpanzee. To Lucy is owed the greatest debt of all, for being her unique self. It is a debt that I can never fully repay.

Notes

1. Marianne was born December 24, 1972 at the Yerkes field station, removed from her mother soon after birth, and reared in the nursery. She was housed for her first two months in a clear-walled incubator in a room which had another infant chimp (Stephanie) and an infant orangutan (Teriang) similarly housed. For the next seven months, Marianne, Stephanie, and Teriang were housed together in a large laboratory cage. From nine months of age until her removal and shipment to us, Marianne was caged with various chimpanzees alike

in age. Standard care in the nursery at Yerkes involved caretakers spending considerable time in contact with each infant every day. Thus Marianne had regular social contacts with both humans and chimpanzees prior to her exposure to Lucy and us.

2. Anthony Kowalski, M.D., personal communication.

References

Brewer, S. 1976. *Chimpanzee rehabilitation.* Special Report of the International Primate Protection League, December.

Brewer, S. 1978. *The Chimps of Mt. Asserik.* New York: Knopf.

Gallup, G. G. 1970. Chimpanzees: Self-recognition. *Science 167,* 86–87.

Gallup, G. G. 1977. Self-recognition in primates: A comparative approach to bidirectional properties of consciousness. *Am. Psychol. 32*(5), 329–339.

Giovacchini, P. 1979. *Treatment of primative mental states.* New York: Aronson, pp 29–31.

Hayes, C. 1951. *The ape in our house.* New York: Harper.

van Lawick-Goodall, J. 1968. The behavior of free living chimpanzees in the Gombe Stream Reserve. *Anim. Behav. Monogr. 1*(3), 161–311.

van Lawick-Goodall, J. 1973. Cultural elements in a chimpanzee community. In *Symposium of the Fourth International Congress of Primatology, Vol. 1: Precultural primate behavior.* Basel: Karger.

Lemmon, W. B. 1971. Deprivation and enrichment in the development of primates. *Proceedings of the 3rd International Congress of Primatology.* Vol. 3. Basel: Karger.

Lemmon, W. B., J. Temerlin, and E. S. Savage. 1975. The development of human-oriented courtship behavior in a human reared chimpanzee (*Pan troglodytes*). In S. Kondo, M. Kawai, and A. Ehara (Eds.), *Contemporary primatology: Proceedings of the 5th International Congress of Primatology, Nagoya 1975.* Basel: Karger.

Mahler, M., F. Pine, and A. Bergman (Eds.) 1975. *The psychological birth of the human infant.* London: Hutchinson.

Mason, W. A. 1965. Social behavior in young chimpanzees. In A. H. Schrier, H. Harlow, and F. Stounitz (Eds.), *Behavior of nonhuman primates.* Vol. 2. New York: Academic Pr.

Mellgren, R., R. Fouts, and W. Lemmon. 1973. *American Sign Language in the chimpanzee: Semantic and conceptual functions of signs.* Paper presented at the Midwestern Psychological Association meeting, Chicago.

Temerlin, M. 1975. My daughter Lucy: *Psychol. Today* 9–6, 59–103. (a)

Temerlin, M. K. 1975. *Lucy: Growing up human—A chimpanzee daughter in a psychotherapist's family.* Palo Alto, Cal., Science and Behavior Books. (b)

Winnicott, D. W. 1953. Transitional objects and transitional phenomena. In *Collected papers.* New York: Basic Books, 1958, pp 229–242.

Chapter 13

Behavior of Great Apes Reared in Zoo Nurseries

Valerie Smith

The great apes hold a unique position in the animal kingdom. They are closely related to humans phylogenetically, and they also attract tremendous human sympathy because of their similar physical and mental abilities. In the past, researchers working in both controlled environments and in the field have often lost their objectivity in analyzing the nature of the great apes. I doubt that anyone will challenge the observation that man still has much to learn about the behavior of the great apes, especially the gorilla and the orang-utan.

Although gorillas (*Gorilla gorilla*) and orang-utans (*Ponga pygmaeus*) have been exhibited in zoos since 1860 and 1640, respectively, their breeding records have been very poor. The first captive-born gorilla was born in 1956 and hand-reared in the zoo's nursery (Haberle, 1963; Thomas, 1958). Although the third gorilla born in captivity was reared by its mother (Lang, 1961), natural mothering has not been the norm. Infant mortality is high (25 percent) in gorillas and only about 10 percent of the 100 or so captive-born infants have been cared for by their mothers (Nadler and Jones, 1974; Martin, 1976). Orang-utans suffer similarly from limited breeding and inadequate parental care. By 1971 there were 152 captive-born orang-utans listed in the "International Zoo Yearbook" (Perry and Horseman, 1972; Perry, 1976). About 25 percent of these infants were cared for by their mothers.

The less-than-desirable breeding record and infant-care behaviors shown by captive great apes no doubt reflect the fact that most specimens were taken from the wild during their early years and reared by zoo attendants. They and now many of their offspring have never been exposed to their natural parents or environment. Thus the zoo condition appears to be a situation in which to conduct an investigation into the development of species-specific behaviors and species-identification processes in any great ape that has been reared in captivity.

The St. Louis Zoo Nursery

Little has been published on the behavior of hand-reared zoo apes. I have been involved with the care of four orang-utans and one lowland gorilla, which were born in captivity and nursery-reared at the St. Louis Zoo.

This zoo nursery is a combination of adapted human nursery equipment and modified zoo-cage fixtures. It offers protection (glass windows) from direct public interaction, yet allows visual contact between the infant and the viewer. Infants are attended 24 hours daily if necessary. Several attendants share the responsibility for the care of the apes, thus preventing the infants becoming overly attached to one individual.

The process of hand-rearing these apes includes several definite stages. The initial stage is incubator isolation with constant attendant care, which lasts from one to 4 weeks. Once past this critical first stage, the infant spends much of its time, at least 6 hours daily, being carried and played with by the attendant. This type of care generally continues until the infant is at least 6 months old, by which time it will have been introduced to playmates of various species, including gibbons, chimpanzees, and gorillas. Only our young gorilla had the opportunity to grow up with one of its own species, a wild-caught female, whereas all the orang-utans had the companionship of species other than their own until they were at least one year old. This nursery facility is some distance from the ape quarters. This situation prevents infant visitation and observation of their adult counterparts until the youngsters are able to withstand the stress of public crowds, cold temperatures, and the chance exposure to infectious disease. Deprived of contact with their natural parents, these young apes develop strong attachments to their human foster parents. Play with humans is also preferred to play with fellow apes in the nursery.

Orang-utan Behavior

The St. Louis Zoo presently has seven orang-utans in its collection. Most have been reared in captivity. Janie, a 20-year-old Sumatran female, has been at the zoo since she was approximately one year old. She has conceived several times and three of her offspring survive: Tom, who is 9 years old; his brother Allen, 7 years old; and their half-sister Merah, now 3 years old. All three were nursery-reared because of Janie's definitely malicious treatment of the newborn infants. Many attempts have been made to induce Janie to motherhood, but she is particularly tempermental.

Kalle is a wild-born Sumatran male approximately 16 years old. He was acquired at the age of 11. Kalle has sired several offspring, and two of his infants (Merah and an infant male) survive. Carmen, an 11-year-old Bornean female, is Tom and Allen's half-sister. She was also raised in the zoo nursery. Carmen had an infant son (born February 28, 1978) sired by Kalle, which she has been caring for expertly, much to the surprise of the zoo staff. (Carmen had a prior infant which died at birth.) Merah is currently housed with Carmen to observe the mothering process.

Merah is now 3 years old. She was, at 17 months, the youngest ape ever introduced to an adult at the Ape House. Merah had observed the other orang-utans through the glass on several occasions and showed either disinterest (focusing her attention on the fencing and glass beside her) or insecurity (tightly gripping the attendant and pushing away from the glass). She displayed no sign of interest in her ape family when caged beside them and screamed constantly while clutching her security rag. This behavior continued for approximately one week, during which the other orang-utans spent much of their time sitting close to the screen partition, staring at her. Within two weeks she was observed sitting by the screen, watching the other orang-utans.

During her third week at the Ape House, Merah was introduced to Allen, her 5-year-old half-brother. Allen had been involved in a then new type of socialization process now being promoted in many zoos (Maple, Zucker, Hope, and Wilson, 1977). Cooperative efforts among zoos have resulted in temporary live-in arrangements through which young animals of the same species are placed in one zoo's facility. Such an arrangement made it possible for Allen (at age one year) to live with another young orang-utan at the Oklahoma City Zoo for approximately one year.

When introduced, Allen and Merah responded with immediate pileal erection, and Merah so frantically avoided Allen that she abraded all the skin on her hands within minutes. It is interesting to note that Allen, although rough with his pulling and poking, never attempted to bite Merah. He lost interest in her within five minutes of the initial introduction and became absorbed in inspecting her toys. Merah's panicky actions and badly torn hands forced us to discontinue our five-minute, twice-daily contacts between the two animals.

The following week Merah was introduced to her 9-year-old half-sister Carmen, who recently had had a baby which died at birth. There were several reasons for these introductions. First, Carmen might have lost her baby because of her improper maternal efforts, and contact with an infant might help to make her more careful. (That idea may have been an influential factor in Carmen's expert maternal care of her second offspring.) Second, Carmen was basically a loner, possibly a result of her early experiences growing up with an aggressive young chimpanzee. Merah benefitted from the quieter, more mature personality of Carmen. During the period of introduction to Carmen, Merah began to repel contact with all humans, especially with those who put her with Carmen. This repulsion may have been a response to an association between the keepers and Carmen in that Merah could have recognized that keepers controlled her periods of visitations with Carmen. These introductions were kept short (approximately 15 minutes daily) and were initiated by the keeper pulling open a shift door between the two orang-utans' adjoining cages. Merah did not enjoy her first encounter with Carmen and tried to stay as far as possible from her. Although by the second week with Carmen she had stopped her constant demand for human attention, Merah was again willing to touch or play with humans. Merah also developed a new response to her orang-utan companion. She now teased Carmen, often hitting, poking, or kicking her, only to be completely ignored. Before the end of the second week, both animals participated in and appeared to initiate play wrestling.

About three months later Kalle, the adult male with a calm, quiet nature, was introduced to Merah and Carmen. Merah demonstrated none of her previous frantic, panicky actions. Kalle ignored Merah until she initiated contact on the second day of the introduction.

The one mature, hand-reared female orang-utan, Carmen, solicits breeding from the male, who breeds acyclically with her. The matings between the two have been observed by Merah, who takes a definite interest in the activity. These observations have been considered critical to the ape's proper social development (Maple, 1977). Carmen and Merah have also observed the male's "rape breedings" with the other female, a wild-caught animal.

Gorilla Behavior

There are four lowland gorillas (*Gorilla gorilla gorilla*) in the St. Louis Zoo collection. One pair, Rudy and Trudy, are each approximately 21 years old. They were acquired in Africa at the age of about one year and have always been kept together. Their only offspring, Mzuri, is a male almost 12 years old. Muke, a 12-year-old female, was taken in the wild as a 6-month-old infant. She was reared in the zoo nursery with Mzuri and is slightly older. They have been constant companions since Mzuri was a few months old.

Our gorilla group illustrates the typical problems involved in great-ape propagation in zoos. Two 21-year-old animals produced a son 10 years ago, who was reared in the zoo nursery. They have not indicated a sexual interest in each other since that birth. Recent introduction of the parents to their son and his female wild-caught companion has failed to stimulate any sexual activity. Incompatibility of the paired individuals, the conditions under which they have been kept, or some elusive combinations of these factors have kept this group from producing off-spring. Mzuri shows species-specific play behavior with both his parents and Muke. He is a playful, active animal and especially enjoys interaction with keepers. Although he has been with both Muke and Trudy when they were in oestrus, he has never shown more than a playful interest in the other gorillas. He masturbates often and has been examined by the veterinary staff, who have found him physically normal. He has been observed to submit to the subordinate role in homosexual behavior with his father, but has never been observed to mount and breed with a female gorilla. Both males are kept with females at all times, yet neither has shown any mounting behavior (other than homosexual) in the past 10 years.

Summary and Conclusions

The purpose of reviewing the complexities of propagating great apes in captivity is to focus on the results of disrupting the natural development of self- or species-identity. One of the San Francisco

Zoo's gorillas, Koko, has been taught American sign language. On two occasions her responses to questioning suggested possible perception of self-awareness. Both mirror reflection and cartoon illustrations of gorillas were identified by the animal as her own image (Hayes, 1977). Chimpanzees and orang-utans presented with mirrors have used them as tools to examine normally unseen parts of their bodies (Gallup, 1977). It appears then that these species have a sense of self-awareness.

Deprived of social contact with adults and peers of their own species, captive-reared apes absorb the behavior of their human foster parents. Prior to forced separation from her human foster parents, Merah exhibited only fear in the presence of her own species. Although possibly aware of their own image, these apes often prefer human companionship when given a choice, and often produce tremendous fear responses upon exposure to their own species. They may be able to differentiate between themselves and humans, since the initial fear of their own species does mellow with time and the lack of choice. Apparently they do not have an innate identification with their own species. (As adults they solicit human contact, yet they all equally seek contact with conspecifics.) The continuing interest in humans on the part of our four nursery-reared orang-utans and one gorilla indicates that several forces may be involved in the young ape's development of self-identity. The pattern does seem to suggest that zoo-reared orang-utans and gorillas have the capacity for a species affinity. Since the living arrangements are more often forced than chosen, it is difficult to analyze a specific species-identity among these animals.

Young apes, captured during their formative years, may lack crucial information about their own species. This may be reflected in their sexual and maternal behavior (Babladelis, 1975). Proper sexual responses often do not mature or may develop abnormally. When apes fail as parents, which is far too often, humans must teach the infants to be apes. Merah did show a possible latent affinity for orang-utans at the age of 18 months, and the other orang-utans have been integrated into group situations at various ages with apparent success. The affinity that infants exhibit for each other may be related to species-similar interests and responses rather than reflecting an innate species identity.

The zoo situation does not offer a hand-raised ape the choice to remain within the human world, nor does it offer a natural social setting (MacKinnon, 1974). In spite of these barriers, some captive-reared apes prefer human companionship, while others breed and successfully rear their young within this environment. One zoo has been able to use the trust of their female gorilla as a teaching aid for maternal care (Joines, 1977). The resulting proper maternal care of an infant may induce the other female gorillas in the enclosure to care for their next infants.

Where human and peer influences have failed to teach proper social responses, including maternal and sexual behavior, there appears to be a gap in species awareness or perhaps the lack of an adequate species identity. More systematic observations are needed on zoo-reared apes before this issue can be clarified.

References

Babladelis, Georgia. 1975. Gorilla births in captivity." *International Zoo News* No. 130, October.

Gallup, G. 1977. Self-recognition in primates: A comparable approach to the bidirectional properties of consciousness. *Am. Psychol. 32* (5), 329–338.

Haberle, and D. V. M. Albert. 1963. Notes on the birth and hand-rearing of a gorilla. *J. Zoo Anim. Med. 4*, 20–27.

Hayes, H. T. P. June 12, 1977. The pursuit of reason. *N.Y. Times Mag. Sec.* 6, 20–23, 73–79.

Joines, S. 1977. A training program designed to induce maternal behavior in a multiparous female lowland gorilla, *Gorilla gorilla,* at the San Diego Wild Animal Park. *Int. Zoo Yearb. 17,* 185–188.

Keiter, M. and P. Pichette. 1977. Surrogate infant prepares a lowland gorilla, *Gorilla* g. *gorilla,* for motherhood. *Int. Zoo Yearb. 17,* 188–189.

Lang, E. M. 1961. Jambo, the second gorilla born at Basle Zoo. *Int. Zoo Yearb. 3,* 84–93.

Mackinnon, John. 1974. The behaviour and ecology of wild orang-utans, *Pongo pygmaeus. Anim. Behav. 22,* 3–74.

Maple, T. 1977. Unusual sexual behavior of nonhuman primates. In J. Money and H. Musaph (Eds.), *Handbook of sexology.* Amsterdam: North Holland Press, 1167–1186.

Maple, T., E. Z. Zucker, M. P. Hoff, and M. E. Wilson. 1977. *Behavioral aspects of reproduction in the great apes.* Paper delivered at the convention of the American Association of Zoological Parks and Aquariums, San Diego.

Martin, R. 1976. Breeding great apes in captivity. *New Sci. 14,* 10–102.

Nadler, R. D. and M. L. Jones. 1974. Breeding of the gorilla in captivity. *Int. Zoo News, 22,* 21–27.

Perry, J. 1976. Orang-utans in captivity. *Oryx,* Feb., 262–264.

Perry, J. and D. L. Horseman. 1972. Captive breeding of orang-utans. *Zoologica 57* (2), 105–108.

Thomas, W. D. 1958. Observations on the breeding in captivity of a pair of lowland gorillas. *Zoologica 34* (3), 95–105.

Chapter 14

Species Identity in Humans: Innate or Not

M. Aaron Roy

> Strange to say, babies born to us need not grow up to be what we think of as human; their humanity seems to be transmitted to them after birth in an as yet ill-understood negotiation or transaction between the babies and chiefly among others their mothers—and its seed may never blossom. . . . (Shotter, 1974, p 215)

The notion that at some point in our lives we all had to learn that we were people is quite intriguing. More than just generating an interest in how this process occurs, it implies that in some individuals such learning may not take place or, at best, is incomplete. This chapter will deal with a human's awareness that he or she is a person, i.e., the phylogeny and ontogeny of a species identity in humans. The following argument will be offered: we are neither born with an innate awareness that we are human and thus like other members of our species, nor does such an awareness automatically develop after birth. Such an identity with our species is primarily dependent upon particular postnatal contacts with other people.

A Species Identity in Humans

The introductory chapter of this volume defined species identity in a way applicable to humans, and the presence of a species identity in many members of our species is easy to detect. We can determine it either by verbal questioning or through inference from a constellation of social behaviors. As will be evident later in this chapter, it is not so easy to determine if a member of *Homo sapiens* does not know he or she is a person.

A scientifically based discussion of a species identity in humans is not possible. Such identifications have received surprisingly little attention to date, and research with humans designed to limit particular postnatal experiences which could impair normal development, as in the nonhuman primate isolation-rearing studies discussed in Chapters 8–11, was (and is) ethically unacceptable. A discussion of the possibility of an acquired species identity must therefore rely extensively on information which was not gathered scientifically. Such information includes theories and speculations of both academicians and clinicians, as well as various reports of children known to have been or suspected of being reared atypically. As a result the plausibility of an acquired species identity in humans can be evaluated in this chapter, although the validity of such a position cannot be ascertained.

The Essence of Being Socialized

People are unquestionably the primary force which socializes the developing child. Their actual function is nevertheless unclear. They may be needed only to provide the necessary protection and nurturance so that the developing child, who does (or will) automatically know it is a person, may grow and eventually behave like other humans. In contrast, people may serve an additional role to nurturing—that of providing models, so that the child who is without innate knowledge that it is a person can form an identity with its biological species.

Count (1973) has said that "a youngster raised without association with age-peers of his own kind may be permanently deflected from the orientations normal to his society . . ." (p 56). In a foreword to Singh and Zingg's (1942) text on wolf children and feral men, Kingsley Davis stated that "man is social because he is made social, not because he is naturally so." Montagu (1950), in an essay entitled "On Being Human," states:

> No organism of the species so . . . named *Homo sapiens* is born with human nature. Being human is not a status *with* which but *to* which one is born. Being human must be learned (p 37).

LaBarre (1954) talks of the child as "a potential culture-bearer [whose] 'Human nature' . . . is not automatically organic, not instinctively spontaneous, but necessarily disciplined and shaped by a long apprenticeship to childhood" (p 219). At issue in these statements is *what* exactly must be shaped or learned by education—is it social behaviors, is it something akin to a species identity, or is it both? The term "socialization" may mean more than just acquiring social skills; it may also imply the acquisition of a "force" or "need" to be like and with other people.

The notion of the infant acquiring its "human nature" and thus becoming socialized is shared by others. Ferguson (1970) speaks of infants "joining the human race," Wrong (1961) refers to the "process of becoming human," and Halliday (1948) discusses the acquisition of a "sense of being a person—the awareness of I'ness." Both Becker (1972) and Richards (1974) believe infants become human beings after interaction with people. Shotter's (1974) position as presented at the beginning of this chapter is relevant here. Schmidt (1973) states that "a child needs a certain minimum of human interaction and care in order to develop the distinguishing biological characteristics of his species in at least a minimal way." (p. 44). Within a more analytic framework, Murray (1938) has said that an affiliation for people is not innate in the child; it is a psychogenic need and thus a function of parental and cultural influences. It should be emphasized that those who have discussed such things as "becoming human" or "joining the human race" may not equate these with acquiring a species identity. It is apparent, however, that many have concerned themselves with something other than the learning of overt social behaviors, and many may in fact agree with a position that one's species identity is postnatally determined.

Some may argue that man innately recognizes members of his species, and they find support in the literature on neonatal responsiveness. Ainsworth, Bell, and Stayton (1974) have stated "that infants are genetically biased toward interactions with other people from the beginning. . . . A child is preadapted to a social world, and in this sense is social from the beginning" (p 99). Brazelton (1969), when discussing various "inner forces" which influence the neonate's development, speaks of "the drive to fit into, to identify with, to please and to become part of his [human] environment" (p. 27). Human beings, according to Jacobson (1964), "are born with a potential capacity for mutually adaptive and mutually gratifying relations with their own species" (p 31).

Just because an infant emits a number of unlearned behaviors which initiate and maintain contact with people does not mean these behaviors are innately directed toward humans. Infants may prefer people because people are the ones who have the most and frequently the only contact with them after birth.

Processes of Acquiring a Species Identity

An analysis of the possibility of an acquired species identity in *Homo sapiens* should begin with a discussion of normal behavioral development. This will both facilitate comparisons with examples of atypical social development, as well as generate some questions pertaining to why certain behaviors which are characteristic of the socialization process do occur.

Some believe that development is a continual and somewhat smooth process of becoming more adultlike, with environmental experiences having a pronounced effect whenever they occur (Bandura and Walters, 1963; Bijou and Baer, 1961, 1965; Miller and Dollard, 1941). This nonstage approach interprets socialization as learning how to react to various social stimuli in order to achieve reinforcing consequences. Consistent with this approach, a species identity could be acquired at any age if the proper socializing contingencies were present.

In contrast, others have emphasized a noncontinuous process of becoming more adultlike (e.g., Erikson, 1950; Freud, 1953; Montessori, 1936; Piaget, 1964; Spitz, 1965). Some have even specified the presence of particular developmental periods or stages (e.g., Caldwell, 1964; Scott, Stewart, and DeGhett, 1974) which seem compatible with the three general developmental periods (critical, susceptible, and optimal) presented by Hess (1973). The infant is believed to pass through stages where the child achieves a particular level of functioning in each one by being differentially responsive to certain types of experiences. Consistent with a stage approach, one would specify that a species identity would be acquired easily during one period and with difficulty, if at all, after that period had passed. Many individuals interested in human development have already suggested specific periods for affiliation and socialization in humans. Most have placed the period within the first year of life: 5–12 weeks (Ambrose, 1963); 6–24 weeks (Gray, 1958); birth to 6 months (Hess, 1959; Rollman-Branch, 1960); 3–6 months (Spitz, 1965); and birth to 10 months (Ferguson, 1970). Halliday (1948) places this stage between two and six years of age.

The issue of whether a stage or nonstage interpretation is most applicable need not be decided here. What is important is the implication that social experiences, especially those early in life, are likely to have considerable influence on the acquisition of a species identity.

Early Social Experiences

Several factors characterize the course of early development. These include neonatal stimulus preferences and behavioral repertoires, attachments between the child and caretaker(s), and the occurrence of separation and stranger anxieties.

The First Six Months

The neonate is capable of emitting behaviors and responding to stimulation from a range of sensory modalities. Infants as young as one day old show preferences for human faces (Fantz, 1965; Kagan, 1970; Lewis, 1969; Maurer and Salapatek, 1976). It has also been reported that neonates are capable of imitating facial and manual gestures of adults (Meltzoff and Moore, 1977). These preferences and abilities, however, do not necessarily reflect an innate awareness that one is human. Research has not determined if these neonatal preferences are for "humans" or for particular stimulus configurations (which we adults with a species identity interpret as being human). It has not been determined, furthermore, if these neonatal preferences are maintained in the absence of contact with humans or if pictures of lower mammals or other primates with eyes, a nose, and a mouth will also generate more viewing time as compared to scrambled faces of humans. Until more research is conducted with varying stimulus configurations, an innate preference for humans remains speculative. Predispositions to interact with humans could merely represent genetic biases which serve to facilitate, but not insure, the acquisition of a species identity.

Though there are differences in basic constitutional characteristics in infants (Thomas and Chess, 1970), normal neonates encourage contact with others, e.g., their human caretakers. These contact-promoting behaviors include crying, smiling, various vocalizations, visual attention and following, clinging, and grasping (Ainsworth, 1967; Bowlby, 1969; Haith, Bergman, and Moore, 1977). These behaviors also serve to sustain contact with the caretaker. What normally evolves is a series of reciprocal interactions between the child and the human caretaker(s) during activities of nursing, hygiene, and play (see Stern, 1977).

The object to which the infant normally attaches is its mother or primary caretaker, but strong attachments can simultaneously occur to other adults or even inanimate objects (Schaffer, 1971; Schaffer and Emerson, 1964). The infant initially comes to recognize a familiar person and differentiates him or her from unfamiliar people; but then as time passes there are signs of an emotionally charged preference for this person whose company the child actively seeks and craves. The child in turn does not solicit and will even reject contact with strangers or nonpreferred acquaintances. This very strong primary attachment between the infant and a stimulus object usually occurs within the first 6 to 8 months (Bowlby, 1969; Gewirtz, 1972; Rutter, 1972). What is needed for formation of this attachment is a sufficient amount of contact between the child and the stimulus object to which it attaches.

There are indications that the attachment process begins within the first few weeks of life. These include visual following of the caretaker and concentration on the eye area (Haith et al., 1977; Robson, Pederson, and Moss, 1969); a preference for the caretaker's face (Fitzgerald, 1968); smiling at the sound of a human's voice (Wolff, 1963); and the development of stable eating and sleeping patterns (Sander, Julia, Stechler, and Burns, 1972). Final indications that the infant has formed this basic attachment include the occurrence of two types of anxiety, neither of which is found in neonates. In Western cultures there are the infant's fear of strangers (stranger anxiety), which emerges at about eight months of age, and the emotional and physical disturbances which occur upon separation from the object of attachment (separation anxiety), which first appears at about seven to ten months of age (Mussen, Conger, and Kagan, 1974; Schaffer, 1971).

Our ability to characterize aspects of the primary attachment phenomenon has not enabled us to understand fully why it evolves. The reinforcing value of the mother in satisfying the child's physiological needs has been emphasized (see Bijou and Baer, 1965); but a strict associative learning interpretation is apparently insufficient, in that strong attachments can form to those who do not provide primary reinforcement (Ainsworth, 1964; Bowlby, 1969; Schaffer, 1963). Contact comfort, as implied by the nonhuman primate studies of Harlow (1960), is apparently important, as is the role of protection offered the infant when it forms an attachment (Ainsworth, 1972; Bowlby, 1976). Many factors are no doubt operative in each child–caretaker attachment. The universality of this phenomenon suggests, nevertheless, that a strong attachment results primarily from a genetic bias to attach to something, rather than it being a sole consequence of the proximity of people. This innate attachment tendency in conjunction with innate contact-promoting behaviors would serve to maximize the probability that the human neonate will form an attachment with another member of *Homo sapiens.* An intriguing question is why there would be an innate tendency to attach in the first place. One may speculate that this innate tendency has evolved, in part, so that exposure to one's biological species is of sufficient duration so that a species identity could form.

An innate tendency to attach is not a novel viewpoint. Halliday (1948) discusses the infant who "is born with active drives to receive love from the Mother—the life giver" (p 93), while Bowlby (1958) has mentioned an innate drive to form a love relationship with a parent figure. An innate "attachment need" to seek the proximity of certain other members of the species has been proposed by Schaffer and Emerson (1964), while Ainsworth and Bell (1970) refer to a "genetic bias" to attach. What may be misleading in these statements is the implication that the infant wants to attach to other people. Since human infants usually are reared solely by members of their biological species, a subsequent preference for people may reflect the lack of postnatal exposure to other species, rather than a preattachment identification with people per se.

Social Experiences after Six Months

Socialization, or the acquisition of species- or cultural-specific behaviors, begins early in life. It is believed to be accomplished by two nonindependent processes. One involves a direct interaction with the environment so that, based upon conditioning principles, behaviors come under the control of various discriminative and reinforcing stimuli. People have an important role in this process in that they either serve as the manipulators of stimuli or, most often and most importantly, they serve as reinforcing stimuli themselves.

Though behavior acquired by means of conditioning is important, assimilation into one's culture is primarily accomplished through a second process, that of copying behaviors and attitudes. Three levels of copying behavior have been discussed (Bandura, 1967; Gewirtz and Stingle, 1968). In order of behavioral complexity and importance in the socialization process, these are observational learning, imitation, and generalized imitation. Just as with associative conditioning, people who interact with the young child have an important influence on the development of copying behaviors. People not only act as important models for the infants, but they also reinforce copying by providing primary and secondary reinforcing stimuli.

Since it is uncommon to find a child who does not exhibit any copying behaviors, three questions arise:

1. Why does observational learning, the antecedent of imitation, occur?
2. Why does imitation occur so readily in real life when specific instances of "imitation training" are infrequent?
3. Why does generalized imitation occur, and why is it so resistant to extinction?

It is obvious that primary and secondary reinforcers are intricately involved in the development and maintenance of copying behavior in humans. A total reliance on reinforcement, however, may not fully explain such behavior. Observational learning could evolve naturally as an attachment to a caretaker occurs and as one's species identity develops. It may in fact evolve independent of reinforcement, since observational learning does occur in nonhuman primates which lack such learning histories with the model (Jouventin, Pasteur, and Cambefort, 1977; Strayer, 1976). The appearance of imitation and generalized imitation can also be discussed by incorporating one's species identity. Mere contact with and stimulation by a model (in the form of smile, praise, etc) may be "rewarding" because the model is a species member with whom the observer identifies. Such an analysis also pertains to the role of intrinsic reinforcement. If one relates to other people and knows one is a person like them, then it may be intrinsically satisfying to behave like those organisms with whom you identify.

In summary, the eventual degree of imitation shown by a child may be based on both the presence and "strength" of its identification with people. Reduced or nonexistent tendencies to learn through observation and imitation may reflect a disturbance in one's species identity.

Atypical Rearing Conditions

Those traits which are crucial to further socialization are present by the time a normal child reaches school age. The child readily interacts with people, it imitates them, and it generally likes to be with them. From these traits it is easy to infer that the child has a species identity. If such an identity determines social preferences, then a distortion in one's species identity should lead to changes in species-typical social behavior. If, furthermore, a species identity in humans is acquired through postnatal contacts with species members, then a disruption in such contacts is likely to be reflected in the child's social preferences and behaviors. More specifically, a child should be less likely to imitate, want to be with, and interact with other people even though it may look normal and engage in many normal nonsocial behaviors.

As has been discussed for nonhuman species in the preceding chapters of this volume, preventing exposure to members of one's biological species early in life can create both subtle and not so subtle behavioral disturbances which are evident when interactions with the biological species eventually occur. These behaviors usually are obtained by intentionally rearing the infant in isolation or with another species. No such data exist for humans. Therefore, one must look at the literature which discusses atypically reared children, i.e., the "experiments of nature."[1] Information in these sources was not scientifically obtained, and it is more likely (though not necessarily) tainted with biases, faulty recollections, and incomplete reporting. The social development of these children, nevertheless, can provide considerable insight into the mechanism(s) whereby an infant's early social interactions with its biological species can influence the possible development of its species identity.

One can expect from the previous discussion of normal developmental characteristics that less contact with people during the first year of life will (1) disrupt the attachment process, (2) interfere with the development of stranger and separation anxieties, (3) decrease the amount of observational and imitation learning, and (4) reduce the tendency to affiliate with other people.

The following review will deal with three types of adverse child-rearing practices: children reared in institutions, those raised in the home but under various degrees of social restriction, and children supposedly nurtured in the wild by animals. Each case will in one form or another encompass those specific deprivation types discussed by Clarke and Clarke (1960) and Rutter (1972).

Early Institutional Upbringing

The effects of social restriction resulting from early institutionalization have been extensively reported by Dennis (1973), Goldfarb (1945), Kohen-Raz (1968), Provence and Lipton (1962), Spitz (1945, 1946), and Tizard and Rees (1975). Additional information is available from reviews by Bowlby (1951) or Casler (1961) and from experiments using institutionalized infants as subjects (e.g., Brossard and Décarie, 1971; Rheingold and Bayley, 1959; Saltz, 1973).

It is generally accepted that institutionalization per se is not the cause of developmental difficulties. Two significant factors are what does or does not happen to the child while in the institution and what the child's experiences were before institutionalization. There are differences among the institutions which have been discussed, yet they were all depriving when one compares them to being raised in a typical family.[2] Total deprivation of human contact is not found. Care is different quantitatively rather than qualitatively from that usually provided in the home (Provence and Lipton, 1962; Rheingold, 1960). In most instances the children who were studied received adequate medical and hygienic attention, and they were believed free of neurological damage. Therefore we are discussing in most cases physically normal infants who, until the time of foster home placement at around three or four years of age, grew up having less physical contact with other members of their species.

The consequences of institutional rearing early in life can be divided into those which are short-term, appearing while the child is still in the institution, and those which are long-term, becoming evident even after successful foster-home placement. Many functional and physical deficits occur prior to placement.

Most cognitive, motor, and physical deficiencies, however, usually dissipate after a child's successful placement (see Clarke and Clarke, 1977), whereas deviant social and personality characteristics are more likely to persist regardless of the desirability of placement.

Spitz (1945) reported that no significant change was noticed in the infants until after the third month. By the end of the first year, however, scores on perceptual, motor, intellectual, social, and environmental responsiveness were significantly lower, and they continued to fall. Stranger anxiety was either completely missing or excessively abnormal. In this ward of children ranging in ages from 18 to 30 months, few spoke words, only two could walk, and only a few could eat alone. A follow-up study was conducted when the children were between two and four years of age and still in the institution (Spitz, 1946). Many children were normal in height and weight, but few could walk unassisted. Speech impairments were prevalent. In 1945, Goldfarb summarized a series of his earlier reports on institutionalized children (1943a, 1943b, 1943c, 1944). As youngsters they exhibited many of the social, motor, and language deficiencies found by Spitz. Even though the children appeared to be very demanding of affection and attention from others, this did not significantly enhance their capacity to form ties. Most were described as removed, emotionally isolated, and affectively impoverished. As adolescents they were socially immature, with only superficial interpersonal relationships. Many showed little emotionality or remorse following acts of aggression, hostility, or cruelty. Changes in foster homes were made with little emotional reaction, suggesting that few, if any, social ties had been formed. Tizard and Rees (1975) also report that the majority of their children who were reared in institutions failed to form close attachments with adults.

Provence and Lipton's (1962) report on the effects of institutionalization supports the previous findings. Not until after the first month was any motor deficiency noted, and this was a decreasing capacity to adjust to their handler while being carried or held. Infants smiled on schedule, but this behavior diminished with age. The children exhibited an emotional "blandness," and with few exceptions stranger anxiety was absent. There seemed to be a very poor sense of self and body image. The children were not likely to initiate social contacts, to seek adults in time of need, to form close ties, or to be attached to any one person. The children reached normal developmental ranges in many areas following successful foster-home placement.

> However, when one looked more closely . . . there were residual impairments of mild to severe degree in their capacity for forming emotional relationships, in aspects of control and modulation of impulse, and in areas of thinking and learning that reflects multiple adaptive and defensive capacities and the development of flexibility in thought and action (p 158).

A ten-year study of children raised in a very depriving institution in Czechoslovakia reported in 1967 by Matejcek (see Langmeier, 1972) also reported long-term impairments in initiating and maintaining social relationships.

Summary and Implications

Children raised in these types of institutions did have regular and somewhat systematic contacts with people early in life, but there were few of these interactions. Many children in the institutions did not exhibit characteristics of the early socialization process: they frequently failed to form primary attachments, their contact-promoting and contact-sustaining behaviors were infrequent, and both stranger and separation anxiety were often missing. Short-term influences of early institutionalization could be found in many functional and physical areas. Successful foster-home placement usually reduced deficits in the physical and cognitive areas, but long-term impairments in social interactions usually remained. All of these studies demonstrate that sustaining the life of a child does not automatically insure normal social development. The impairments in social functioning which were evident in many of the children reared in institutions do not imply that any or all of them did not have a species identity. These children obviously identified with humans, but their ability to form attachments with conspecifics and their desire to be with other people were often less than normal. One can conclude that social development and functioning are both most vulnerable to "less-than-normal" early contact with conspecifics and least modifiable with "normal" social contacts.

Social Restriction in the Home

There are numerous references in the literature to children whose history included fewer than normal interactions with people, but who were not brought up in institutions. Their social "isolation" was created in the home by caretakers. It is not always clear what social experiences each child had prior to or during these periods of limited social interactions; but consistent with both Gewirtz's (1961) and Rutter's (1972) differentiation, there were degrees of social deprivation (the loss of attention and contact) and social privation (the lack of attention and contact). There was also less nonsocial stimulation in general. For some, their contact with people was reported to be even less than that normally found in institutionalized children; thus, one would expect to find more severe social deficits.

Six cases are particularly relevant in that they involved seven infants who were separated from normal social activities for periods ranging from two to 12 years. Although the duration of atypical rearing and the quality of postdiscovery care was related to recovery, the most important variable was the type and degree of social contacts the child had both before and during its period of social restriction. Some of these children showed considerable improvement in social interactions after recovery (e.g., Isabelle, Genie, Carl); others did not (e.g., Anna, Albert, Anne).

Anna

This child was raised in seven different foster-home locations from birth to 5½ months of age (Davis, 1940). Anna's repeated placements were attributed to maternal indifference and insufficient funds rather than to any physical disorders or deformities. She was finally returned to her mother who, because of Anna's illegitimate status, was forced by Anna's disapproving grandfather to confine her in an out-of-the-way attic room. Here Anna remained tied to a chair for 5½ years. This attic room was also used by her mother and older brother as a dormitory, but they spent little time in it other than to sleep. Anna was routinely fed a substandard diet—only milk until age 5, then both milk and oatmeal until age 6. Other than feeding, neither her mother nor her brother took "the trouble to bathe, train, supervise, or caress her" (Davis, 1940, p 557). Anna was generally ignored by all.

Anna was removed from her confinement by authorities at the age of 6 and placed in a local county home. At admission she was described as apathetic, expressionless, indifferent, immobile, and incontinent. So severe was Anna's state that she was considered an "unsocialized individual" who was in an "animal-like" stage. This child improved during her nine-month stay in this home in that she became more responsive to her environment and she showed signs of learning a few basic personal and social skills. Anna, however, remained socially and linguistically retarded. This child was then placed in a foster home for 7½ months, where she continued to improve: she learned to walk a few steps, she understood and responded to some verbal commands, she acquired some toilet habits, she began to feed herself, and familiar people were recognized with some pleasure. These changes, however, should not be construed as normality, for she was still far from being rehabilitated. "She said nothing, hardly played when alone, she had little curiosity, little initiative; it seemed still impossible to establish any communicative contact with her" (Davis, 1940, p 562). Anna was finally assigned to a school for retarded children, where she lived for 2½ years before succumbing to hemorrhagic jaundice at the age of ten. At the time of her death she was fully continent. Other achievements included calling attendants by name, expressing her needs in complete sentences, and conforming to group activities, but only as a follower.

This child lived for four years after her discovery. During this time she acquired some social skills, but these (e.g., self-feeding, continence, obeying commands) did not represent a recovery. In spite of regular contacts with people, Anna remained retarded in social interactions. She did not appear to "want to be like and with" other people.

Davis's interpretation of Anna's poor recovery varied from 1940 to 1947. Her continued retardation, especially in language and social skills, was originally attributed to her deprived rearing. Davis felt that the early social restriction without an intimate, primary relationship with another human made it "almost impossible for any child to learn to speak, think, and act like a normal person after a long period of isolation" (Davis, 1940, p 564). Anna was viewed as missing a critical period for socialization. Davis (1947) later reversed his position when Anna was compared to another atypically reared child, Isabelle (see next case). Anna's continued retardation was now attributed to a suspected feeblemindedness which was compounded by her inferior treatment. (This change was apparently made to account for the discrepancy in recovery between Anna and Isabelle, given Davis's conclusion that they had very similar prediscovery rearing situations.)

Isabelle

In 1947 Mason reported a case involving an illegitimate girl named Isabelle. She had lived with her deaf-mute mother isolated in a darkened room of a house for 6½ years. Neither mother nor daughter was ever allowed to leave the room. Isabelle and her mother were found after their escape from confinement. At this time Isabelle made a strange croaking sound, but no speech was evident. She was attached to her mother and especially fearful of strange men. Isabelle received a comparatively thorough rehabilitative training program. By the age of 8, only 1½ years after discovery, she was considered an average child. Isabelle was bright, cheerful, energetic, and had a good vocabulary. There were no noticeable social or motivational deficiencies. By the age of 14 years, Isabelle was attending sixth grade in public school and she was considered normal by both peers and teachers.

Genie

Genie is the third example of a child reared under unusual conditions within a family setting (Curtiss, 1977; Fromkin, Krashen, Curtiss, Rigler, and Rigler, 1974). She was born without known abnormalities other than a congenital dislocated hip. This child apparently received adequate care from her family until the approximate age of 20 months. From then until the time of her removal from her parents' house around 14 years of age, she was confined alone in a small room, usually restrained in a crib or potty chair. The door to this room was closed and the windows were covered. Genie's contact with people while she was confined was limited to routine feeding (cereal and baby food) by a near-blind mother, physical punishments by her father when she made noise, and periodic harassment by her brother and father, who barked at her like dogs. The mother was only permitted to spend a few minutes with Genie during feeding, but unlike the father and brother she did appear to treat her kindly.

Genie was first placed in the rehabilitation center of a large city hospital for nine months and then relocated in a foster home. On admission to the hospital she was described as "pale, thin, ghost-like, apathetic, mute, and socially unresponsive" (Fromkin et al., 1974, p 86). Rigler (1972) described her as being partly socialized. By the second day she was imitating verbalizations of the staff, communicating her needs nonverbally, and showing signs of frustration when not satisfied. These were behaviors not seen in Anna so soon after discovery. A report made only four weeks after admission said that "now she had become alert, bright-eyed, engaged readily in simple social play with balloons, flashlight, and toys, with familiar and unfamiliar adults . . . and

[had] ample latent affect and responses" (Fromkin et al., 1974, p 86). Rapid progress continued to occur in the foster home. Genie's social and linguistic skills, as well as her general interest in the environment, became much more normal.

Carl

A fourth instance of atypical rearing was reported by Balikov (1960). This case involved a family with two blind parents and three physiologically normal children. There was an older girl and two brothers, all of whom were born approximately 20 months apart. The report focused on Carl, the middle child, because he was said to be the most verbal and his symptoms paralleled those of his siblings.

Carl lived in a dirty, barren, and darkened apartment for the first two years of his life. He was kept in either his crib or playpen with the door to the bedroom frequently closed if he cried. Nevertheless, he did have some social interactions with both his parents and siblings. Carl was discovered when he started to attend a nursery school at the age of two. On his first day he was a clumsy and fearful child; however, he did relate to people, and he was quite verbal. Not surprisingly, Carl's behavior closely resembled that of a blind person. When observed accompanying his father on walks, Carl imitated the father so completely that he acted equally blind. With only seven months of attendance at the nursery school, Carl had made considerable progress. He was adjusting quite well to the daily routine, the staff, and his peers.

Albert and Anne

Freedman and Brown (1968) discuss the early rearing conditions of two siblings, Anne and Albert. Anne was confined to a bedroom with a normal older brother who had full access to family and friends. She slept on a straw pallet and was fed with a bottle. Excluding a few visits to the doctor, Anne remained in this room for 6 years until a relative initiated legal proceedings to see her. She (with Albert) was then removed from the home, hospitalized for 8 weeks, returned to the home for 9 months, placed in a foster home for 18 months, and finally adopted by another family.

At recovery Anne was indifferent and lacking affect with people, and she uttered some words in an echolalic manner. Her height and weight were well below average. While in the hospital she became continent, learned the name of some objects, began to feed herself, and showed some affect for staff. Anne continued to expand her expressive abilities while in the foster home, and she began to respond more affectionately toward the family members. This child, however, remained abnormal in her ability to form close attachments with people, in her discomfort following periods of separation from familiar people or surroundings, and in her spontaneous interchanges with others. These residual impairments in social interactions parallel those of many of the institutionalized children.

Albert was born two years after Anne and for four years was kept in a small room by himself. He stayed in his crib or on a potty chair. The mother was only present to give him a bottle or to change his diapers. Albert was retarded physically, mentally, and verbally when removed from his room and, like his sister, he continued to improve in these areas. He was, however, more deficient than Anne, especially in his social interactions. He had no speech when discovered nor even a social awareness of people around him, i.e., he did not react to or play with people as Anne did. Albert's postdiscovery development, furthermore, was much slower than Anne's. Two years after final removal from the home, both children showed "very poorly differentiated ego boundaries and little or no capacity to enter into mutual affectional relations with another person" (Freedman and Brown, 1968, p 435).

A Boy Named M.

The final case is that of a boy, M., who was brought in for treatment by relatives at the age of seven (Morrison and Brubakken, in press). There is no early developmental history on M. until the age of 3, but relatives said that he was frequently left unattended by both his alcoholic mother and his absent, working father. When he was three, M. was confined to a crib for nine months without many social contacts. For the first four years of life M. was only bottle-fed and, when old enough, only when he said "eat." From the age of 5 to 7, this boy received better care in the home of relatives, but they too had marital and alcoholic problems.

M. was scored as deficient in a range of sensorimotor learning tasks when admitted to a state mental hospital at the age of 7. Furthermore, he was socially inept: emotional relations with others were impaired, there was no sign of a personal identity, and he was preoccupied with inanimate objects. Language development was at the 2-year-old level. M. received four years of multidisciplinary therapy in the home of relatives, in schools, and in a state hospital. This intensive remedial program succeeded in changing many facets of his behavior, especially in the motor and cognitive areas. Socially, however, M. was still tainted by his early rearing.

> Attachment behaviors remain clearly illusory . . . he continues to remain affectively unattached. It is noted that he evidences a minimal capacity for self–object differentiation, play ability with other children, and "purposeful" activity in learning but has not developed object relations of greater than a transient nature. All of the reported intellectual advances have not been sufficient in and of themselves in the socialization of this child (Morrison and Brubakken, in press).

Summary and Implications

Important differences as to the age when socially restricted, the length and the degree of restriction, and the postdiscovery experiences limit any generalizations about all seven children without qualifications. Some facets are worth mentioning, however. All of the children showed postdiscovery improvement in sensorimotor learning activities. The three children (Isabelle, Carl, Genie) who recovered socially had the opportunity to form close ties with people before or during their periods of restriction. Isabelle was attached to and communicated effectively with her mother, who shared her isolated life. Carl interacted with all family members while living in darkness, even identifying with his father to the point of imitating the latter's blind characteristics. Genie not only had social contact with family members for over 1½ years prior to her restriction, but the limited social interactions with her parents during the following 12 years no doubt enabled her to maintain an orientation toward people. This was demonstrated by her awareness and imitation of people so soon after discovery, as well as by Genie's subsequent signs of maternal attachment.[3] When evaluated by professionals, all three of these children seemed to possess a number of traits which are necessary for social functioning. These included an orientation toward people, a desire to be with and like these people, and an ability to learn many social behaviors without specific instruction (e.g., by imitation and copying). Their recovery is consistent with the outcome of those examples of social restriction where siblings have been confined together (Koluchová, 1972, 1976) or where a sibling has acted as the primary caretaker of a confined child (Kagan, 1976).

In contrast, there were fewer postnatal interactions with people for Anna, M., Albert, and Anne. In addition to less social contact, these children apparently never had an opportunity to form a strong attachment to someone during or before their confinement. (See Schmidt, 1973, for a similar interpretation.) These four children showed pronounced and lasting socialization deficits in spite of seeing people around them early in life. This finding parallels the research with nonhuman primates, which has demonstrated that seeing peers without receiving tactile–kinesthetic stimulation from them will not prevent the development of self- and peer-directed abnormalities (Mitchell, 1975; Prescott, 1975). These four children spent no more time under social restriction than did Isabelle or Genie, yet they did not exhibit the type of sociality when discovered. The basic and "natural" affiliation for people was not present in Anna, M., Albert, or Anne.

These seven cases of social deprivation and privation imply that early rearing with reduced contacts with people can disrupt a child's affiliation for, desire to be like, and generalized imitation of its biological species. To say that any of the children raised under social restriction were without a species identity would be too speculative, given the information available. Yet the behavior shown by four of them (Anna, M., Albert, Anne) suggests that their identification with their biological species was not normal when they were first seen by professionals, nor did it become normal through the period of postdiscovery contact with people whose goal was to foster normal social interactions in these children.

Feral Children

An analysis of the developmental histories of feral (or wolf) children is potentially most relevant to this discussion of the ontogeny of a species identity in humans, in that such children are supposedly raised without any human contacts. They would be expected to lack an identification with humans and therefore be even more deviant socially than either children raised in institutions or those who experienced social restriction at home. It is unfortunate that these cases are not documented scientifically. This, plus their very essence of supposedly rearing human infants by another species or in the wild, tends to raise sufficient suspicion in many professionals to have the cases automatically discounted as being exaggerated or fabricated.

There is reason to doubt many of the cases which have been reported. (See those presented by Singh and Zingg, 1942.) Some feral children suffered from enough subnormal intelligence or other functional–physical anomalies to have made them less sophisticated socially even if they had been reared by humans. Other cases strongly suggest that, like Genie, there was considerable contact with humans prior to or during their prediscovery periods of social restriction. There is one child, however, about which there is enough information to conclude that mental retardation was not a major factor and for which there were few regular contacts with people prior to capture. This is the case of the girl called Kamala, which was described by Reverend Singh, an Indian missionary (Singh and Zingg, 1942). It is important to detail Kamala's development after capture because many of us have only read secondary, brief summaries which could not effectively convey her condition.

Singh was the rector of an orphanage in Midnapore, Bengal, India, who had no apparent reason to "manufacture" this child and her behavior. No professional or monetary gain resulted as far as is known. On the contrary, Kamala's postcapture existence was hidden from the local people for many months. The extensive diary on which the report was eventually written was very specific and time-consuming to prepare, thus adding credibility to the description of this child. There were 22 photos of Kamala living in the orphanage.

Whether or not Kamala was raised by wolves is an unanswerable question. Many would say that it was (and is) not possible and, in turn, reject Kamala's entire case history. Singh's explanation of wolf-rearing is difficult to believe. Yet it is even more difficult to believe that she had been raised by humans to any extent. At her capture and for many years thereafter, Kamala's behavior was so animal-like that it is incomprehensible to accept the notion that humans had reared her prior to her jungle living. It is difficult to believe that interacting with the wolves or other jungle animals after weaning from her parents could so completely erase her orientation toward people. One is left with accepting Singh's explanation or with creating an alternative explanation. One could hypothesize that Kamala had been kept alive early in life by humans, but confined in a hut or pen with dogs or pet wolves (possibly with the "mother" wolf with which she was found) and away from regular human contacts. In this way her reference was canids, a species she identified with and imitated. Then, not many months prior to her capture (the natives had reported this "man-ghost" in the vicinity for about three or four months), she may have escaped or been released into the jungle with her canine peer(s). Kamala's original rejection by her parent(s) may have resulted from facial (e.g., high jawbones) or eye (i.e., a peculiar "blue glare" at night) characteristics later described by Singh which made her unwanted, yet not so much that someone would kill her or desert her in the jungle to be eaten by tigers.

Kamala was captured on one of Singh's regular missionary tours into the dense jungle of Bengal, where aboriginal people were so uncivilized that they ran from strangers who were dressed in modern garments. In this region there were small, remote villages connected by footpaths too narrow for a cart to traverse. It was about seven miles from one of these villages that Kamala was found living with a family of wolves in a deserted white-ant mound.[4] The date was October 17, 1920. Singh judged Kamala to be about eight years old and speculated that she was originally brought to the den as food when she was an infant.

Immediately after her capture, Kamala was caged and fed raw milk and raw meat, which she ate readily. At the orphanage she was placed under limited confinement and constant supervision to prevent her from running away, which she nevertheless did accomplish a few times. Her social state was regularly equated with that of a lower creature, e.g., "she was virtually an animal" (p 116), and her "nature was wholly that of an animal to all intents and purposes" (p 43). She was aloof and indifferent toward people, but she had a very strong affinity for other mammals such as dogs, cats, and a hyena cub. The animals in turn treated her as a peer. An exception was Kamala's early liking for a small preverbal boy who was just learning to crawl. Singh felt Kamala was attracted to this boy because his mode of movement was like hers and that of other animals. This playful relationship only lasted a few weeks, and then the boy was bitten and roughly scratched by Kamala for no apparent reason. Singh felt this aggression occurred when Kamala learned the boy was "different in nature" from herself.

Mrs. Singh was Kamala's primary and almost sole caretaker. In spite of this, nine months passed before Kamala took a biscuit from Mrs. Singh's hand. Others in the orphanage had difficulty feeding and attending to Kamala in Mrs. Singh's absence. Even though over the years Kamala developed some liking for others in the orphanage, only Mrs. Singh was noticeably missed when absent or greeted with concern upon her return. It appears, furthermore, that Mrs. Singh was the only one who could consistently control Kamala when it was needed. For example, Kamala remained incontinent for five years. After that, if and when she did retreat to void in the bathroom, it was only if Mrs. Singh was present.

It is easy to accept Kamala's "animal nature" when one considers how she behaved. She only began liking to have clothes on after four years; until then she had a loin cloth sewn on so that it could not be easily removed. She ate, slept, and moved like a dog. Kamala only gradually began to eat food prepared for humans. Yet she was found eating the entrails of a dead fowl two years after capture and chasing about to find raw meat hidden in the garden as long as three years after capture. She lapped liquids like a dog. This child's only mode of movement for years was a crawl using arms and legs. In spite of this she was quite mobile and, at times, could "run" faster than the other children. It took five years before she began to walk like other humans, but she continued to crawl as well. Kamala did not like being bathed, but whether she primarily disliked the water or the tactile contact with Mrs. Singh is not clear. After six years she was described at being "somewhat tamed" during her baths.

Kamala was not verbal when captured. The only sound she made was a peculiar cry, or howling, which usually was emitted during the night at periodic intervals. Her first wordlike utterances did not occur until three years after her capture. From then until her death about nine years after removal from the forest, Kamala acquired a limited number of words which were used in simple sentences. Kamala was 17 years old when she died from uremia.

Summary and Implications

There was considerable improvement in Kamala's condition during the last nine years of her life. In spite of these changes, Kamala apparently did not identify with humans. She did not imitate others and did not show signs of either wanting to be with or like other humans. This prevented Kamala from socially developing at a rate consistent with the supportive care and stimulation she received. Many of her social behaviors occurred on the periphery of the group, primarily in Mrs. Singh's presence, and frequently "out of synch" with the other children. Too often it appears that the social behaviors she emitted during the first 6 to 7 years of her stay in the orphanage were merely conditioned actions, just as a trained chimp would emit a number of "social behaviors" such as shaking hands and smiling when greeted. It appears Kamala acquired behaviors like those around her without knowing that she was a person. Walking, for example, did not originate through imitation or training with social rewards. It was shaped with physical constraints and primary reinforcements.

The desire to be like other species members was not present for many years. Yet it is erroneous to conclude that she remained an unattached being. Something about Kamala changed near her death, and it apparently was more than a change in overt behaviors per se. As late as October 10, 1926, Singh wrote "it was hoped that someday in the near future Kamala would be reclaimed as a human child" (p 96). Later he wrote that "Kamala became a new person in the year 1928" (p 118). May we speculate that the *new person* was more than one who had learned some new behaviors—perhaps a basic change in her personality had finally occurred, one which was more "natural" and which had been expected of her for eight years.

The most striking aspect of Kamala's stay in the orphanage was her slow socialization. Even the basic skills which were acquired in a comparatively short time by Anna and M. were not learned by Kamala without systematic instruction over many years. Some may argue that Kamala's continued social retardation could be attributed to her lack of professional care. This may be, but it is unlikely when one considers the constant maternal-like attention and peer support Kamala did receive. Another interpretation could be that Kamala had her affiliation for people extinguished. This is unlikely. Her postcapture behaviors did not hint of a history of harsh treatment, and nonhuman-primate research does not support the notion that affiliations only develop with a nonpunitive caretaker (Arling and Harlow, 1967; Seay, Alexander, and Harlow, 1964).

Apparently there was some foundational obstacle in Kamala's personality which, other than being "tamed" by humans, prevented her from fully benefiting from social interactions with people. If Kamala innately knew that she was a person, one wonders why she did not behave after her capture as if she were pleased to be finally with her own kind. She should have been more receptive to interactions with humans as soon as her fear of them dissipated, but this did not happen as it did with Isabelle and Genie. If Kamala were merely without any species identity, one would think that she would have become a "new person" in less than eight years. A normal child shows copying behaviors and affiliation for other people well before the age of three; thus Kamala should have acquired her species identity in the same or even less amount of time. Kamala's behavior toward people and animals, as well as the long period after capture before she was considered a "person," suggests that she identified with another species (i.e., canids) and this identification had to be dissolved before an identity with her biological species could evolve.

Species Identification and Early Infantile Autism—A Hypothesis

The implication that an identification with the biological species may not exist in some people has more than theoretical significance, and it is more than a convenient though speculative explanation for Kamala's behavior. Certain abnormal individuals in our society may not possess a species identity. As a result, they may not benefit from therapy which involves interactions with people and which is frequently predicated on the notion that the patient wants to please the therapist, get better, and be more like other people.

There is a small group of disturbed children who appear to have inadequately developed or nonexistent species identities. They are diagnosed as suffering from Kanner's syndrome, or the more frequently used Early Infantile Autism (EIA). This syndrome should not be confused, as happens regularly, with autistic children in general.[5] Leo Kanner originally wrote about the EIA child in a 1943 article entitled "Autistic Disturbances of Affective Contact."

The EIA syndrome is unique among other childhood disorders both in the cluster and range of symptoms. Important characteristics include (1) obsessive tendencies and stereotypic movements, (2) unusual cognitive abilities and interests, (3) linguistic problems (e.g., muteness, echolalia, pronominal reversal), and (4) a reduction in spontaneous activities. The children appear normal both physically and intellectually. Socially, however, they are markedly different, and it is this area which sets them apart from all other normal and abnormal children. In essence they do not relate to or affiliate with people. Any social behaviors which are present are usually emitted without affect. Generalized imitation is lacking. Kanner (1943, 1971) has maintained that the fundamental disorder of these children is their inability to relate to people in a normal way. Early signs that social interactions are not progressing as expected include a sparsity of contact-promoting and contact-sustaining behaviors, infrequent primary attachments, and the absence of stranger and separation anxieties. Problems with social functioning magnify as the child matures. An account of a 4-year-old's behavior on a crowded beach follows:

> He would walk straight toward his goal irrespective of whether this involved walking over newspapers, hands, feet, or torsos, much to the discomfort of their owners. The mother was careful to point out that he did not intentionally deviate from his course in order to walk on others, but neither did he make the slightest attempt to avoid them. It was as if he did not distinguish people from things, or at least did not concern himself about the distinction (Kanner and Eisenberg, 1956, p 559).

Social behaviors are learned as the child matures, but they are frequently emitted without the proper meaning.

The prognosis for the EIA child is poor, regardless of what is attempted therapeutically. In a follow-up report on 42 EIA children with a mean age of 14 years (range 8–24 years), only one had shown any real recovery to normality (Kanner and Eisenberg, 1955). The 12 who did show improvement sufficient to live at home and function in the community on a limited basis had schizoid tendencies, disturbed interpersonal relationships, and a "tenuous contact with reality." As adolescents they had retained the primary characteristic of EIA but had lost many of the secondary characteristics. In a 1971 report on the original 11 children, Kanner cited two cases who showed near-normal adjustment and another with marginal adjustment who was living in a sheltered environment. The two "real successes" (ages 34 and 36) were single, living at home, holding nondemanding jobs, and engaging in some community activities. However, in one case (and possibly in the other, although the report was insufficiently detailed), there was a lack of initiative, little social conversation, and no interest in the opposite sex.

The cause of the EIA child's abnormal behavior, especially the neutrality toward people, is not understood. Kanner (1943) originally felt that this social disinterest from birth implied an innate inability to form affective relationships with other people. He later modified this position and stated that the child's neutrality toward people resulted from detached, mechanical care provided by the parents (Kanner, 1954). There is support for this belief in that some parents intentionally minimized social interactions with their child, e.g., Brian (Kanner and Eisenberg, 1956) and Patricia (Kanner, 1949).

There are various reasons why I believe a child with Early Infantile Autism has a faulty or nonexistent species identity. First, there is Kanner's use of an innate factor. This implies that the few children he designated as autistic exhibited some type of uncommon and pervasive deficiency which could not be explained by using any current physiochemical or psychogenic explanation. Second, one would expect the situations that could impair or prevent the development of a species identity not to occur very frequently. This is consistent with the rarity of Kanner's syndrome. Finally, there are my personal experiences with many children diagnosed as either autistic-like or suffering from EIA. Though it is difficult to convey the Gestalt I experienced, my impressions were that the few EIA children did not know that they were people.

Suspicions arise when an unusual position is presented, especially one which is based on nonscientific literature and is not readily testable. Thus it is satisfying when a lay person independently reaches conclusions about her autistic son's species identity which are consistent with my interpretation.

Joan Hundley wrote about her son and other similar children in 1971. The title of her book, "The Small Outsider," is highly suggestive. It is important to note that I first read this book in 1977, *after* I had returned from participating in a symposium which addressed the issue of species identifications (see Preface). At this symposium I had discussed the lack of a species identity in some EIA children (Roy, 1978). It is as difficult to summarize this book as it was to review the works of Kanner or the development of Kamala. A few passages where Mrs. Hundley discusses her son's personal identity and his basic relationship with people are presented.

> I had the feeling that Richard didn't know what I was, that to him I was a thing to be investigated. I remember thinking at the time that neither he nor David knew they were people, knew what they were (p 69).
>
> If the autistic child seems strange to other people, other people seem strange to the autistic child. We at least have the advantage of knowing that he is one of us . . . but he often doesn't know what he is and can't see himself in relation to his environment (p 106).
>
> He gives the impression of being quite remote, as though he is living separately from the rest of us, just looking in on the world (p 139).
>
> His sense of his own identity is still not clear. Just because he can see other people doesn't necessarily mean that he knows he is the same (p 157).

The similarities between my position and that of Mrs. Hundley are unquestionable and need not be elaborated.

Conclusions and Interpretations

There is little disagreement as to whether normal people with whom we interact are actually aware that they are people. They know it and we know it. Our conclusion that other humans possess an identity with their biological species is based on how they behave, rather than on their physical characteristics. Not every member of *Homo sapiens* appears to possess an identification with its species, however. The social behavior of some members of our species leaves little doubt that their basic affiliative preference for humans is severely distorted or nonexistent.

The literature on institutional rearing of children and on social restriction in raising children at home generates one important conclusion—infants need social contacts with other people early in life to insure the normal development of affiliative tendencies for, and subsequent social interactions with, other humans. A positive relationship seems to exist between the degree of postnatal social interactions and the adequacy or normality of one's affiliative tendencies. Though the terms "social contacts" and "social interactions" as used in this chapter are undefined, they need not remain so.

The social inadequacies of many institutionalized children (as well as M., Anna, Albert, and Anne) suggest that visual and auditory contact with the biological species is not sufficient to insure social development without deficiencies. Some degree of tactile–kinesthetic and/or motion-induced stimulation of the child while it is in the presence of another person is necessary as well (see Hofer, 1978). Such stimulation no doubt contributed to Carl's and Isabelle's normality in spite of their privation of distal stimulation. It is interesting to note, however, that visual and auditory contact may be sufficient in maintaining an identity with people, which apparently was the case with Genie.

If we accept the notion of an acquired species identity in humans, and the importance of social stimulation in modalities other than vision and audition, there must be some point (all other things being equal) where insufficient stimulation will prevent the infant's species identity from developing. This point cannot be determined from the available literature, yet to create this state the social contacts must surely be more restrictive than those found in the children who have been reared in institutions. Social contacts in such a situation are also likely to be more restrictive than the experiences of M. and Anna, but not by much, for these children showed long-lasting deficits in their basic affinity for people and in their ability to form social ties with humans. The point at which social stimulation is insufficient to establish a species identity may never be reached if the average child is cared for by humans. Yet there may be instances where neurological inadequacies prevent the nonaverage neonate from benefitting from the necessary stimulation even when it is present. This appears to be the case for some children who later are diagnosed as suffering from EIA.

Most importantly, if a species identity is not an all-or-none state but present in degree, a distorted identity could be reflected in fewer affiliative behaviors and a "weaker" preference for people. It would help if an agreed-upon set of behaviors could be used to determine the presence and/or strength of a species identity, for then one could group these individuals and study the social stimulation they received early in life. Unfortunately such a criterion does not exist.

The literature suggests that the acquisition of a species identity in humans is best described as being optimal-period dependent (i.e., acquired during a particular stage of development), and this period appears to occur sometime within the first year of life. A species identity may eventually be acquired by those organisms without one, but only at a slower rate and possibly only with social experiences which are accentuated in type, degree, or frequency. A normal species identity may eventually be acquired by individuals like Kamala and Anna, but only through an extended process. It even appears that some EIA children acquire a clear understanding of who their conspecifics are, but not as readily as it is learned by children during the optimal period.

It is unclear if people have an innate predisposition to identify with their biological species or whether they possess an innate predisposition to identify with any species which may be available. The case of Kamala with her many canine characteristics would support a generalized tendency. This is not to imply that the proximity-promoting behaviors and stimulus preferences exhibited by neonates are not effective in fostering a situation (i.e., an attachment to caretakers) which maximizes the development of a species identity in humans. It is likely that these behaviors actually reflect this innate predisposition. Regardless of what predisposition exists, it is likely that postnatal social experiences determine in large part the species with which human infants eventually identify.

The literature reviewed in this chapter suggests two things. A species identity in humans is not innate, and a species identity in humans does not automatically evolve postnatally without particular social experiences which, as yet, are poorly understood.

Notes

1. Normality is a relative term, and rearing practices among and within cultures vary significantly. The atypical rearing conditions to be discussed, however, differ sufficiently from those in the reference culture so that most people would classify them as being considerably different.
2. Institutional rearing practices are frequently more socially restrictive than one expects. The reader who is unfamiliar with these will find excellent detailed descriptions by Dennis (1973) and Spitz (1965).
3. This relationship was evident by the picture Genie drew of her mother (see p 325, this volume) and from Kent's report which appears in Appendix III of the text by Curtiss (1977). It thus appears that the use of "wild child" to describe Genie is inappropriate and misleading.
4. A younger girl called Amala was also found in this den. This child, then judged to be about one and a half years old, will not be discussed because her death 11 months after capture limits an evaluation of long-term socialization changes.

5. The term "autistic child" has become a category for any child who cannot be readily diagnosed or to whom authorities, professionals, and parents do not want to assign another label because of stigmas associated with other diagnostic classifications. Facilitating this trend has been the National Society for Autistic Children, whose definition of an autistic child is almost without limitation. A brochure dated 1975 from the society states:

> The term "Autistic Children" . . . shall include persons, regardless of age, with severe disorders of communication and behavior whose disability became manifest during the early developmental stages of childhood. "Autistic Children" includes, but is not limited to, those afflicted with infantile autism (Kanner's syndrome), childhood psychosis, childhood schizophrenia or any other condition characterized by severe defects in language ability (such as profound aphasia) and behavior and by a failure to relate appropriately to others. The autistic child appears to suffer primarily from profound central processing disorders, i.e., a selective impairment of his cognitive and/or perceptual functioning, the consequences of which are manifested by sensory-motor, cognitive, social and language developments; impediments which reduce the ability to understand, communicate, learn and participate in social relationships.

I must again emphasize that I am writing about children with infantile autism (Kanner's syndrome) and not any child loosely called autistic. The reader who fails to make this crucial distinction will find my discussion incomprehensible.

References

Ainsworth, M. D. S. 1964. Patterns of attachment behavior shown by the infant in interaction with his mother. *Merrill-Palmer Q. 10*, 51–58.

Ainsworth, M. D. S. 1967. *Infancy in Uganda: Infant care and the growth of love.* Baltimore: Johns Hopkins Univ. Pr.

Ainsworth, M. D. S. 1972. Attachment and dependency: A comparison. In J. L. Gewirtz (Ed.), *Attachment and dependency.* New York: Wiley.

Ainsworth, M. D. S. and S. M. Bell. 1970. Attachment, exploration, and separation: Illustrated by the behavior of one-year-olds in a strange situation. *Child Dev. 41*, 49–67.

Ainsworth, M. D. S., S. M. Bell, and D. J. Stayton. 1974. Infant–mother attachment and social development: 'Socialisation' as a product of reciprocal responsiveness to signals. In M. P. M. Richards (Ed.), *The integration of a child into a social world.* London: Cambridge Univ. Pr.; Cox and Wyman.

Ambrose, J. A. 1963. The concept of a critical period for the development of social responsiveness in early human infancy. In B. M. Foss (Ed.), *Determinants of infant behavior,* New York: Wiley.

Arling, G. L. and H. F. Harlow. 1967. Effects of social deprivation on maternal behavior of rhesus monkeys. *J. Comp. Physiol. Psychol. 64*(3), 371–377.

Balikov, H. 1960. Functional impairment of the sensorium as a result of normal adaptive processes. *Psychoanal. Study Child 15,* 235–242.

Bandura, A. 1967. The role of modeling processes in personality development. In W. W. Hartup and N. L. Smothergill (Eds.), *The young child: Reviews of research.* Washington, D. C.: National Association for the Education of Young Children.

Bandura, A. and R. H. Walters. 1963. *Social learning and personality development.* New York: Holt, Rinehart, and Winston.

Becker, E. 1972. *The birth and death of meaning.* 2nd ed. London: Penguin.

Bijou, S. W. and D. M. Baer. 1961. *Child development.* Vol. 1. New York: Appleton-Century-Crofts.

Bijou, S. W. and D. M. Baer. 1965. *Child development.* Vol. 2. New York: Appleton-Century-Crofts.

Bowlby, J. 1951. *Maternal care and mental health.* New York: Columbia Univ. Pr.

Bowlby, J. 1958. The nature of the child's tie to his mother. *Int. J. Psychoanal. 39,* 350–373.

Bowlby, J. 1969. *Attachment and loss.* Vol. 1. New York: Basic Books.

Bowlby, J. 1976. Human personality development in an ethological light. In G. Serban and A. Kling (Eds.), *Animal models in human psychobiology.* New York: Plenum Pr.

Brazelton, T. B. 1969. *Infants and mothers: Differences in development.* New York: Delacorte Pr.

Brossard, M. and T. G. Décarie. 1971. The effects of three kinds of perceptual–social stimulation on the development of institutionalized infants: Preliminary report of a longitudinal study. *Early Child Dev. Care 1* (1), 111–130.

Caldwell, B. M. 1964. The effects of infant care. In M. L. Hoffman and L. W. Hoffman (Eds.), *Review of child development research.* New York: Russell Sage Found.

Casler, L. 1961. Maternal deprivation: A critical review of the literature. *Monogr. Soc. Res. Child Dev. 26* (80, 2).

Clarke, A. D. B. and A. M. Clarke. 1960. Recent advances in the study of deprivation. *J. Child Psychol. Psychiatry 1,* 26–36.

Clarke, A. M. and A. D. B. Clarke. 1977. *Early experience: Myth and evidence.* New York: Free Press.

Count, E. W. 1977. *Being and becoming human*. New York: Van Nostrand-Reinhold.

Curtiss, S. 1977. *Genie*. New York: Academic Pr.

Davis, K. 1940. Extreme social isolation of a child. *Am. J. Soc. 45*, 554–565.

Davis, K. 1947. Final note on a case of extreme isolation. *Am. J. Soc. 52*, 432–437.

Dennis, W. 1973. *Children of the creche*. New York: Appleton-Century-Crofts.

Erikson, E. H. 1950. *Childhood and society*. New York: Norton.

Fantz, R. L. 1965. Visual perception from birth as shown by pattern selectivity. *Ann. N.Y. Acad. Sci. 118*, 793–814.

Ferguson, L. R. 1970. *Personality development*. Belmont, Cal.: Brooks Cole.

Fitzgerald, H. E. 1968. Autonomic pupillary reflex activity during early infancy and its relation to social and nonsocial visual stimuli. *J. Exp. Child Psychol. 6*, 470–482.

Freedman, D. A. and S. L. Brown. 1968. On the role of coenesthetic stimulation in the development of psychic structure. *Psychoanal. Q. 37*, 418–438.

Freud, S. 1953. *Three essays on the theory of sexuality*. London: Hogarth Pr.

Fromkin, V., S. Krashen, S. Curtiss, D. Rigler, and M. Rigler. 1974. The development of language in Genie: A case of language acquisition beyond the "critical period." *Brain and Language 1*, 81–107.

Gewirtz, J. L. 1961. A learning analysis of the effects of normal stimulation, privation, and deprivation on the acquisition of social motivation and attachment. In B. M. Foss (Ed.). *Determinants of infant behavior*. New York: Wiley.

Gewirtz, J. L. 1972. *Attachment and dependency*. New York: Wiley.

Gewirtz, J. L. and K. G. Stingle. 1968. Learning of generalized imitation as the basis for identification. *Psychol. Rev. 75*, 374–379.

Goldfarb, W. 1943. The effects of early institutional care on adolescent personality (graphic Rorschach data). *Child Dev. 14*, 213–223. (a)

Goldfarb, W. 1943. The effects of early institutional care on adolescent personality. *J. Exp. Educ. 12*, 106–129. (b)

Goldfarb, W. 1943. Infant rearing and problem behavior. *Am. J. Orthopsychiatry, 13*, 249–265. (c)

Goldfarb, W. 1944. Effects of early institutional care on adolescent personality: Rorschach data. *Am. J. Orthopsychiatry, 14*, 441–447.

Goldfarb, W. 1945. Psychological privation in infancy and subsequent adjustment. *Am. J. Orthopsychiatry, 15*, 247–255.

Gray, P. H. 1958. Theory and evidence of imprinting in human infants. *J. Psychol. 46*, 156–166.

Haith, M. M., T. Bergman, and M. J. Moors. 1977. Eye contact and face scanning in early infancy. *Science 198* (4319), 853–855.

Halliday, J. L. 1948. *Psychosocial medicine.* New York: Norton.

Hampson, J. L. and J. C. Hampson. 1961. The ontogenesis of sexual behavior in man. In W. C. Young (Ed.), *Sex and internal secretions.* Vol. 2. Baltimore: Williams and Wilkins.

Harlow, H. F. 1960. Primary affectional patterns in primates. *Am. J. Orthopsychiatry, 30,* 676–684.

Hess, E. H. 1959. Imprinting. *Science 130,* 133–141.

Hess, E. H. 1973. *Imprinting.* New York: Van Nostrand.

Hofer, M. A. 1978. Hidden regulatory processes in early social relationships. In P. P. G. Bateson and P. H. Klopfer (Eds.), *Perspectives in ethology.* Vol. 3. New York: Plenum Pr.

Hundley, J. M. 1971. *The small outsider.* New York: Ballantine Books.

Jacobson, E. 1964. *The self and the object world.* New York: International Univ. Pr.

Jouventin, P., G. Pasteur, and J. P. Cambefort. 1977. Observational learning of baboons and avoidance of mimics: Exploratory tests. *Evolution 31* (1), 214–218.

Kagan, J. 1970. The determinants of attention in the infant. *Am. Sci. 58,* 298–306.

Kagan, J. 1976. Resilience and continuity in psychological development. In A. M. Clarke and A. D. B. Clarke (Eds.), *Early experience: Myth and evidence.* New York: Free Press.

Kanner, L. 1943. Autistic disturbances of affective contact. *Nerv. Child 2,* 217–250.

Kanner, L. 1949. Problems of nosology and psychodynamics in Early Infantile Autism. *Am. J. Orthopsychiatry 19,* 416–426.

Kanner, L. 1954. To what extent is Early Infantile Autism determined by constitutional inadequacies? *Proc. Assoc. Nerv. Ment. Dis. 33,* 378–385.

Kanner, L. 1971. Follow-up study of eleven autistic children originally reported in 1943. *J. Autism Child. Schizophr. 1,* 119–145.

Kanner, L. and L. Eisenberg. 1955. Notes on the follow-up studies of autistic children. In P. Hoch and J. Zubin (Eds.), *Psychopathology of childhood.* New York: Grune and Stratton.

Kanner, L. and L. Eisenberg. 1956. Early Infantile Autism. *Am. J. Orthopsychiatry, 26,* 556–566.

Kohen-Raz, R. 1968. Mental and motor development of kibbutz, institutionalized and home-reared infants in Israel. *Child Dev. 39,* 489–504.

Koluchová, J. 1972. Severe deprivation in twins: A case study. *J. Child Psychol. Psychiatry, 13,* 107–114.

Koluchová, J. 1976. A report on the further development of twins after severe and prolonged deprivation. In A. M. Clarke and A. D. B. Clarke (Eds.), *Early experience: Myth and evidence.* New York: Free Press.

LaBarre, W. 1954. *The human animal.* Chicago: Univ. Chicago Pr.

Langmeier, J. 1972. Personalities of deprived children. In F. J. Monks, W. W. Hartup, and J. deWit (Eds.), *Determinants of behavioral development.* New York: Academic Pr.

Lewis, M. 1969. Infants' responses to facial stimuli during the first year of life. *Dev. Psychol. 1,* 75–86.

Mason, M. K. 1947. Learning to speak after six and one-half years of silence. *J. Speech Disord. 7,* 295–304.

Matejcek, A. 1967. Personality development of institutionalized children. *Psychol. patopsychol. dietata 3,* 17–31.

Maurer, D. and P. Salapatek. 1976. Developmental changes in the scanning of faces by young infants. *Child Dev. 47,* 523–527.

Meltzoff, A. N. and M. R. Moore. 1977. Imitation of facial and manual gestures by human neonates. *Science 198*(4312), 75–78.

Miller, N. E. and J. Dollard. 1941. *Social learning and imitation.* New Haven, Conn.: Yale Univ. Pr.

Mitchell, G. 1975. What monkeys can tell us about human violence. *The Futurist* (April) 75–80.

Money, J. and A. A. Ehrhardt. 1972. *Man and woman, boy and girl: The differentiation and dimorphism of gender identity from conception to maturity.* Baltimore: Johns Hopkins Univ. Pr.

Montagu, M. F. A. 1950. *On being human.* New York: Abelard-Schuman.

Montessori, M. 1936. *The secret of childhood.* New York: Longmans, Green.

Morrison, H. L. and D. M. Brubakken. In press. Social isolation and deprivation: An environment of rehabilitation. In J. Money and G. Williams (Eds.), *Traumatic neglect and child abuse.* Baltimore: Johns Hopkins Pr.

Murray, H. A. 1938. *Explorations in personality.* New York: Oxford Univ. Pr.

Mussen, P. H., J. J. Conger, and J. Kagan. 1974. *Child development and personality,* 3rd ed. New York: Harper and Row.

Piaget, J. 1964. Development and learning. In R. Ripple and V. Rockcastle (Eds.), *Piaget rediscovered.* Ithaca, N.Y.: Cornell Univ. Pr.

Prescott, J. W. 1975. Body pleasure and the origins of violence. *The Futurist* (April) 64–74.

Provence, S. and R. C. Lipton. 1962. *Infants in institutions.* New York: International Univ. Pr.

Rheingold, H. 1956. The modification of social responsiveness in institutional babies. *Monogr. Soc. Res. Child Dev. 21* (2).

Rheingold, H. 1960. The measurement of maternal care. *Child Dev. 31,* 565–575.

Rheingold, H. and N. Bayley. 1959. The later effects of an experimental modification of mothering. *Child Dev. 9,* 1–40.

Richards, M. P. M. 1974. First steps in becoming social. In M. P. M. Richards (Ed.), *The integration of a child into a social world.* London: Cambridge Univ. Pr.; Cox and Wyman.

Rigler, M. 1972. Adventure: At home with Genie. Paper presented at the 80th Annual Convention of the American Psychological Association, Honolulu, Hawaii.

Robson, K. S., F. A. Pederson, and H. A. Moss. 1969. Developmental observations of dyadic gazing in relation to the fear of strangers and social approach behavior. *Child. Dev. 40,* 619–627.

Rollman-Branch, H. S. 1960. On the question of primary need. *J. Am. Psychoanal. Assoc. 8,* 686–702.

Roy, M. A. 1978. Consequences of atypical rearing experiences in humans. *Resources Educ.* (July) *150,* 761.

Rutter, M. 1972. *Maternal deprivation reassessed.* New York: Penguin Books.

Saltz, R. 1973. Effects of part-time "mothering" on IQ and SQ of young institutionalized children. *Child Dev. 44,* 166–170.

Sander, L. W., H. L. Julia, G. Stechler, and P. Burns. 1972. Continuous 24-hour interactional monitoring in infants reared in two different caretaking environments. *Psychosom. Med. 34,* 270–282.

Schaffer, H. R. 1963. Some issues for research in the study of attachment behavior. In B. M. Foss (Ed.), *Determinants of infant behaviour II.* New York: Wiley.

Schaffer, H. R. 1971. *The growth of sociability.* London: Penguin.

Schaffer, H. R. and P. E. Emerson. 1964. The development of social attachments in infancy. *Monogr. Soc. Res. Child Dev. 29* (3, Serial no. 94).

Schmidt, W. H. O. 1973. *Child development.* New York: Harper and Row.

Scott, J. P., J. M. Stewart, and V. J. DeGhett. 1974. Critical periods in the organization of systems. *Dev. Psychobiol. 7*(6), 489–513.

Sears, R. R. 1965. Development of gender role. In F. A. Beach (Ed.), *Sex and behavior.* New York: Wiley.

Seay, B. M., B. K. Alexander, and H. F. Harlow. 1964. Maternal behavior of socially deprived rhesus monkeys. *J. Abnorm. Soc. Psychol. 69,* 345–354.

Shotter, J. 1974. The development of personal powers. In M. P. M. Richards (Ed.), *The integration of a child into a social world.* London: Cambridge Univ. Pr.; Cox and Wyman.

Singh, J. A. L. and R. M. Zingg. 1942. *Wolf children and feral man.* Hamden, Conn.: Shoe String Pr. (Reprinted in 1966 by Harper and Row.)

Spitz, R. A. 1945. Hospitalism: An inquiry into the genesis of psychiatric conditions in early childhood. *Psychoanal. Study Child 1,* 53–74.

Spitz, R. A. 1946. Hospitalism: A follow-up report. *Psychoanal. Study Child 2,* 113–117.

Spitz, R. A. 1965. *The first year of life.* New York: International Univ. Pr.

Stern, D. 1977. *The first relationship.* Cambridge, Mass.: Harvard Univ. Pr.

Strayer, F. F. 1976. Learning and imitation as a function of social status in macaque monkeys (*Macaca nemistrinia*). *Anim. Behav. 24*, 835–848.

Taketoma, Y. 1968. The application of imprinting to psychodynamics. *Sci. Psychoanal. 12*, 166–183.

Thomas, A. and S. Chess. 1970. Behavior individuality in childhood. In L. R. Aronson, E. Tobach, D. S. Lehrman, and J. S. Rosenblatt (Eds.), *Development and evolution of behavior.* San Francisco: Freeman.

Tizard, B. and J. Rees. 1975. The effect of early institutional rearing on the behaviour problems and affectional relationships of four year old children. *J. Child Psychol. Psychiatry 16*, 61–73.

Wolff, P. H. 1963. Observations on the early development of smiling. In B. M. Foss (Ed.), *Determinants of infant behaviour II.* New York: Wiley.

Wrong, D. H. 1961. The oversocialized conception of man in modern sociology. *Am. Soc. Rev. 26*, 183–193.

Part III

Commentaries

Chapter 15

Becoming Human: An Epigenetic View

Inge Bretherton
Mary D. S. Ainsworth

Is species identity a term that can usefully be applied to humans? Before we can discuss that question we must consider how the concept has been used by the various investigators who have contributed chapters to this volume. The term has, in fact, been used in two quite distinct ways: first, as operationally defined, and second, to refer to an affective-cognitive process.

Definitions of Species Identity

An operational definition of species identity merely requires a list of criterial behaviors, i.e., the organism is said to have an appropriate species identity if it chooses the correct partner for mating and displays the appropriate reproductive, parenting, and other species-typical social behaviors. Although this is sometimes described as "a fish knowing that it is a fish," nothing is implied about cognitive processes other than that the organism has the capacity to direct the appropriate behavior to the appropriate conspecific at the appropriate time. The operational definition of species identity allows us to compare species along the whole phylogenetic scale without concerning ourselves, for the time being, with how species-typical behavior and species recognition are achieved. It also allows us to compare a variety of species in terms of the disruption of species identity without concerning ourselves with the underlying cognitive–affective processes. We may, using the concept in this way, distinguish the following ways in which species identity may be distorted:

1. The animal directs its own species-typical social behaviors to an inappropriate species. Examples:
   a. A jackdaw courts a human being with jackdaw courting behavior, i.e., attempts to feed the human partner with worm pulp (Lorenz, 1952, quoted in Chapter 1, this volume).
   b. Male rhesus monkeys reared with a cloth surrogate display the complete mating sequence toward the surrogate when exposed to it in adulthood, although attempts at mating with conspecifics are nonexistent or inappropriate (Deutsch and Larsson, 1974, as quoted in Chapter 1, this volume).
   c. The chimpanzee Lucy (Chapter 12, this volume) vocalizes to her human mother with chimpanzee-specific vocalizations.
2. The animal directs its social behavior to an inappropriate species, using the social behavior typical of the target species, not that of its natal species. Examples:
   a. The chimpanzee Lucy converses with her human family in American Sign Language.
   b. The human child Kamala (Singh and Zingg, 1942) not only directs all her social behavior to dogs and doglike creatures, but eats foods which dogs eat, laps water like a dog, walks on all fours like a dog, and howls like a dog at night; she prefers the company of dogs and behaves like them.
3. The animal directs social behavior to an inappropriate species even though it has neither acquired the social behavior typical of the target species, nor knows how to perform the behavior in its own species-typical way. There is, however, strong evidence that the animal is motivated to perform the behavior. Example: the chimpanzee Lucy (Chapter 12) had not developed chimpanzee-specific sexual behavior (e.g., presenting), but gave ample evidence of being attracted to nonfamily human males when in estrus by seeking genital contact in a manner typical of neither chimp nor human.
4. The organism fails to develop any species identity whatsoever and remains asocial (the term asocial here is not to be confused with solitarization in cats, which is in fact a form of

socialization typical of felines). Examples: some isolate-reared rhesus monkeys who have not been rehabilitated; some of the severely isolated children described in Chapter 14.

It is already obvious that the behavior of the chimpanzee Lucy falls into several of these categories, i.e., the disturbance of her species identity can be described in a number of different ways according to the particular behavior under study. Similarly, failing to develop species identity may also apply to some behavioral systems, but not others (e.g., a cat which fails to become solitarized fails to develop one aspect of its species identity; see Chapter 7, this volume). This leads one to consider the second definition of species identity, namely the use of the term to refer to an affective–cognitive process. It has only been used in this way by authors who describe the behavior of higher vertebrates and especially that of primates.

Species identity viewed as an affective–cognitive process implies that the animal will *not* direct one class of social behavior to one species and another type of social behavior to a second species, but rather that something like imprinting has taken place so that the target species is, as it were, fixed. One could speak, for example, of a young bird as having species identity when both filial and sexual imprinting are completed. In other words, when we are interested in whether an individual organism has or has not achieved species identity in the affective–cognitive sense, we are asking if the animal knows with which species it should interact and associate.

We do not believe that the animal must have a self-concept in order to identify with a target species. What is necessary is that the organism have a concept of the target species. For example, Scott (Chapter 6, this volume) reports that dogs raised exclusively with humans seem to lose all desire to associate with dogs, even to the point of not performing innate behavior patterns with conspecifics. Do dogs who are raised with humans "know" that *they* are human or do they merely have the concept of the *target species* as being human? We are inclined to think the latter. Questions about self-awareness in animals have only been raised about primates (Chapters 11 and 12, both in this volume).

Species identity becomes more complex when we look at organisms that have self-awareness. Gallup (Chapter 11, this volume) has demonstrated convincingly that chimpanzees at the very least have an awareness of their body-self. Temerlin (in Chapter 12) describes a similar self-awareness in Lucy, yet Lucy's non-human physical appearance, of which she was very much aware, did not prevent her from identifying with humans and from showing fearful behavior toward the first chimpanzee to whom she was exposed. Lucy seemed fully at home in a human household, using human utensils, eating like a human, and respecting human possessions. Another telling example of Lucy's human identification was her "embarrassment" at having begun to eat monkey chow, a fact which she tried to hide from her human companions when she was provided with a chimpanzee companion at the age of 4½.

Having a sense of self is clearly not to be equated with having a biologically appropriate species identity. Lucy and Gallup's chimpanzees showed very similar mirror behavior, but Lucy did not let her discrepant appearance deter her from identifying with human companions. In higher primates the emergence of a sense of self may nevertheless be a prerequisite for developing some species identity.

When thinking about species identity as an affective–cognitive process, it becomes obvious that how the animal behaves and how the animal comes to know its conspecifics are two separate questions. This is not possible if we are content with an operational definition of species identity. The distinction becomes important in interpreting the behavior of the isolated and feral children described in Chapter 14, this volume. Before discussing this point further, we would like to devote some time to considering how normal species identity emerges in human beings.

The Development of Species Identity in Humans

A useful framework is that provided by Erikson (1968) in his book *Identity, Youth and Crisis,* in which he outlines how early psychosocial development contributes to achieving a human identity. We do not claim that Erikson conceived of his theory in terms of species identity, but merely that it can be so viewed.

According to Erikson, the child negotiates a number of crises ("crisis" is used not in the sense of catastrophe, but in the sense of turning point or in the sense of some issue to be resolved) during the first five years of life.

Infancy and the Mutuality of Recognition

During the first year (corresponding to Freud's oral stage) the child's mode of interacting with the world is "taking in" in a number of different ways, not only through the mouth but with all the senses. At this time the infant is dependent on the mothering person to furnish what it needs in terms of affection, food, stimulation, and interaction. This is not to say that the child is a passive recipient of what the world has to offer, but that the child's ability to influence the social world is very much dependent on the cooperativeness of the child's caregiver. The child learns to accept the ministrations of others and at the same time learns that it can cooperate with and influence these others (mainly the mother). Through this mutual regulation of mother and child, the child develops not only a sense of trust (as opposed to a predominance of mistrust), it also develops a sense of self and other:

> What would be considered to be the earliest and most undifferentiated sense of identity? I would suggest that it arises out of the encounter of maternal person and small infant, an encounter which is one of mutual trustworthiness and mutual recognition (p 105).

Early Childhood and the Will to Be Oneself

During the next stage (corresponding to Freud's anal stage) the issue to be resolved is the development of a sense of autonomy, a sense of being able to do something on one's own, a sense of mastery within the framework of parental support and discipline. This issue is precipitated by the child's rapid muscular maturation and by the rapid development of language and cognitive abilities. If the child does not learn to exercise self-control because of too rigid or too early training or because of too little parental control, the child may be burdened with a lasting capacity for self-doubt. The social environment must back the child up in the will to be himself or herself.

Childhood and the Anticipation of Roles

Having become aware that he is an autonomous person, the child must now decide what sort of person to become. New achievements in locomotion and language allow the child to play out and imagine himself or herself in many different roles. The hallmark of this period is a sense of initiative, but an initiative which begins to be governed by conscience. The child hears the inner voice of self-observation, self-guidance, and self-punishment, a necessary part of human identity.

> A comparative view of child training, however, suggests a fact most important for identity development, namely, that adults by their own example and by the stories they tell of life and what to them is the great past, offer children of this age an eagerly absorbed ethos of action in the form of ideal types and techniques, fascinating enough to replace the heroes of picture book and fairy tale (Erikson, 1968, p 121).

At the end of this stage (at about five years of age), as Erikson describes it, one could say that human children have attained species identity. At this point the child is set not only as a member of the species here and now, but as someone who will one day perpetuate it.

Erikson's view is an epigenetic one, where the concept of epigenesis is borrowed from embryology. It implies that an organism has some basic ground plan of development in which different issues gain ascendancy at particular stages in development. An epigenetic view does not view innate factors and learning as opposites, but rather proposes that for healthy development the right input must be present at more or less the right point in time. Later stages of development will be adversely affected if some of the earlier stages have not been successfully negotiated, although the organism can sometimes be put back on the right course later on. The epigenetic point of view is very compatible with the ethological concepts of "sensitive period" and "learning propensity." As Erikson phrases it: "Personality therefore can be said to develop according to steps predetermined in the human organism's readiness to be driven toward, to be aware of, and to interact with a widening radius of significant individuals and institutions" (Erikson, 1968, p 93).

How does Erikson's clinically derived framework correspond to the findings of developmental psychology? Evidence (summarized by Bowlby, 1969) suggests that human infants, though they do not "know" their conspecifics, have at their disposal a repertoire of behaviors and propensities which, in the ordinary, expectable environment, serve to channel them into interaction with a human caregiver whose behavior meshes with their own.

Human caregivers tend to pick up, hold, and rock infants, and infants, in turn, are particularly easily soothed by the vestibular stimulation inherent in picking up and rocking, and by the ventroventral contact that often occurs while they are being held; caregivers talk to their infants, and infants are particularly responsive to human voices; infants cry, and human caregivers attempt to alleviate their distress; infants orient to human faces and voices, which leads the caregiver to continue interacting with the infant. This does not, of course, imply that the baby's behaviors are intentionally directed toward an adult. It merely means that the baby is "prepared" for interaction by being especially responsive to the package of stimuli normally associated with the behavior and appearance of human partners.

> A neonate may be an essentially asocial creature, in the sense of not being capable as yet of truly reciprocal social relationships and of not yet having the concept of a person. However, the nature of his early interactive behavior is such that it is increasingly difficult to avoid the conclusion that in some sense the infant is already prepared for social intercourse. Not that this should surprise us: if an infant arrives in the world with a digestive system to cope with food and a breathing apparatus attuned to the air around him, why should he not also be prepared to deal with that other essential attribute of his environment, people? (Schaffer, 1977a, p 5)

In addition to protosocial behavior and propensities human infants also appear to have a remarkable capacity for adaptation and reciprocal regulation, which is evident even within the first ten days of life, as Sander (1977) demonstrated. The sleeping and feeding rhythms of newborns (given up for adoption), who spent the first ten days of their life rooming in with a nurse who acted as a substitute mother, were severely disrupted upon being transferred to a second substitute mother at the age of 10 days. In another study Sander (1977) showed that 7-day-old infants have acquired not only rhythms that are shared with the mother figure, but rules of interaction as well. Although the infants did not pay attention to the fact that the mother was wearing a mask when she walked about the room and when she bent over the crib, they appeared startled at the moment the mother put them in the feeding position and they looked up at her masked face. Papoušek and Papoušek (1975) showed that at 4 months of age, infants responded with distress to gross violations of interaction rules. The infants began to turn away from their mothers upon her return after a brief absence if the leave-taking took place in an unusual or abrupt fashion (by sneaking out of the room while the light was turned off) instead of disengaging gradually, as would have been their normal practice. There is now a whole literature in which early mother–infant interaction is described in terms of dialogue or protoconversation and in which the mutual regulation of infant by the mother and mother by the infant is investigated in minute detail (see Schaffer, 1977b; Lewis and Rosenblum, 1977, for reviews).

We would not claim that the infant abilities which have just been described are evidence for the attainment of species identity, but they are evidence for a growing *capacity* for species identity. During the course of mutual adaptation of mother and infant, infants come to prefer interaction with the mothering person: they greet her more enthusiastically upon her return after a brief absence from the room (Stayton, Ainsworth, and Main, 1973), they tend to protest when she (but not a familiar visitor) leaves the room, and they tend to engage in longer and livelier face-to-face interaction with her (Blehar, Lieberman, and Ainsworth, 1977). Ainsworth and her colleagues (1978) have found great variations in the goodness of fit which mothers and infants achieve and, in fact, have discovered that the preference for the mother is much less evident in those mother–infant pairs where interaction is not harmonious. The reciprocal meshing of mother–infant interaction has been studied by Ainsworth et al. with a view to describing the infant's growing attachment to the mother, but the same behavior can also be understood in terms of identity formation. In order to achieve a smooth interaction with another person, one has to be able to predict the partner's behavior, and this in turn requires that one have an internal representation (at least in a primitive form) of the partner's behavioral programs. Present evidence does not suggest that young infants are capable of symbolic representation, i.e., they do not appear capable of using internal representations of mother independently of her actual presence. Schaffer (1971) makes this same point. Yet one must surely claim that infants have what one of us has called a "sensorimotor representation" of the mother (Bretherton, in press) which enables them to make short-term predictions of her behavior and thus to mesh their behavior with hers. To have an internal representation of the partner (even if it is not yet symbolic) is surely the first step to identification and self–other differentiation. The infant's growing attachment to the mother goes hand in hand with a growing capacity for species identification, for developing a sense of trust (or a sense of security), and for "knowing" the other through interaction.

A turning point is reached around six to nine months when infants begin to be able to conceive of objects as separate from themselves (when they can find an object which has been completely hidden under a cloth), when infants start to grieve for an absent mother instead of merely protesting her departure, and when they tend to become somewhat wary of strangers. This turning point is marked not only by the first beginnings of a capacity for symbolic representation but also by the first appearance of independent locomotion. The infant who, for the first time, can voluntarily separate from the caregiver can now also carry around an internal representation or working model of the caregiver.

Although Erikson (1968) would put the beginning of the stage of autonomy somewhat later, when the child can launch out into the world by independent walking and when he becomes ready for toilet training, we believe that there is some justification for conceiving of it as beginning with the onset of crawling, of object permanence, and of an attachment which endures over time. The mother now becomes the secure base (Ainsworth, 1973) from which the child can explore the environment somewhat on his own while still being protected by the mother. Only a short time after the discovery of object permanence (usually preceded by the concept of person permanence; Bell, 1970) comes the dim realization that persons are independent agents with wills of their own who can be activated by intentional signals. No longer do infants have to take the mother's hand and physically move it to perform the desired action; instead, they convey messages by looking up at the partner expectantly and by giving intentional and ritualized signals, such as pointing to the desired object while looking up at the adult, or making opening and closing motions of the hand instead of merely grabbing the object from the hand of an adult (Bretherton and Bates, 1978). The realization of the other as an independent agent also brings the realization that one can refuse to cooperate with an adult by saying no (Spitz, 1965). Thus self-assertion comes to the forefront, but only in the framework of a secure base (without it there is no self-assertion but merely depression and withdrawal, as the studies on isolated children show). Imitation of those persons with whom the infant interacts daily is yet another facilitator of species identity, especially the growing capacity to imitate actions which are novel and which the child cannot see himself perform (Piaget, 1962). It is interesting, in this connection, that infants just under a year of age prefer to imitate "meaningful actions"; that is, actions which they have seen others perform, such as drinking from a cup but not from a toy car, or driving a toy car around on the high-chair table but not driving a cup (Killen and Uzgiris, 1978). Already imitation is socialized and not merely a generalized capacity to copy just any behavior. The increased capacity to imitate and the motivation to imitate the actions of others also lead to more elaborate and precise representation of the social and nonsocial environment.

What do internal, symbolic representations (or working models of the environment) have to do with species identification? As Barkow (1977) suggests, man can be considered the mapmaking species. In constructing cognitive maps (or models) of the environment which can be used to predict future events and plan future actions, those aspects which are most salient become most elaborate, including in early childhood working models of attachment figures and later of playmates. The ability to generate cognitive models of significant others is part and parcel of the general human symbolic capacity. In humans cognition is not just superimposed as an extra ability upon other already existing capacities, but is an integral part of their social–emotional being—to build internal models of the social world is the way in which humans achieve species identification.

At the end of the second year the human child usually has a firmly established sense of self separate from others and is now beginning to understand different roles. Money and Ehrhardt (1972) suggest that gender identity is already established at this time, although cognitive gender constancy is a much later attainment (Kohlberg, 1966). This early establishment of gender identity would put the beginning of the Erikson stage of initiative earlier than he suggests, i.e., before the age of three. In the third year children also become aware of the different roles of adults and children (the roles of playmate, partner, worker, and parent). Their ideas of the sex role are very stereotypic, even in communities where adult sex roles are flexible (Katz, 1978), but they are increasingly looking forward to performing adult roles, such as mommy (less often daddy), policeman, nurse, and doctor, that is, those adult roles whose performance is most evident to them. In addition, children now begin to master the rules of social interaction at a much more sophisticated level, i.e., they become aware of polite speech (Bates, 1976) and even address baby talk to younger children (Shatz and Gelman, 1973). They also begin to acquire internalized morality, which Freud attributed to the internalization of the parent qua authority figure. Barkow (1977), in an effort to cast Freud in modern terms, links this internalization not with the Oedipus complex but with the child's capacity to create symbolic representations of the parents. If the child possesses internal models of the parents which are available for consultation even in the absence of the actual parents, then it would not be unreasonable to suppose that these internal models are sometimes perceived to give orders as to right or wrong behavior—hence internalized morality and a new rendering of Freud in Piagetian terms.

The material just presented is compatible with Erikson's outline of the stage of initiative. It signals the attainment of species identity in the sense of an individual not only preferring the company of particular conspecifics, but also in the sense of having acquired the behavior and the behavioral rules which will allow for reasonably smooth interaction with those conspecifics in many different situations and roles, and in the sense of seeing oneself as a future member of adult society. During the latter stages of development of identity as proposed by Erikson, the emphasis is not so much on species (human) identity as on individual identity: "Where do I fit in?"

After this brief account of what is more usually described as the development of attachment, as social development, or as socialization, but which could also justifiably be regarded as the acquisition of human species identity, let us examine which of these processes go awry in children who have been atypically reared.

Distorted Species Identity in Humans

Several categories of atypical rearing are discussed in Chapter 14 of this volume: institutional rearing, rearing in isolation, and rearing with an inappropriate species. In all except the very worst instances of isolation there was at least some partial socialization or (in the case of Kamala) resocialization, depending on the degree to which rearing in the first few years was atypical and the age at which it came to an end.

Evidence from Tizard and Rees (1975) would suggest that some development of basic trust and of the capacity for attachment is possible in an institution which offers cognitive stimulation and has a high staff-to-child ratio, but which does not especially encourage the formation of close emotional ties between the children and their caregivers. "We don't encourage the children to become too attached, it isn't fair to them, it isn't fair to us," as the matron of one residential nursery was reported to have said. If children reared in such an institution were adopted before age 4½, they tended to be normal in terms of cognitive development and also tended to form strong attachments to their adoptive parents, although one-third were described as being somewhat overfriendly with strangers and exceptionally affectionate toward their adoptive parents. These abnormalities are very slight when compared to those of the children described in earlier studies by Spitz (1945) and Goldfarb (1945). Clearly these children were socialized beings, even though they were somewhat withdrawn, shy, and insecure as long as they still lived in the institution.

More surprising is the story of Isabelle (reported in Chapter 14, this volume). Her environment was unstimulating in the extreme. The only thing that Isabelle had going for her was a good relationship with her deaf-mute mother. Within 18 months of her discovery at the age of 6½, Isabelle was indistinguishable from normal children at her elementary school. In this way she was quite unlike Genie (Curtiss, 1977).

Genie

Genie is presumed to have developed an attachment to her mother during the first 20 months of her life, which were not optimal but bore no resemblance to the miserable years which were to follow. We may speculate that Genie's supply of early basic trust was sufficient to carry her through the next 11 years of incarceration in a bleak room until she was discovered as an adolescent. Genie resembles Isabelle in that she had one kindly person, her mother, to whom she was attached. Unlike Isabelle Genie did not have continuous access to that mother and, in addition, suffered much abuse from her father and brother. Still, her moving drawing which is reproduced here (p 325) attests to the fact that Genie had species identity and she saw herself as a human being.

It is in Genie that we can clearly see that for human beings a division between species identity and species-typical social behavior is clearly possible. As Curtiss (1977) said—and this refers of course to Genie at the time of writing about her—Genie remained in many ways an unsocialized being. She had great difficulty in learning the rules of social interaction, even though her intelligence was excellent in many other areas. She had acquired language, but would generally take her partner's hand and place it in the position for action rather than making a requesting gesture—a behavior normal infants outgrow before the end of the first year (Bates et al., in press). It was also difficult for Genie to conceive of what was socially proper and what was taboo. She was reported to walk up to unfamiliar persons who had possessions she desired and take them away or hang on to them if the desired object was part of the person's clothing. She was also reported to have removed a sanitary napkin in a supermarket and until 1977 to have masturbated almost constantly, both indoors and out.

Genie acquired language at a time assumed to be beyond the sensitive period for its development, but she learned to speak in a manner peculiar both syntactically and socially. She often failed to acknowledge requests and summons (as if not spoken to), and conversation with her had to be kept going by the partner, not by Genie, who not only seemed to be deficient in initiative but also could not produce correct pitch and intonation except when she was imitating directly. She spoke as little as possible. Her grammar was deficient in a number of ways, but on the other hand she had acquired an extensive vocabulary.

Unlike many autistic children, Genie was able to describe her emotional states—angry, mad, sad, and happy—and to talk about "loving" "hating," and "thinking about." She could talk, furthermore, about events which occurred long before she acquired language. Although it took five years from her discovery, she mastered the distinction between "I," "me," "mine" and "you" and "yours," unlike many autistic children (Schuler, 1978).

Her cognition in terms of holistic processing and perceiving was excellent, but poor when it came to sequential analysis. This, as well as the peculiar speech and syntax, led Curtiss to the conclusion that Genie was a right-hemisphere thinker (which was supported also by the results of dichotic listening tasks). This may have occurred because Genie acquired language when left-hemisphere differentiation for language was no longer possible.

The poor sociolinguistic abilities which Genie displayed, however, in conjunction with her other unsocialized behavior, may be attributed to her social rather than linguistic deprivation. Is there a sensitive period for the acquisition of socialized behavior, i.e., for the acquisition of social rules of interaction on which she missed out? One comes away with the impression that this is a possibility, because Genie's unsocialized behavior never seemed naughty or mean, just the result of an inability to understand what was required. In fact, Genie's difficulty with socialization reminded us of the lab-reared chimp Marianne, who became Lucy's companion in attempts to reorient Lucy to her own species (Chapter 12, this volume). Marianne interacted with humans, but could not be socialized as a member of a human household at the age of 4½ years, as was possible with Lucy, who lived with the Temerlins almost from birth.

Despite her incomplete socialization, Genie clearly saw herself as a member of the human race (see Figure 1). She was a person who had developed a capacity for attachment to conspecifics and who, moreover, also had the capacity for eliciting deep attachments from others despite her sometimes repulsive behavior and the missed opportunity for normal language. As Fraiberg's (1977) work with blind children shows, a seemingly less severe deprivation (e.g., deprivation in only one sensory modality) can wreak havoc with a child's developing identity in quite unforeseen ways.

Blind Children and Autism

Fraiberg (1977) reports that 20 percent of all congenitally blind children with no other abnormalities come eventually to be labeled "autistic," i.e., show symptoms usually associated with early infantile autism. Roy in Chapter 14 of this volume makes the case that children with Early Infantile Autism have neither species identity nor self-identity (they do not think of themselves as human), but it is still a mystery what underlying deficits lead to the autistic syndrome in sighted children. Arguments rage as to whether autism is due to faulty interaction with the caregiver (but only 5 percent of mothers who have one autistic child have a second one also) or some biological deficit or weakness. An incapacity to symbolize and the inability to formulate intentional behavior are among cognitive deficits which have been postulated to lead to an inability for social behavior (Schuler, 1978). It has been alternatively suggested that autism is due to some deficit in social motivation which may be innate and/or environmentally engendered. The question is more easily answered for blind autistic children than for those who are sighted.

When Fraiberg first became involved with research into the special problems of congenitally blind children, she assumed (as

Figure 1. "This drawing is testimony to the importance and strength of the mother–child relationship for all human beings, and to Genie's need for a sense of her own history. Early in 1977, filled with loneliness and longing, Genie drew this picture. At first she drew only the picture of her mother and then labeled it 'I miss Mama.' She then suddenly began to draw more. The moment she finished she took my hand, placed it next to what she had just drawn, motioning me to write, and said 'Baby Genie.' Then she pointed under her drawing and said 'Mama hand.' I dictated all the letters. Satisfied, she sat back and stared at the picture. There she was, a baby in her mother's arms. She had created her own reality." (From Curtiss, S., *Genie: A psycholinguistic study of modern-day "wild child."* 1977. New York: Academic Press. Reprinted with permission.)

we would have assumed) that blindness can be compensated for by using the hands as a substitute for the eyes. This, she found, was far from correct; congenitally blind children, unless special intervention is provided, will not reach out into space and explore their environment. Without help, the hands of congenitally blind children remain mouth-centered. If something is touched by the hands it is grasped and brought to the mouth, but the hand will not move outward in search for an object which has made a sound, even if the sound is familiar. Sound–hand coordination is not automatically substituted for eye–hand coordination. As Fraiberg puts it, "A biological sequence, usually facilitated by vision, has been derailed." Congenitally blind children will hold their hands, babylike, up near shoulder level and engage in stereotypic movements until an object is grasped which can be brought to the mouth for exploration. Fraiberg observed such behavior in one boy who was nine years old. Later, when she began to engage in therapy with congenitally blind infants, Fraiberg was able to help the parents help their child toward success in the task of reaching for sound. But what has reaching for sound got to do with identity development? As it turns out, a lot.

Fraiberg found that reaching into space for sound leads the child also to crawl after sound. In her earlier studies she had found blind children who were extremely delayed in the achievement of independent locomotion. In her later intervention study, on the other hand, Fraiberg discovered that once the idea of "out there" was established, the baby would not merely reach into space but was also motivated to move out into and about in space. Without special encouragement the congenitally blind infants reached all the other postural and motor milestones on schedule (when compared with sighted infants), i.e., they rolled over and sat up on time and achieved the creeping position on all fours in time, but then seemed unable to move forward. Being able to explore space, Fraiberg found, is closely linked for the congenitally blind child to comprehending the object as something separate from one's own actions, as something which has solidity even though one can only hear it. Without the concept of the object, i.e., without the concept of a permanent world "out there," both the development of attachment and self–other differentiation (and therefore species identity) are severely impeded.

When babies do not overcome these barriers at the scheduled time, they begin autisticlike stereotypic behavior and also become arrested in their social–cognitive development. If they can surmount these difficulties, then normal social and language development ensues and the stereotypies cease; in other words, the difficulties result from the lack of a vital input, vision, at a specific time and are not due to inability in social behavior.

Fraiberg found in congenitally blind children a second barrier to social–cognitive development and to the establishment of a representational self (and therefore to human identity). Many of the blind children in Fraiberg's intervention study had achieved adequate to superior language production in the second year. They had neither the syntactic nor sociolinguistic deficits which Genie displayed. However, they did share one difficulty with Genie, namely their late attainment of the correct use of "I" and "you." What is known as the syncretic "I" (as in "I wanna") made its appearance at the expected time, but then no further progress occurred for several years. Katie, one little girl described by Fraiberg, would up to the age of 3½ say "give it to her" whenever she wanted something.

Related to this developmental lag in the proper use of personal pronouns was a delay, in comparison with sighted children, in symbolic play with dolls, in the ability to listen to stories, and in the ability to invent imaginary events. Not only did Fraiberg's blind children not play (in fact refused to play) with dolls at the time when such play was commonplace in sighted children, but they also could not point to a doll's nose or mouth, even though they were perfectly capable of pointing to the observer's or to their own facial features. The first appropriate use of "I" and "you" coincided with the beginnings of doll play; using the self as an object and taking an object to represent the self seemed to be based on the same underlying achievement. Without special help of the type Fraiberg provided for these children and their parents, species identification for a congenitally blind child is, it seems, fraught with as much danger as the total deprivation suffered by the most isolated and maltreated children previously described in Chapter 14, this volume.

Of course, mother–infant interaction in many of the families Fraiberg observed and helped was deficient in some ways and muted; but it probably never even approached the deprivation endured by the isolated children like Genie. Muted social interaction is, of course, a further impediment to self–other differentiation and to symbolic representation of the self and others. One reason for the low level of interaction in some of Fraiberg's dyads was the baby's seeming unresponsiveness. Congenitally blind babies do smile at the mother's voice; but only a few mothers discovered that the social smile could be much more reliably elicited if they also augmented their voices with much tactile stimulation. Much of the time a blind baby's face seems bland and does not provide the eye-to-eye contact which is so important in mother–infant interaction (Robson, 1967). Fraiberg taught the mothers of blind babies to watch their infant's hands instead of their faces for clues to the baby's wants and intentions. The face of an infant may seem to stare into space and look bored; but the hands may be seen to make grasping motions when, for example, a familiar bell is sounded. The baby's inability to use visual feedback seems to let the innate ability for facial expression atrophy. Some other modality (hand watching) has to be substituted if the mother is to find a way to communicate with her infant and thus help her child develop an identity as a member of the human species.

Conclusions

It must be evident by now that we agree with the statement "A species identity in humans is not innate," which concludes Chapter 14. We differ from Roy in taking a much more explicitly epigenetic stance. To reiterate what an epigenetic viewpoint, derived from embryology by Erikson (1968), implies: organisms have a ground plan which can only develop in a particular environment. For some organisms the tolerable environmental variation may be large; for others, small. If, however, an organism cannot obtain, either by its own efforts or with the cooperation of conspecifics, the necessary input at the appropriate time (i.e., during sensitive periods), the developmental process may go awry to a greater or lesser extent, although for some capacities, e.g., langauge or socialization, the sensitive periods in humans appear to be very long. It is also plausible to suppose that some organisms, including humans, have ground plans which are suited to a number, but not an infinite number, of environmental variations, so that if the organism encounters environmental input A at time 1, the eventual outcome at time 2 will be different from (but not necessarily worse than) the organism encountering environmental input B.

This epigenetic viewpoint leads us to believe that the human neonate's preference for conspecific faces over plain, highly colored stimuli and over newsprint (Fantz, 1961) is not due to coincidence. The preference evolved in an environment where learning to interact with human partners had survival value, whereas being able to scan newspaper print did not. When we previously stated that human infants seem to be "prepared" for interaction with human caregivers we did *not* mean to suggest that human babies necessarily have innate recognition of the schematic human face, but merely that they are especially responsive to stimuli which *share* important features with the human face. Roy's suggestion that one compare human infants' responses to human mammalian faces is, in our opinion, a good one. Since the newborn human infant is most likely to encounter *human* faces rather than those of other mammals, the operative "key stimulus" need not be all that precisely specified. We similarly believe that the 3-to-4-month-old babies' entrancement with mobiles which turn in response to the head movements (Watson, 1972) is related to their propensity for reciprocal interaction ("the Game"). Watson himself claims that "the Game" is not important to the infant because people play it, but rather that people become important to the infant because they play "the Game." We would interpret the baby's responsiveness differently: to be able to interact with caregivers is of survival value, so that the capacity for responding to reciprocal games (e.g., back-and-forth cooing) and for mutual regulation can be demonstrated in any situation where the infant's and the partner's behavior are contingent upon one another, even when the partner is only a mobile. In other words, "the Game" is attractive to human infants because they are "prepared" for such interaction, but the partner need not necessarily be human. The human baby's responsiveness to close bodily contact with conspecifics may even have some antecedents in prenatal experiences (such that prenatal input leads to seemingly innate preferences postnatally). Blehar et al. (in preparation) have suggested that the preference for a warm, encompassing environment with vestibular stimulation (e.g., being held) may be the consequence of prior experience in the womb. Their argument was based on Salzen's (1970) suggestion that precocial birds' preference for nestling close to the warm body of the mother bird may result from a neuronal model of the environment built up during its life in the egg, an environment which the infant bird tries to reestablish after hatching. While we can probably never demonstrate the correctness of this claim, it is at least a plausible epigenetic alternative to pure nativism, i.e., that "preparedness" for certain behaviors may involve necessary prior learning.

We have much to discover about what kinds of inputs or experiences are particularly important or effective at what developmental period, and at what periods that very same input may no longer be quite so vital for further development. Ainsworth (1973) has suggested that a mother's ability to read her baby's signals and to respond both promptly and appropriately to them are highly important during the infant's first year. At later periods such promptness in response may no longer be of equal importance. Fraiberg (1977) suggests that visual input is necessary for self–other differentiation and for the development of species identity; if a person is blinded after these milestones have been achieved, the lack of vision, although a hindrance, does not interfere with the sense of self. Gender identification similarly seems to occur early with great ease (around age two) and is difficult to establish later (Money and Ehrhardt, 1972). On the other hand, the capacity for language appears, on the basis of evidence from Isabelle (quoted in Chapter 14), to persist for many years, provided an initial parent–child attachment was established.

The evidence assembled here does not, of course, prove that an epigenetic approach to the development of a human species identity is correct or ultimately the most fruitful one. We do suggest, however, that it is a plausible and promising working hypothesis which permits more precise questions, even if the answers in the end are not the expected ones. Fraiberg's work suggests that we begin to look for much more narrowly defined experiential deficits (rather than global ones, such as "social deprivation") in working out which interventive efforts will bring the greatest rewards.

References

Ainsworth, M. D. S. 1973. The development of infant-mother attachment. In B. M. Caldwell and H. N. Ricciuti (Eds.), *Review of child development research 3.* New York: Russell Sage Foundation, 1–94.

Ainsworth, M. D. S., M. C. Blehar, E. Waters, and S. M. Wall. 1978. *The strange situation: Observing patterns of attachment.* Hillsdale, N.J.: Lawrence Earlbaum Associates.

Barkow, J. H. 1977. Human ethology and intra-individual systems. *Sci. Inf. 16,* 133–145.

Bates, E. 1976. Acquisition of polite forms: Experimental evidence. In E. Bates (Ed.), *Language and context: The acquisition of pragmatics.* New York: Academic Press.

Bates, E., L. Benigni, I. Bretherton, L. Camaioni, and V. Volterra. In press. Cognition and communication from 9–13 months: Correlational findings. In E. Bates et al., *The emergence of symbols: Cognition and communication in infancy.* New York: Academic Press.

Bell, S. M. 1970. The development of the concept of the object as related to infant–mother attachment. *Child Dev. 41,* 291–313.

Blehar, M. C., M. D. S. Ainsworth, and M. Main. Monograph in preparation. *Mother-infant interaction relevant to close bodily contact: A longitudinal study.*

Blehar, M. C., M. D. S. Ainsworth, and S. M. Bell. Monograph in preparation. *Developmental changes in the behavior of infants and their mother relevant to close bodily contact.*

Blehar, M. C., A. F. Liebermann and M. D. S. Ainsworth. 1977. Early face-to-face interaction and its relation to later infant-mother attachment. *Child Dev. 48,* 182–194.

Bowlby, J. 1969. *Attachment and loss, Vol. 1: Attachment.* New York: Basic Books.

Bretherton, I. In press. Young children in stressful situations: The role of attachment figures and unfamiliar caregivers. In G. V. Coelho and P. Ahmed (Eds.), *Uprooting.* New York: Plenum Press.

Bretherton, I., and E. Bates. 1979. The emergence of intentional communication. In I. Uzgiris (Ed.), *New directions for child development, Number 4: Social interaction and communication in infancy.* San Francisco: Jossey-Bass.

Curtiss, S. 1977. *Genie: A psycholinquistic study of a modern-day "wild child."* New York: Academic Press.

Deutsch, J. and K. Larsson. 1974. Model-oriented sexual behavior in surrogate-reared rhesus monkeys. In W. Riss (Ed.), *Brain, behavior, and evolution.* Basel, Switzerland: Karger.

Erikson, E. H. 1968. *Identity, youth and crisis.* New York: Norton.

Fantz, R. L. 1961. The origin of form perception. *Sci. Am. 204,* 66–72.

Fraiberg, S. 1977. *Insights from the blind.* New York: Basic Books.

Goldfarb, W. 1945. Psychological privation in infancy and subsequent adjustment. *Am. J. Orthopsychiatry 15,* 247–255.

Killen, M. and I. Uzgiris. 1978. *Imitation of actions with objects: The role of social meaning.* Paper presented at the International Conference on Infant Studies, Providence, Rhode Island, March.

Katz, P. May, 1978. Personal communication.

Kohlberg, L. 1966. A cognitive-developmental analysis of children's sex-role concepts and attitudes. In E. E. Maccoby (Ed.), *The development of sex differences.* Stanford, Cal: Stanford University Press.

Lewis, M. and L. A. Rosenblum (Eds.). 1977. *Interaction, conversation and the development of language.* New York: Academic Pr.

Lorenz, K. 1952. *King Solomon's Ring*. New York: Crowell.
Money, J. and A. Ehrhardt. 1972. *Man and woman, boy and girl*. Baltimore: Johns Hopkins University Press.
Papoušek, H. and M. Papoušek. 1975. Cognitive aspects of preverbal infant–adult interaction. In CIBA Foundation Symposium 33, *Parent–infant interaction*. New York: Associated Scientific Publishers, 241–269.
Piaget, J. 1962. *Play, dreams and imitation in childhood*. New York: Norton.
Robson, K. S. 1967. The role of eye-to-eye contact in maternal-infant attachment. *J. Child Psychol. Psychiatry 8*, 13–25.
Salzen, E. A. 1970. Imprinting and environmental learning. In L. R. Aronson (Ed.), *Development and the evolution of behavior*. San Francisco: Freeman, 158–178.
Sander, L. W. 1977. The regulation of exchange in the infant caregiver system and some aspects of the context–content relationship. In M. Lewis and L. A. Rosenblum (Eds.), *Interaction, conversation and the development of language*. New York: Wiley.
Schaffer, H. R. 1971. *The growth of sociability*. London: Penguin.
Schaffer, H. R. 1977. Early interactive development. In H. R. Schaffer (Ed.), *Studies in mother–infant interaction*. New York: Academic Pr. (a)
Schaffer, H. R. (Ed.). 1977. *Studies in mother–infant interaction*. New York: Academic Press. (b)
Schuler, A. 1978. The interaction of social, linguistic and cognitive development in childhood autism. In W. H. Fay and A. L. Schuler (Eds.), *Emerging language in autistic children*. Baltimore: University Park Press.
Shatz, M. and R. Gelman. 1973. The development of communication skills: Modifications in the speech of young children as a function of the listener. *Monog. Soc. Res. Child Dev. 38*, Serial 152.
Singh, J. A. L. and R. M. Zingg. 1942. *Wolf children and feral man*. Hamden, Conn.: Shoe String Press. (Reprinted in 1966 by Harper and Row.)
Spitz, R. A. 1945. Hospitalism: An inquiry into the genesis of psychiatric conditions in early childhood. *Psychoanal. Study Child 1*, 53–74.
Spitz, R. 1965. *The first year of life*. New York: International Universities Press.
Stayton, D. J., M. D. S. Ainsworth, and M. Main. 1973. Development of separation behavior in the first year of life: Protest, following and greeting. *Dev. Psychol. 9*, 213–225.
Tizard, B. and T. Rees. 1975. The effect of early institutional rearing on the behavior problems and affectional relationships of four year old children. *J. Child Psychol. Psychiatry 16*, 61–73.
Watson, J. S. 1972. Smiling, cooing and "the Game." *Merrill-Palmer Q. 18*, 323–339.

Chapter 16

An Evolutionary Perspective

David P. Barash

In Dr. Seuss's delightful children's book *Horton Hatches The Egg*, Horton, the lovable elephant hero, agrees to care for an egg laid by the Mazie Bird ("lazy bird") who then departs on a rather long vacation. True to his word, Horton perseveres in his task through many hilarious adventures, only to be displaced by the Mazie Bird just before the egg hatches. Did Horton commit an error in species identity? It certainly appears so, in that he spent time and energy rearing young that were not his own; indeed, the Mazie Bird took parasitic advantage of him. But wait! When the egg finally hatches, we find that the young bird has an elephantine trunk and large ears; somehow the reader feels that justice has been served.

The case of Horton provides a microcosm for understanding the difference between psychological and biological approaches to species identity. Thus a psychologist might wonder about Horton's early experience (perhaps Horton himself had been reared by a Mazie Bird), about his endocrine state (perhaps he has a pathologic excess of avian-type hormones), or maybe about his own "self-concept." Psychologists, in any event, have traditionally emphasized "proximal" factors in explaining behavior—the immediate causative mechanisms, whether they lie in anatomy, physiology, or in prior experience. The evolutionary biologist, in contrast, is most likely to view behavior from the perspective of natural selection and accordingly to be concerned primarily with "ultimate" or "distal" causation—the adaptive significance of the behavior in question (Barash, 1977). This approach treats proximal mechanisms as the handmaidens of distal designs. In the case of Horton the apparent introgression of elephant genes into the Mazie Bird's offspring suggests an ultimate adaptive significance to Horton's otherwise inappropriate behavior; by incubating the egg he was not only keeping his word (apparently an important consideration to elephants), he was also enhancing his fitness, since some of his genes appeared in the offspring. More serious analyses of this sort are the cornerstone of sociobiology (Wilson, 1975), a discipline that is largely concerned with the evolutionary biology (hence adaptive significance) of behavior.

Of course, interest in distal causation does not preclude concern for proximal mechanisms as well, and, indeed, sometimes it may suggest some novel approaches. In the present example, elephant genes perhaps can diffuse through Mazie Bird eggs, displacing some of the existing genotype. There is also an even more intriguing explanation, one that not surprisingly did not appear in the children's book: perhaps Horton and the Mazie Bird had been secret lovers before she laid her egg. This would render Horton's behavior adaptive and supercede considerations of his species identity, although psychologists might still wish to investigate the proximal factors that led him to make so unusual a mate selection. (As will be shown later, they should be even more concerned with the Mazie Bird.)

Fanciful examples aside, there are real and important differences between the approaches of social science and of sociobiology, although the two ideally are complementary rather than antagonistic. Despite its avowedly phylogenetic perspective, the present volume is heavily inclined toward the former approach. A notable exception is Cooke's work (Chapter 5, this volume) on avian species identity, especially his research on the snow goose–blue goose complex. His work is especially laudable for its integration of laboratory and field studies and because it is concerned with both the evolutionary causes of mating preferences—specifically narrowly and broadly selective behavioral morphs—and its consequences for the breeding population. In the introductory chapter of this volume, Roy aptly points out that most researchers in human behavior have never asked themselves why humans turn almost exclusively to other humans, i.e., why *Homo sapiens* species identity occurs as it does. However, his own discussion as well as those in most of the other chapters, goes on to consider "how" without really coming to grips with the distal "why."

From an evolutionary perspective, "why" concerns the evolutionary benefit accruing to each individual as a result of making a species identity "decision." And there are many such decisions, each with associated costs and benefits (all measured in the currency of fitness). One such decision, previously thought to be unique to *Homo sapiens*, is in response to the question, "Who am I?" Nonhuman animals can behave as though they have answered this potentially profound question, without having necessarily ever posed it to themselves at all. Thus a chickadee may act like a chickadee and even show preferences for other chickadees, without having announced to itself, "I am a chickadee." Gallup's work (Chapter 11, this volume) is especially fascinating in this regard, since it comes closer to evaluating this potential among nonhuman animals than any has before. It is suggestive and certainly consistent with an evolutionary perspective on animal behavior which suggests an underlying biologic unity in all living things, with no dramatic discontinuities between human and nonhuman animals.

The view of this chapter, however, is that a phylogenetic perspective is quite different from a truly evolutionary one; the former comes closer to the *scala naturae* still often implicitly assumed in comparative psychology, as well as to the static conception of behavioral phylogenies that characterizes much of Lorenzian ethology (e.g., Eibl-Eibesfeldt, 1975). A dynamic use of evolution, in contrast, would emphasize the adaptive consequences of various behavioral "choices" with regard to mating, social grouping, defense, and foraging. Alternatively (or additionally) it may focus on the various information channels available to organisms involved in behavior related to species identity, perhaps with a special eye to phylogenetic inertia (Wilson, 1975) and the role of structural aspects of the environment in predisposing different communicative modes. Such an approach, unlike that of the present volume, would not exclude the social insects (Wilson, 1971) and would necessarily give deserved attention to audition and pheromones (Sebeok, 1977; Thiessen, 1977). Humans are highly visual animals, whereas most other species are very much less so.

Furthermore, it is conceptually barren to ask whether or not a species has a species identity just as it has teeth, feathers, or mammary glands. No sexually reproducing species lacks it, or else reproduction would not occur. Thus species identity of one sort or another characterizes successful individuals of virtually every natural population, and one only sees the relative successes. This is not to deny the possibility of error, and certainly my intention is not to denigrate research into experiential mechanisms that achieve species recognition. As many of the contributions to this volume attest, we can profitably inquire into the modifiability of species recognition, and such inquiry may even be relevant to human clinical experience such as early infantile autism (see Roy, Chapter 14, this volume). An approach more consistent with evolutionary considerations, however, is to treat species identity as a complex information pool with many relevant dimensions—recognition of an appropriate mate, identification of offspring, choice of foraging or grooming partners, selection of models for allelomimetic behavior, and so on. Each of these behaviors has important consequences for evolution, thereby selecting for optimum decision mechanisms consistent with information acquisition and retrieval systems that in each case maximize the difference between benefits and costs (in fitness) to the organism in question. The sum total of this potentially heterogeneous assemblage may be considered the species identity of the individual in question, but we should recognize that both functionally and operationally it is a "grab bag" term.

Every individual member of every species eventually acquires information, on which basis it makes numerous social decisions (cf. Lorenz, 1965). These may be as simple as whether or not gametes should be extruded (clams) or as complex as who to invite to a cocktail party (humans). The information may accordingly vary dramatically, not only in type ("yes–no" versus "Would Barbara and Sue get along together?"), but also in source (largely genetically influenced, versus largely acquired by experience). However, both genotype and environment are involved in both, and in all such cases valuable intellectual leverage is gained by seeing such behavior as dependent on information, with different strategies existing for the acquisition of different types of information, depending on the organism and the behavior in question. In this regard it should be emphasized that the terms "genetic" or "instinctive" are not necessarily synonymous with "biological," as opposed to "learned" or "strongly influenced by experience." Learning is not unbiological; it is a perfectly good (i.e., adaptive) mechanism for the acquisition of information that generally has great significance, both proximally and distally. Depending on the species and the behavior in question, we can expect relatively more reliance on one mechanism or the other. In evolutionary perspective, the important thing is that the necessary information be obtained, not how it is obtained.

In Chapter 9 of this volume, Meier and Dudley-Meier point out the remarkable persistence of basic behavior necessary for reproduction, at least in primates. Thus much reproductive behavior in particular is retained despite the often dramatic modification of naturalistic rearing conditions. A new perspective on isolate-rearing accordingly suggests that living things are frequently very stable in their behaviors, insofar as those behaviors directly subserve fitness. They are often highly "canalized" (Waddington, 1957) in that such phenotypes tend to resist environmental perturbation. It is clearly adaptive that they are resistant since individuals whose sexual and reproductive behaviors did not subserve fitness would be less successful reproductively than other individuals whose behaviors were more "fit." As a consequence, alleles influencing such behaviors would be less successful than their counterparts and would be replaced by less "modifiable" alternatives.

On the other hand, it is not necessarily optimum to encode maximum information in the genotype, especially since there is only a finite amount of room available in the DNA of any species, such that codons devoted to information on species identity, for example, must preclude coding for some other parameter of information, whether structural, physiological, or behavioral. Given the finite information-bearing capacity of the genotype (and accordingly the cost in fitness of coding one bit of information versus another), we can envision a cost–benefit analysis carried out by natural selection during the evolutionary history of any population. If the organism has sufficient neural complexity (i.e., appropriate phylogenetic inertia) to acquire such information through experience rather than prewired neural connections, and if the rearing experiences of the organism normally result in a sufficiently high probability of predictable identification being achieved, then selection should favor a more flexible system for acquiring species identification. Accordingly there is nothing inappropriate about malimprinting, for example, when we recall that geese in nature do not normally encounter Konrad Lorenz squatting on his haunches and quacking vigorously. If such experiences were common to the species, we can be sure that selection would favor those individuals whose program for acquiring that type of species identity was substantially less susceptible to experience.

Similarly, we can predict that females ought to be more discriminating than males in choosing a sexual partner (Williams, 1966) insofar as they make a greater parental investment than males (Trivers, 1972). The consequences of a bad decision accordingly fall more heavily on females than on males (at least for most birds and mammals), so individuals who make such decisions should be more heavily penalized in terms of their fitness. As a consequence of such a penalty, greater selectivity would be expected among females than among males, i.e., a reduced likelihood of evincing "incorrect" species identity regarding choice of a mating partner. This is consistent with the observation that at least among dimorphic bucks, males are much more susceptible to sexual malimprinting than females (Schutz, 1965).

Considerations of species identity are also of evolutionary significance insofar as they contribute to isolating mechanisms, influencing the degree of hybridization among species. It is generally agreed among evolutionary biologists that new animal species arise primarily through allopatric speciation, where some physical barrier prevents gene flow among isolated populations (Mayr, 1963). If representatives of these populations subsequently meet, hybridization is rendered less likely by the genetic differentiation that occurred while the two populations were separated and subjected to differential selection causing incompatible gene pools.

From the viewpoint of each individual, interspecific hybridization is almost invariably maladaptive, since the products of such mismatings are less likely to be fit than offspring produced by intraspecies matings. Hence parents that produced such offspring are less fit, and selection should therefore operate strongly against alleles that permit such errors in species identification. Thus, given a degree of genetic differentiation passively occurring as a result of adaptation to different environments, behavioral ability to discriminate among individuals as a function of similarities in genotype should spread via selection and express itself as an enhanced capacity for species identification, regardless of the mechanism whereby it is achieved.

In general, we would expect that reliance upon genotypic, as opposed to experiential, factors in shaping most behaviors classed as species identity should vary directly with the likelihood of making a mistake in nature and directly with the consequences of so doing. No less than mating, behaviors related to parenting should therefore also be especially amenable to evolutionary analyses, since parenting is a major route to individual fitness. In general this area does seem to be especially productive of such analysis (Barash, 1976). Since there is no evolutionary payoff in caring for a stranger's offspring, individuals should be resistant to caring for another's children. Examples of parents caring for offspring of another species can generally be discounted as the result of artificial rearing conditions (with some important exceptions to be discussed later). Regardless of the proportion in which the relevant information is acquired genotypically and experientially, the simple fact is that organisms end up identifying their own young and caring for them, and not those of another species. Indeed, the discrimination is generally much more precise.

Not only do parents correctly identify their young as to species, they also tend to discern their own young from those of another conspecific. To be more accurate, they are able to discriminate their young in proportion to the cost associated with not being able to do so. Thus the eggs of ground-nesting gulls do not normally roll out of their nest and into that of neighboring gulls; accordingly, parent gulls do not recognize their eggs individually, since they can safely rely on the permanent location of their nests to correlate with the placement of their eggs (i.e., their genes). The same situation arises shortly after the chicks hatch, since they do not leave the nest. Within a few days, however, their young begin wandering about, and the parents' risks of mistaking a neighbor's offspring for their own become high. Accordingly parent gulls rapidly develop the capacity for individually recognizing their offspring at just this time (Tinbergen, 1953).

By contrast, kittiwake gulls nest on narrow cliff edges, and not surprisingly kittiwake chicks do not wander from their nests even as they mature. Also not surprisingly, kittiwake parents do not develop the capacity for identifying their own offspring (Cullen, 1957); the nest location is a sufficiently reliable predictor. A similarly adaptive pattern occurs across all species—the ability of parents to recognize their own young correlating with the importance (to their fitness) of making a correct decision. It is well known, for example, that many rodents such as rats and mice will readily accept cross-fostering. In contrast, many ungulates such as wildebeests and goats will reject strange young. Rats and mice typically give birth in natal nests constructed in isolated burrows;

furthermore, their young are altricial, thus making it doubly unlikely that in a free-living system the parents will be faced with the necessity to discriminate their offspring from strangers. On the other hand, ungulates typically produce precocial young that can easily become mixed with strangers and, in addition, they often travel in large herds, giving birth synchronously as part of an adaptive strategy to "swamp" predators and thus minimizing individual loss through predation. As a result such species are typically quite precise about recognizing their offspring, their behavior generally mediated by a form of olfactory imprinting achieved within a few minutes of birth (e.g., Klopfer and Gamble, 1966).

In general and as expected, parental care varies directly with the certainty of genetic relatedness between parent and offspring. Teleost fish which broadcast their gametes, for example, typically do not engage in postzygotic investment by either sex. Postzygotic biparental investment is found exclusively in strictly monogamous species, in which both parents are very probably the biological parents of the young in question (most passerine birds, columbiform and psittaciform birds, certain foxes, beavers, and gibbons). Males typically provide postzygotic parental investment when they are more likely than females to be the biological progenitors (e.g., rheas, tinamous, and most damselfish). Finally, females are the primary postzygotic parents when they have a significantly greater confidence in genetic relationship with their offspring than males (most reptiles, polygamous and promiscuous birds, and most mammals). The mechanism of offspring identification is not necessarily specified in this type of analysis; the ends are predicted, not the means by which they are accomplished.

Among normally monogamous species such as the mountain bluebird, in which males typically participate in the care of the offspring, experimentally removed mated males were replaced by new males who did not provision the young or give alarm calls at the approach of danger (Power, 1975). They accordingly may well have "identified" the existing offspring as not theirs, and their failure to invest in them supports sociobiologic expectations that true altruism does not occur in nature. Infanticide following male takeover, furthermore, has been increasingly reported as a common strategy by which male mammals eliminate infants which have been sired by previous males. This strategy, of course, also induces the lactating females to recycle, thereby enabling them to be inseminated by the newly ascendant male (Steiner, 1972; Schaller, 1972; Hrdy, 1977).

Parental investment, of course, is not always adaptively correlated with the likelihood of genetic relatedness; nest parasitism (see Chapters 1 and 4) is a prominent case of parents "misidentifying" offspring and investing in a manner that decreases their fitness. Such cases invariably result from conflicts between fitness maximization of the host and of the nest parasite; the parasite is selected for behaviors that enhance the chances of the host being deceived (e.g., Rothstein, 1975). Such systems are responsible for the remarkable mimetic resemblances between parasite and host, since natural selection by hosts favors mimics that are as accurate as possible in resembling (and thus deceiving) their hosts. The consequences of being deceived into making an error of "species identity," on the other hand, select for acute discrimination by the hosts (see Wickler, 1970 for numerous examples).

In addition to the adaptive orienting of mate selection and parental care, species-identity decisions relate to choice of a social group. Social groups may be advantageous in providing defense against predators (Alexander, 1974), in facilitating mate selection, in acquiring otherwise unavailable resources (e.g., Kruuk, 1972), and in providing information concerning available food resources (e.g., Ward and Zahavi, 1973). Social grouping may also be adaptive in reducing the threat of predation by providing increased fighting ability, by mimicking a larger individual (as has been occasionally suggested for fish schools), by increasing the probability that a neighbor rather than oneself will be taken by the predator (Hamilton, 1971), or by the increased watchfulness of many alert individuals. This listing, of course, is certainly not exhaustive, but it provides interesting implications for species identity.

Whereas a conspecific species identity is valuable, for example, insofar as social groups subserve mate selection, it is somewhat disadvantageous because it increases the likelihood of misdirected parental care (see Hoogland and Sherman, 1976, for an excellent evaluation of the pro's and con's of sociality in one species). Mixed species groups among wintering and tropical birds, furthermore, may be especially advantageous in increasing foraging efficiency (Morse, 1970). Mutually beneficial antipredator associations between baboons and antelope, for example, are well known in Africa, where baboons profit by the greater olfaction and hearing of the antelope, which in turn benefit from the keener eyesight of the baboons. In such cases can one justifiably say that an "error" in species identity has occurred? It seems more appropriate to consider that individuals of each species are selected for making adaptive responses to all features of their environment, conspecifics forming an important but not exclusive part of that environment. The specific response pattern will vary with the species and with the behavior in question. The only cross-species consistency is the adaptive nature of the patterns themselves.

Often the interest of evolutionary biologists in species identity does not concern the gross malfunctions which we can produce experimentally as much as the fine tuning of individual recognition that occurs in nature. Mountain sheep, for example, not only "identify themselves" as such, they also respond differently to each other as a function of horn development, which in turn correlates with success in aggressive competition (Geist, 1971). A major component of such interindividual fine tuning concerns the recognition by an individual of the proportion of genotype shared with another through common descent, their coefficient of relationship. This value, identified as "r," can also be seen as the probability that two individuals share the same allele at one locus, due to their common ancestry. Concern with coefficients of relationship has developed following Hamilton's (1964) identification of inclusive fitness, a concept that has become a cornerstone of modern sociobiology (Wilson, 1975; Barash, 1977). Thus individuals are perceived as maximizing their inclusive fitness by directing behavior toward related individuals so as to maximize the probability that relevant genes will be passed to future generations. Such kin selection (Maynard, 1964) induces individuals to behave altruistically toward others in proportion to those who share genes with the altruist. In other words, a major thrust of the sociobiology of altruism is that it actually represents genetic selfishness, with each potential beneficiary devalued in proportion as it is more distantly related.

There are several proximal mechanisms whereby such identification can be achieved; spatial proximity may well correlate with relatedness, so selection could in some cases favor individuals who behave more altruistically toward neighbors than toward strangers. Ability to discriminate as a function of proximity has certainly been well demonstrated in a variety of supposedly "solitary" species (Weeden and Falls, 1959; Emlen, 1972; Barash,

1974). In addition individuals can identify each other and respond differentially as a function of phenotypic resemblance, an adaptive strategy insofar as such resemblance correlates with genotypic similarity, a reasonable assumption (Barash, Holmes, and Greene, 1978). Among many species, early experience can also provide reliable information for the identification of siblings, an important class of individuals, since they share an *r* of one-half. Thus sensitivity to one or several of the phenotypic characteristics of littermates could identify a sibling after dispersal occurs, and the opportunity for maximizing inclusive fitness via differential behavior toward kin and nonkin presents itself. Even among primates with a litter size of one, the long period of juvenile dependence enables sibs to identify each other; a semiindependent juvenile can identify its sib (or at least its half-sib) as "the infant who is nursing from mother." Indeed, Japanese monkeys perform sophisticated kinship analyses (Kurland, 1977), and kinship considerations provide considerable insight into alarm calling among Belding ground squirrels (Sherman, 1977). The adaptive significance of such behavior is clear, but the analysis of mechanisms used to achieve such fitness maximization via kin selection has lagged. That is, it is one thing to say that kin recognition could be achieved in such-and-such a manner, it is another to show that it really is.

Research by Holmes and his colleagues at the University of Washington has recently provided surprising insights into the mechanism by which individuals identify their kin. Thus in a recent study, pigtail macaques (*Macaca nemestrina*) separated from their mothers at birth show a significant preference for their half-sibs when given a choice between an unrelated individual and a half-sib in a laboratory situation (Wu, Holmes, Medina, and Sackett, 1978). This is all the more striking since the individuals involved had no experience with each other or with their parents prior to the test situation. Although these results do not identify the precise modality by which identification of kin is achieved, they demonstrate that early experience with half-siblings is not necessary for individuals to distinguish them from unrelated individuals of equal age. That is, both the ability to discriminate kin from nonkin and a preference for the former appear largely to be genetically influenced. These results have paralleled recent studies of kin identification in the Arctic ground squirrel, *Spermophilus undulatus*. Cross-fostering of laboratory-born offspring resulted in four experimental types of individual pairings: biological sibs reared together, biological sibs reared apart, nonrelatives reared apart, and nonrelatives reared together. Individuals of these sorts were then paired in an open, neutral arena (13 pairs of each type). Data on the mean rate of agonistic behavior per minute showed the following: biological sibs reared together scored the lowest, followed in increasing order by biological sibs reared apart, nonrelatives reared together, and nonrelatives reared apart. It is especially striking that biological sibs reared *apart* showed significantly less agonistic behavior than nonrelatives reared *together*, thereby once again demonstrating a powerful inherited capability for both identifying and responding differently to relatives versus nonrelatives.

These results are significant for an evolutionary perspective, in that they demonstrate a capacity in each individual for obtaining information allowing it to behave in a manner that maximizes its exclusive fitness. They are also significant for an "experiential" perspective, since they suggest numerous hypotheses that need testing concerning the proximal mechanisms involved in this discrimination among conspecifics. Indeed, Hamilton's theory of inclusive fitness maximization does not require an innate recognition of kin, simply that kin be differentially treated regardless of the proximal mechanism. But findings on the pigtail macaque and Arctic ground squirrel highlight the potentially fruitful interaction between proximal and ultimate considerations in eventually suggesting a well-rounded picture of behavior, including species identity. Indeed, the time now seems ripe for a comprehensive theory of preferential associations, based on the underlying principle that individuals will associate with each other in a manner that maximizes their inclusive fitness. Such optimum association requires striking an optimum balance between the benefits and costs associated with kin interactions, as well as those alternatively available through reciprocity from unrelated individuals (Trivers, 1971). Efforts to approach this difficult topic have already begun (Popp and DeVore, 1978; Wasser, 1978). They will doubtless continue, as will the complementary and perhaps even mutualistic activities of evolutionary biologists and social scientists.

References

Alexander, R. D. 1974. The evolution of social behavior. *Ann. Rev. Ecol. Syst.* 5, 325–383.

Barash, D. P. 1974. Neighbor recognition in two "solitary" carnivores: The raccoon (*Procyon lotor*) and the red fox (*Vulpes fulva*). *Science 185*, 794–796.

Barash, D. P. 1976. Some evolutionary aspects of parental behavior in animals and man. *Am. Psychol. 89*, 195–217.

Barash, D. P. 1977. *Sociobiology and behavior*. New York: Elsevier.

Barash, D. P., W. Holmes, and P. Greene. In press. Exact versus probabilistic coefficients of relationship: Some indications for sociobiology. *Am. Nat.*

Cullen, E. 1957. Adaptations in the kittiwake to cliff-nesting. *Ibis 99*, 275–302.

Eibl-Eibesfeldt, I. 1975. *Ethology, the biology of behavior*. New York: Holt, Rinehart and Winston.

Emlen, T. 1972. An experimental analysis of the parameters of bird song eliciting species recognition. *Behaviour 41*, 130–171.

Geist, V. 1971. *Mountain sheep*. Chicago: Univ. Chicago Pr.

Hamilton, D. 1964. The genetical theory of social behaviour: I and II. *J. Theor. Biol. 7*, 1–52.

Hamilton, D. 1971. Geometry for the selfish. *J. Theor. Biol. 31*, 295–311.

Hoogland, L. and P. W. Sherman. 1976. Advantages and disadvantages of bank swallow (*Riparia riparia*) coloniality. *Ecol. Monogr. 46*, 33–58.

Hrdy, S. B. 1977. *The langurs of Abu*. Cambridge, Mass.: Harvard Univ. Pr.

Klopfer, R. and J. Gamble. 1966. Maternal "imprinting" in goats: The role of chemical senses. *Z. Tierpsychol. 23*, 588–592.

Kruuk, H. 1972. *The spotted hyena*. Chicago: Univ. Chicago Pr.

Kurland, A. 1977. Kin selection in the Japanese monkey. *Primatology 12*, 1–145.

Maynard, S. J. 1964. Group selection and kin selection. *Nature 201*, 1145–1147.

Mayr, E. 1963. *Animal species and evolution*. Cambridge, Mass.: Harvard Univ. Pr.

Morse, D. H. 1970. Ecological aspects of some mixed species foraging flocks of birds. *Ecol. Monogr. 40*, 49–168.

Popp, J. L. and I. DeVore. 1978. Aggressive competition and social dominance theory. In D. O. Hamburg and J. Goodall (Eds.), *Perspectives on human evolution*. Menlo Park, N.J.: Benjamin.

Power, H. W. 1975. Mountain bluebirds: Experimental evidence against altruism. *Science 189*, 142–143.

Rothstein, S. I. 1975. Evolutionary rates and host defenses against avian brood parasitism. *Am. Nat. 109*, 161–176.

Schaller, G. B. 1972. *The Serengeti lion*. Chicago: Univ. Chicago Pr.

Schutz, F. 1965. Sexuelle Prägung bei Anatiden. *Z. Tierphyschol. 22*, 50–103.

Sebeok, T. 1977. How animals communicate. Bloomington: Indiana Univ. Pr.

Sherman, W. 1977. Nepotism and the evolution of alarm calls. *Science 197*, 1246–1253.

Steiner, A. L. 1972. Mortality resulting from intraspecific fighting in some ground squirrel populations. *J. Mammol.* 53, 601–603.

Thiessen, D. D. 1977. Thermo energetics and the evolution of pheromone communication. *Progr. Psychobiol. Presidential Psychol.* 7, 91–191.

Tinbergen, N. 1953. *The herring gull's world.* London: Collins.

Trivers, R. L. 1971. The evolution of reciprocal altruism. *Q. Rev. Biol.* 46, 35–57.

Trivers, R. L. 1972. Parental investment and sexual selection. In C. B. Campbell (Ed.), *Sexual selection and the descent of man.* Chicago: Aldine.

Waddington, C. H. 1957. *The strategy of the genes.* New York: Macmillan.

Ward, P. and A. Zahavi. 1973. The importance of certain assemblages of birds as "information-centres" for food finding. *Ibis* 115, 517–534.

Wasser, S. 1978. *Preferential associations: A function of complementarity and relatedness.* Paper presented at the annual meeting of the Animal Behavior Society, Seattle, Wash.

Weeden, J. S. and J. B. Falls. 1959. Differential responses of male ovenbirds to recorded songs of neighboring and more distant individuals. *Auk* 76, 343–351.

Wickler, W. 1970. *Mimicry.* New York: McGraw-Hill.

Williams, G. C. 1966. *Natural selection and adaptation.* Princeton, N.J.: Princeton Univ. Pr.

Wilson, E. O. 1971. *The insect societies.* Cambridge, Mass.: Harvard Univ. Pr.

Wilson, E. O. 1975. *Sociobiology: The new synthesis.* Cambridge, Mass.: Harvard Univ. Pr.

Wu, H., W. Holmes, S. Medina, and G. P. Sackett. In preparation. Kin preference in infant *Macaca nemistrina.*

Chapter 17

Species Identity and Self-awareness: Some Ethical and Philosophical Issues

Michael W. Fox

This chapter focuses on a philosophical/legal and ethical issue that has surfaced in recent years (Morris and Fox, 1978; Ryder, 1975; Singer, 1975), namely the issue of animal rights in relation to the phenomena of species identification and self-identity awareness. Griffin (1976) has cautiously explored the question of animal awareness. Under the impetus of such writings and also this particular volume of studies and discussions, it is hoped that we will develop the tools and conceptual abilities to examine with greater scientific objectivity and precision what we may already feel intuitively, that many sentient creatures possess not only species identity but also self-identity awareness.

Implications of the Trend toward Quantitative Ethology

Even the most unprejudiced appraisal of research studies of animal behavior leaves one with the impression that the descriptive, objective terms which are used reflect more than scientific rigor and tradition and also perhaps more than the necessary avoidance of subjective anthropomorphic terms such as "she," or "he." (The animal is always an "it.") The reference or inference to analogous human states of joy, sorrow, jealousy, depression, and the like is generally avoided. The study might even be on some aspect of canine behavior, yet curiously the dog in the laboratory or field seems qualitatively different from the investigator's dog at home, whom the investigator knows very well to manifest these emotions.

Griffin (1976) observes that "behavioral scientists have grown highly uncomfortable at the very thought of mental states or subjective qualities in animals. When they intrude on our scientific discourse, many of us feel sheepish, and when we find ourselves using such words as fear, pain, pleasure, or the like, we tend to shield our reductionist egos behind a respectability blanket of quotation marks" (p 47). Has something been lost in our quest for scientific objectivity and descriptive accuracy in the writing of research studies? Or is the style of writing a sterile reflection of a subjective consensus that animals are unfeeling machines and should be described as such in scientific literature?

I do not wish to give the impression that I am soft-hearted (which I am, since life without empathy would be limbo). However, the hard nose of science may also have hardened the scientist's heart, and worse, contracted his conceptual abilities by restricting the words that are available to describe research findings. The trend from descriptive to quantitative ethology has accelerated over the past decade, again reflecting a major limitation in the scientific consensus that ostensibly attempts and purports to study life, yet reduces life to a mechanistic terminology and then further reduces it to a matrix of numbers, statistical probabilities, Markoff chains, and so on. This tells an outsider to the discipline little about the inner state of an animal and yet surely this is just as important a project to undertake as, for example, the study of animals' predatory behavior or socio-ecological adaptations.

The outsider may want to know something about the "inscape " of animals because he is a poet, artist, philosopher, or lawyer. Yet what can we ethologists, psychologists, and developmental psychobiologists offer them? Numbers and probabilities? Not entirely, but we do need a new vocabulary, new tools and a new conceptual framework in which to operate. In my opinion, a major cultural contaminant still biases our approach to the *umwelt* and inscape of the animal, and that is the legacy of Descartes, i.e., animals are nothing more than unfeeling machines.

The Cartesian Influence in Ethological Studies

This Cartesian view is one of the major concerns of philosophers and others (Morris and Fox, 1978) since it not only affects the ways in which animals are utilized in biomedical research, but also may influence the way in which ethologists and others study and describe the behavior of animals. The ultimate conceptual pathology of the mechanistic view is (post reductio absurdum) that purpose or consciousness have to be put somewhere, so where else (when there is nothing left) but in the genes? Dawkins's (1976) and Wilson's (1975) and Triver's recapitulation of the nature–nurture dichotomy under the guise of genetically deterministic sociobiology attests to the reality of this conceptual pathology.

Some may debate the validity of the inference that such conceptual pathologies are derived from Cartesianism, but they should remember that the earlier Aristotelian concept that animals do not have divine/immortal souls is still being taught in some universities in this country. How we think and perceive is in part determined by past cultural influences. This may be disturbing yet acceptable; but then why is it unacceptable to suggest that those same influences affect our scientific thinking and modes of investigation and description? This should not be the case.

A text dealing with such difficult questions as species identity and self-identity awareness is surely a positive sign of a transmutation of our own awareness from the rigid confines of an enculturated, anthropocentic, and outwardly mechanistic (often also dominionistic; Fox, 1975) world view.

Determining the Existence of Species Identity among Nonhumans

The concept of species per se is phenomenologically valid, yet the degree to which animals conceptualize and classify (in regard to Linnaeus) remains to be determined. A preverbal child systematically will place kitchen utensils (knives, forks, spoons) in "species" categories, his response being based upon simple discrimination of like and unlike. Conceptualization presumably comes after discrimination, followed by verbalization. (Verbal recognition occurs before verbal exposition.)

Species identity may be defined operationally as a phenomenon of attraction and affiliation which is based upon instinctual and/or learned responses and specific cue stimuli—chemical, tactile, visual, or auditory. Chemical affinity precedes but may later reinforce emotional affinity. Teleologically, species identity has many adaptive and survival functions, and the underlying motivations for fission/fusion span the continuum from physiological homeostasis to socio-emotional homeostasis and socio-ecological adaptation.

Phylogenetically (and ontogenetically in the more continuously evolved carnivores and primates), does species identity evolve prior to the ego formation and conceptualization of "me" and "us"? In Chapter 11, Gallup argues that it does. In the ontogeny of the human infant, the phylogenetically earlier symbiotic stage of attachment to the mother may represent a chemophysiological and emotionally homeostatic affiliation facilitating species identity. It is later superceded by self-awareness as the ego develops and individuates from the symbiotic stage of early maternal attachment.

But for a mature organism to behave autonomously, an ego need not be present. This is not to imply that an autonomous amoeba has an ego, but rather raises the question of when the ego (an awareness of self in relation to others) appears phylogenetically.

The behavior of animals towards conspecies vivifies and verifies our concept of species identity, but does not ipso facto imply that they share such a concept. When does the organism attain sufficient neuronal complexity to acquire this conceptual ability and become self-reflective (i.e., aware) instead of instinctually other-directed? The phenomena of personal space, social distance, and territoriality suggest that operationally, if not always conceptually, an animal can respond with reference to place and to others that are not self. This is true for all sentient life, but while an invertebrate may detect the chemical difference of its own slime trail or territory from another invertebrate, such discrimination may be based more on habituation (self-generated stimuli) and the rudimentary ability to recognize familiar from unfamiliar than on concept formation of self versus other. Since receptors and effectors are centered within the individual, the nexus of relationships creating species recognition, affinity, and preference entails some sensory ability to discriminate same from different species, and possibly self from other.

Returning to the development of the human infant, around 7 to 8 months of age, the child goes through a phase of fearfulness around strangers. This occurs after the infant has started to individuate from the mother. Although there may be exceptions, this sequence of infant development may be an accurate phyletic mirror of the evolution of consciousness in other vertebrates. Fear of strangers implies the ability to recognize familiar from unfamiliar on the basis of exposure learning and primary socialization to the family. Where there is an emotional component associated with the recognition of and attachment to kin and emotional reactions to separation, species identity takes on a new dimension—family identity and, at a later age, peer group, sexual, and social role identity. All of these phenomena can be wholly other-directed (i.e., determined by the individual's experiences and interactions with various age and sex classes of conspecies) and be totally independent of any reflective or reflexive self-identity. The existential crisis of late adolescence, which may be extended to a much later age in more dependent other-directed individuals, is a phenomenon recognized by humanistic psychologists. This is the "who am I" crisis, the consciously reflexive phase of development toward inner-directedness and self-actualization. To attempt to determine the existence of such a state of consciousness in a human being without using multiple indices and verbal communication would be impossible.

The verbal barrier is now being broken between humans and primates and cetaceans. With nonhuman primates, growing evidence supports the possibility that we are not the only "intelligent" (i.e., reflective) beings on earth. Multiple indices, however, must be used. Unitary measures, such as those employed by Gallup (Chapter 11) using response to a change in mirror image as a cue for self-recognition, may lead to erroneous or premature conclusions.

A Word of Caution and a Note of Optimism

In order to make a case for "animal rights," some philosophers, lawyers, ethologists, and others are eager to demonstrate that other nonhuman species have some degree of humanlike self-awareness. Such an approach in the final analysis is humanocentric and "species-bound" since it assumes that only humanlike (or suprahuman) beings are worthy of being accorded rights. Surely all living creatures of creation, by virtue of their existence and being an integral part of the interdependent whole which constitutes the biosphere, have the basic right to exist, live, reproduce, and fully actualize their natural potentials (within the natural constraints of ecological harmony rather than under the constraining forces of human dominion). If they are sentient, the right not to have to suffer as a direct or indirect consequence of human actions is essential both for our own well being and for the benefit of all life.

The degree of self-identity which we as a species have evolved may qualitatively and quantitatively be very different from all other life forms on earth. The ancient Indian Brihadaranyaka (Nikilananda, 1949) states that "the inorganic is life that sleeps, the plant is life that feels, the animal is life that knows and man is life that knows that it knows."

Although our closer animal kin may be primarily directed by purposive actions and manifest reason, they may not exhibit rationality, empathy, and altruism. Some (but not all) humans ask the question *why*? This ability to separate the mind from the action or controlling purpose, or from the constraining and directing social and environmental stimuli and contingencies of reinforcement, places us in a very different mode of being. Humans can change their minds, change actions, and conceptually and operationally change the environment with symbols and tools. Self-awareness consists of being able to separate the mind in this way. Without self-awareness, the mind and the environment remain one. With self-awareness, we can change—not physically, but conceptually—and we can change the physical environment. This ecology of mind in which ideas act like genes and where we can consciously change the environment is a new dimension to evolution, for in the past, things have worked the other way. The environment primarily shaped the mind. Stability between the environment and self-awareness arrests evolutionary change, and instability promotes evolutionary change—for better or worse. In the past, the instabilities were natural, environmental, and ecological changes; but today environmental, ecological, and social instabilities are created by man himself. Is this the price of self-awareness? Is the Zen state of equipoise (oneness with the environment) a regression to a more primitive state or a transcending of the human mind from self-centered ego consciousness to transpersonal eco-consciousness?

Conclusion

There are two important points to be learned from the studies on species and self-identity in animals. First, we must realize that we are not the only vertebrates who have some degree of awareness. Other vertebrates are not Cartesian machines. They should be accorded some rights as we do to our own mentally retarded and brain-damaged conspecifics. Second, we should realize that species identity alone is illusory; it only gains significance in its proper perspective in relation to the ecological totality of all interrelated and interdependent life on earth. The less we identify with our own kin and instead identify with all of creation, the more our narrow, egocentric world view will be broadened and such pathological aberrations as racism, "specie-ism," and dominionism become exclusive to past generations. Researchers working with animals today can do much to implement such changes, which are far greater priorities than satisfying scientific curiosity. We must examine our own attitudes and values, and then develop the tools and acquire the conceptual freedom and skills to explore the question of animal awareness more fully.

References

Dawkins, R. 1976. *The selfish gene*. New York: Oxford University Press.

Fox, M. W. 1971. *Integrative development of brain and behavior in the dog*. Chicago: University of Chicago Press.

Fox, M. W. 1976. *Between animal and man*. New York: Coward, McCann & Geoghegan, Inc.

Griffin, D. 1976. *The question of animal awareness*. New York: Rockefeller University Press.

Morris, R. K., and Fox, M. W., Eds. 1978. *On the fifth day: Animal rights and human obligations*. Washington, D. C.: Acropolis Press.

Nikilananda, Swami. 1949. *Upanishads*. New York: Harper & Brothers.

Ryder, R. 1975. *Victims of science*. London: Davis and Poynter.

Singer, P. 1975. *Animal liberation*. New York: Random House.

Wilson, E. O. 1975. *Sociobiology: The new synthesis*. Cambridge, Ma.: Harvard University Press.

Chapter 18

Species Identity

Peter H. Klopfer

"Species identity" is as intuitively appealing but operationally elusive a concept as Lewis Carroll's Cheshire cat. The notion that we (and other organisms) "know our own kind" is consistent with a wide range of human social behavior. It is also consistent with a considerable body of literature that reveals organisms to be capable of complex discrimination of familiar from alien animals or of similar-to-self from dissimilar-to-self stimulus patterns (sounds, smells, colors, etc). Certain male bowerbirds, for instance, when decorating their bowers, preferentially select objects whose color matches that of their eyes (Marshall, 1954). As several contributors to this volume have indicated, however, discriminative ability and "species identity" are not synonomous but merely overlapping concepts (e.g., in Chapter 2 and Chapter 3). A cryptographic decoder or a simple acoustic band-pass filter, for that matter, can "discriminate" one signal from another. If a system is appropriately wired, any particular signal when discriminated can release or direct a complex behavioral response. The mating behavior of many insects, for instance, may be evoked by a particular olfactant (a pheromone) and need not involve principles different from those of a band-pass filter.

In short there is little doubt that animals respond selectively to the world around them. They do so, furthermore, in a manner that generally assures the efficient production of progeny. There exist a multitude of mechanisms or processes that minimize inappropriate efforts at reproduction. Bars to hybridization, which is wasted reproductive effort, are as varied as the animal kingdom itself. The question is whether, given procreative fealty, one can say more about "species identity."

One contributor, Gallup (Chapter 11), has reworded this query: Can one demonstrate self-awareness? The argument that implicitly follows is that organisms with an interior map of their own appearance are most likely to be suitable subjects for speculations on broader "specific" mapping. Gallup's ingenious mirror studies allow an affirmation. Some animals, even if they chase their own tails, can be shown to be aware that the tail is attached. Where the experiment fails, as it apparently does for most beasts, we are left where we began, ignorant of the existence of a self-image.

Let us begin again, however, by considering a case where evidence for self-recognition is at hand. Does this, in fact, afford a basis for a useful definition of "species identity"? There is, of course, no problem in stating a definition, per se. Definitions are often arbitrary: We may follow Lotka's (1956) lead and call a circle "a plane figure bounded by four sides and with four angles." The definition itself is unassailably self-consistent and operationally sound. It is merely not useful in this case because of a clash with existing conventions.

Roy (Chapter 1, this volume) has proposed a definition that would require the demonstration of one or more responses (courtship, territorial defense, communication, or allelomimetic behavior) to conspecifics. Assuredly, among Gallup's self-recognizing primates, these responses are prevalent in the appropriate contexts. Just as assuredly they are present in other species for whom no evidence of self-recognition exists. Some observations (see below), furthermore, suggest that self-recognition and species identity apparently are largely independent constructs.

One should reconsider, however, the problem of species identity from a functional point of view. Should one expect animals to evolve cognitive schemata of their own kind?

Ecologists generally agree that the significance of specific differences is that they reduce interspecific competition for scarce resources. If two warblers in the same tree differ in bill size or in which part of the tree they forage, they are less likely to impinge upon one another's food supply or nesting space than would two more similar individuals (MacArthur, 1955). Interspecific hybridization would reduce the existing partitioning of resources, adversely affecting the fitness of both hybridizing partners, because the hybrid offspring would have similar ecologic habits.

This appears to be a principal factor in the evolutionary development of bars to hybridization. Such bars may include gametic or chromosomal incompatibilities, structural differences precluding copulation (body size or shape of penis), and differences in the timing of reproductive activities or in the courtship displays that may precede breeding. These last generally waste least time and gametes, since they can abort relationships early in their formation. The effectiveness of behavioral differences depends, however, on the recognition of the appropriate display, whether it be the dance frequency of a fruit fly or the colors of the speculum of a duck's wing. Where recognition of the appropriate display or plumage, as with Cooke's snow geese (Chapter 5), does occur, does it then signal the existence of "species identity"? It need not. Consider the dilemma of a duck. In many species of the Anatidae, the sexes differ greatly in appearance and in their courtship displays. Indeed, the females of several species may be more similar to each another than to males of their own species, e.g., mallards (*Anas platyrhyncos*), pintails (*A. acuta*), or gadwalls (*A. strepera*). Whatever the basis on which a male selects an appropriate female, it must differ from that on which other males are identified. It is possible, of course, that male ducks, when mating, are unselective—any female will do—and that the hens exercise choice (note Schutz, 1965). The paradox remains—the hen must select among males, none of which resemble her. In short, recognition of an appropriate mate involves cues different from those likely to be involved in self-recognition.

A comparable paradox is provided by certain Pomacentrid fish. *Pomacentrus flavicauda,* for example, defends a territory against both its own kind and certain alien species, many of which do not resemble it closely (Low, 1971). This is not the only instance of interspecific territoriality. If the responses that define recognition of one's own kind are not exclusively directed to one's own kind, they cannot be used as criteria for a "species identity."

In many species communal feeding groups form where the food is especially superabundant and transitory, such as a school of fish herded into shallows by organized flocks of pelicans. Pelicans clearly benefit by recognizing members of their own species, so that they can join in the cooperative fishing which is of such benefit to pelicankind. Other feeding groups may be composed of individuals from several species, however. Multispecies flocks are particularly common among tropical birds or among overwintering birds of the temperate zones. The existence of communal feeding groups must therefore be viewed as independent of the ability to recognize their own kind.

The answer to the question of function can be hazarded: there are indeed occasions where the ability to identify conspecifics must pay. But one cannot generalize about when the appropriate occasions will arise. This precludes general statements about "species identity" and limits the concept to particular species in which a clear empirical basis for asserting self-recognition does exist. Neither proper mate selection nor the inducements to formation of feeding groups, two of the most important aspects of sociality, affords a basis for a useful concept of species identity.

The inquiry into the existence of a "species identity" began with a comment on the lonely and troubled seeking solace from their own kind (Chapter 1, this volume). There is an opposite side to that coin—privacy—the avoidance of one's own kind. It is as ubiquitous a trait as sociality. The dying elephant and the depressed man may avoid contact as often as they seek it. "I want to be alone" is not so rare a cry.

Privacy is best viewed as "a regulatory process that serves to selectively control access of external stimulation to one's self or the flow of information to others" (Klopfer and Rubenstein, 1977). It is so clearly related to identifying one's self to another as a conspecific as to allow us to consider it a complementary process. It may even depend upon precisely the same processes as socialization. Territorial behavior may either control or limit interactions with conspecifics or, by providing a periphery where interactions can occur, it may maximize social contacts (Armstrong, 1947).

Even highly social animals display "privacy" in the sense that they withhold information about themselves that might otherwise benefit a conspecific competitor. A fatigued fish, for instance, having to defend its breeding territory, may overtly signal "I will attack unless you retreat," even though in reality it will withdraw when challenged. As Klopfer and Rubenstein (1977) put it, "privacy should be considered an essential element in the struggle for survival, and its existence underscores the claim of Wallace (1973) and Otte (1974) that animals actively deceive others." Humans do this, too, of course. We are experts in withholding information for the purpose of achieving personal, political, sexual, and social goals.

Like sociality, privacy also entails costs as well as benefits (see Klopfer and Rubenstein, 1977). The equilibrium, where a marginal gain in fitness due to increased privacy is just offset by a loss in fitness, depends upon the specific features of the organism's social and physical environment. This follows from the fact that the shape of the fitness functions is determined by environmental variables. Hence generalizations respecting the desired level of privacy or sociality are only possible at a very abstract level.

The evolution of privacy or sociality, in short, is readily interpretable in economic terms; where the cost margin is profitable, it will persist. The variables that define the cost–benefit equation will vary for each species, however, so for every case the details must be individually considered. This argument could apply as well to the notion of "species identity." Whether and when it is important and how it is mediated are questions that defy generalizations beyond the level of governing principles. The behavior of some species may indeed be usefully viewed as incorporating a "species identity," but the challenge to discover *which* species (and why) has yet to be met.

Acknowledgment

I am indebted to M. Aaron Roy for his stimulating critique. This work was supported by NIMH Grant MH 04453.

References

Armstrong, E. A. 1947. *Bird display and behavior.* Fair Lawn, N.J.: Oxford University Press.

Klopfer, P. H. and D. Rubenstein. 1977. The concept "privacy" and its biological basis. *J. Soc. Issues 33,* 52–65.

Lotka, A. J. 1956. *Elements of mathematical biology.* New York: Dover.

Low, R. 1971. Interspecific territoriality in a Pomacentrid reef fish, *Pomacentrus flavicauda. Ecology 52,* 648–654.

MacArthur, R. H. 1955. Population ecology of some warblers of Northeastern coniferous forests. *Ecology 39,* 599–619.

Marshall, H. J. 1954. *Bower birds.* Fair Lawn, N.J.: Oxford University Press.

Otte, D. 1974. Effects and functions in the evolution of signalling systems. *Ann. Rev. Ecol. Syst. 5,* 385–418.

Schutz, F. 1965. Sexuelle Prägung bei Anatiden. *Z. Tierpsychol. 22,* 50–103.

Wallace, B. 1973. Misinformation, fitness and selection. *Am. Nat. 107,* 1–7.

Chapter 19

Species Image, Self-image, and the Origin of Persons

Lawrence B. Slobodkin

The questions asked in the chapters of this volume range from those concerned with behavioristic examination of such things as the movements of organisms towards or away from various objects in their environment (including other members of their own species) to those in which the organisms have names as individuals and are assumed to be capable of giving and receiving love. The forms in which the questions are asked begin to sound like those usually considered in literary and poetic contexts.

That is, perhaps without intending to, the investigators themselves become involved with their animals on much the same level at which the animals become involved with each other. The purely objective and manipulative role dissolves into that of a personal involvement that transcends the scientific–maternal or pet–owner relation. This is itself a phenomenon of great interest. What must "I" and "thou" be before an "I–thou" relationship is possible?

One could infer that in the invertebrates we would find neither new insights into the nature of self nor models for clinical association. The mechanism of species recognition in hydra, protozoa,

Contribution No. 282, Ecology and Evolution Dept., SUNY, Stony Brook. This paper reflects study at the Smithsonian Institution, where the author was a Guggenheim fellow and Senior Visiting Scientist and, research support from the National Science Foundation General Ecology Program.

and bacteria melts into problems of surface chemistry and protein specificity. Clinical problems cast their shadow on the discussions of humans and of dogs, who are man's emotional if not phylogenetic relatives.

One regrets the absence of any discussion of the Cephalopoda, since they may have complex self-related behavior. That is, octopus brains are large and complex, and learning is well demonstrated (Wells and Young, 1975). The Arthropoda also present such a startling array of behaviors that one would have wished for a discussion of them. It might have been of interest to include the complexities of the ants' nest; the curious eroticisms of the spiders, in which mating involves the male pumping sperm into a web rather then into a female (Meglitsch, 1972); and fiddler crabs, in which elaborate courtship dances occur. The readers should refer to the rather vast literature on invertebrates most recently summarized by Wilson (1975).

The questions at issue change among chapters. For reptiles (Chapter 3) the central issue seems to be how interspecific miscegenation is avoided. In fish behavior (Chapter 2) the focus shifts slightly, to how the early experience of a fish alters the likelihood of its associations in the future. In fish there is already some evidence that associations among animals are to some degree self-reinforcing, so that the fact of having associated with a particular type of fish alters the probability of associating with that type of fish in the future. In general, early-association types tend to persist as the animals get older.

In the analysis of birds (Chapter 4), and particularly snow geese (Chapter 5), the phenotypic polymorphism between blue and white geese requires an explanation in terms of mating preferences of individuals, which in turn are based on their sibling associations in the nest. This is of critical importance, because here memory has itself become an evolutionary force. That is, since the geese have somehow developed the biological property of having their present behavior contingent on the "memory" of who their siblings were, the gene frequency distribution in the populations is now in part a result of that memory. In population genetics, genetic frequencies are usually considered a resultant of the interplay between various selectional and mutational forces. Memory in the snow goose modulates those forces to some degree. Even if the blue-color gene has no selective meaning one way or the other, it is possible to describe the changes in gene frequency with time in the goose population by the model based on relative abundances of the different sibling groupings and the effect of these sibling groupings on later mating partners. A model of gene-frequency change which did not depend on selective alterations imposed from outside the organism, but was contingent on the preexisting properties of the organism, would then have been constructed.

While this simple concept is of enormous importance, since it underlines the necessity of bearing in mind the preexisting biology of organisms when one considers the actual evolutionary changes that may occur in them. This becomes of increasing importance in the discussion of mammals and in particular of primates.

In the chapters on dogs and primates, the nature of the evidence changes in another way. There seems to be an implicit assumption that in the mammals the degree of variation in behavior among individuals cannot be understood without describing in detail the individual life history of the animal itself. Rather than in statistical arrays of animals, the evidence is presented as case histories in which the animals are identified individually. For each animal the details of its early life are discussed at fair length. This is an even further departure from the paradigm of population genetics. We will return to this problem later.

Further, the nature of the questions being asked alters. In terms of dogs and apes, the examined behavior is rather delicate, and we do not simply ask whether or not a dog lies down near another dog rather than near a wolf, a lion, or a tree, but rather if it is a "well-adjusted" dog that can carry out its full and rather complex social role in a social context which may be that of dogs, of people, or of both.

Scott (Chapter 6) showed that there are specific age periods required for socialization of a puppy and that the process of socializing to humans is distinct from socializing to dogs. A well-adjusted puppy in a purely canine society behaves as a wild animal in a human society, and a puppy deprived of contact with other dogs during a critical period does not socialize with dogs at all. In addition he cited evidence that a certain amount of social enrichment is in fact required if the dog is to behave in a generally normal way.

At this point it is as if we are discussing the personality development of a human where environmental richness, forms of early socialization, and early experience are generally assumed to alter drastically future behavior patterns and apparent future norms.

Anyone who has lived with a dog will not doubt the complexity of its personality; but the inference that it really is a personality, that it really is a person thinking of itself as a person, is not unequivocally supportable. That is, dogs seem to show guilt, pride, deceit, hatred, and affection, and I have evidence from my own dog that they do have explicit dreams. My *sense* of personality in dogs is not evidence, however. It is at best a merely plausible assertion generally accepted by dog owners, but not demonstrable by normal scientific means. It is generally possible to find other, perhaps more farfetched, explanations for the "personlike" behavior of dogs.

It is not until we get to research on primates that one deals with organisms in which the fact of personality is objectively demonstrable. The brilliant experiments of Gallup (Chapter 11) are in effect unequivocal demonstrations of the presence of self-image in the primates. Gallup's and other mirror studies have unequivocally demonstrated that the chimpanzee, gorilla, and orangutan have overt behavior which can be most parsimoniously explained only in terms of an animal having an internal image of itself which somehow has a strong visual complement.

The fact that mirror studies have been negative in this sense for other primates and for other mammals does not demonstrate the absence of a self-image in these species. Gallup suggests that the mirror seems to be showing the chimpanzees something they had always known; that is, the mirror provides a mechanism for making overt behavior out of the subjective sense of identity. Is it too much to suggest an analogy with a musical instrument that permits explicit demonstration of creativity that is otherwise latent? The analogy is not perfect. Bach in a noninstrumental world would perhaps have had his children shout the trumpet parts. It is certainly possible and even plausible to consider that, in many animals which have been subjected to mirror tests and have not shown this response, there are at least some who have a sense of personal identity which is effectively the same as or very similar to that of the great apes, but does not manifest itself in visual terms. Consider, for example, the behavior of preadolescent boys and elderly professors who do not seem to use mirrors in the same way as other people.

The correspondence between the behavioral anomalies in great apes and monkeys reared in the absence of normal social context and that of humans reared in abnormal social contexts is striking, as demonstrated in Part II of this volume.

What we have seen, then, is a general survey of behavior related to conspecifics in many kinds of vertebrates. Had we continued the invertebrates, we would have found phenomena not dissimilar to those in reptiles and fish, including behavior related to specific odors and associations between adult behavior and early development. Insects reared on the same host plants are likely to return to those plants and mate with other individuals of similar rearing, for example (Dethier, 1954).

Stepping back from the problem for a moment, however, it is clear that the level of miscegenation in the animal kingdom is, by definition, not excessive. Certain populations may have interbred. They are either extinct or one now thinks of them as single populations. In any case what one sees now is a collection of solutions to the problems of biology in general and of species identity in particular.

There is no *a priori* reason to believe that the rather complex socialization processes of dogs and apes have as their primary function species identity, since protozoans and herrings and salmon do quite well in species-identity choices without all this difficulty or subtlety. It seems best to turn the argument around and consider that the problems of species identity and the elaborateness of the species-identity mechanism in dogs and primates arise because the animals are capable of learning in general and of solving certain problems, and that the complexity of their normal life and the simple development of species identity are incompatible.

The development of the capacity to learn seems to be fairly general. Certain organisms can learn odors but cannot learn individual identities, etc. It would be impossible at the moment to deny that learning capacity in many animals may be channelled or restricted in this way. Recent analysis of dyslexia indicates that some children learn material presented verbally but cannot learn the same substantive material presented in written form (Lerner, 1976). To develop the capacity for learning while maintaining the automaticity and simplicity of interspecific recognition, however, may be a very difficult thing.

In the abstract this can be circumvented by evolutionarily preserving some engram which excepts interspecific recognition from the general loosening of stimulus–response relationships present in the learning process. While this cannot be unequivocally denied, the evidence for it in animals capable of extensive learning is not apparent. The simultaneous development of complex learning capacity may be incompatible with the maintenance of simple species-identity mechanisms, in the same sense that the development of flight is incompatible with the development of very large body size and excessive body weight. That is, there always seem to be constraints on what evolution can accomplish and which paths evolution will take, depending on the preexisting properties of the organism. Most of these constraints involve grossly physical attributes and were already apparent to Galileo (Thompson, 1948). A recent development in evolutionary speculation focuses on "optimality" considerations, i.e., the incompatibilities among different physiological and behavioral activities, each of which contributes to the survival and well-being of the organism but interferes to some degree with the others. The sophistication and the usefulness of this line of thought differ among authors. In the present context, a general incompatibility is suggested between the degree to which learning capacity is developed and the degree to which decisions of a relatively subtle kind can be behaviorally fixed. We might also expect that the selective circumstances which promote learning capacity are the same as those that promote social intimacy and flexibility.

In general it can be demonstrated (but only to the degree that other things are equal) that a greater reproductive output per animal has selective advantage (Slobodkin and Rapoport, 1974; Stearns, 1976). The restriction that other things be equal sets constraints on how that reproductive advantage may be attained. In particular, for animals with a low reproductive rate and with a physiologically demanding reproductive process, as in female primates, it is "easier" to enhance total reproductive capacity by increasing longevity rather than by increasing the number of young in a litter. We would expect, in most animals that have just a few young in a litter, that there is an ongoing pressure toward greater longevity. The greater the longevity, the more meaningful the learning process.

If an animal's life span is only one year in a sharply seasonal environment, there is no selective advantage to its learning the properties of springtime, since it will never see another spring. Learning in general would seem to have its advantageous properties if there is a recurrence of reasonably similar events as stimuli during the life of the organism, with the added stipulation that these stimuli and the appropriate response to them differ somewhat from generation to generation. If this were not the case, stereotyped behavior would seem advantageous.

As a female primate grows older one would expect her to be surrounded by her progeny of various ages and conceivably even by her grandchildren. The female is emphasized in this discussion, since there is more variation that may be expected in male behavior among species, but the nursing system may be expected to forge some kind of reasonably permanent bond between mothers and their offspring.

As animals grow older, their present reproductive value is reduced, and the ratio of the progeny they have contributed to the progeny they will contribute is reversed. This can be portrayed in a formal theory, which will not be developed here. In compensation, the animals' value as repositories of information may have increased (Fisher, 1930). If learning continues throughout life, the amount of information in an old animal will be so great that, by following such an animal, the young one can short-circuit some of the pain and misery of stimulus–response learning. Even if learning does not continue into adult life or if patterns of behavior are fixed with age, the older animals' behavior will embody information relating to problems that occur relatively rarely; if these problems are important, the behavior of these old animals may be of enormous value should young animals imitate them.

Rowell (1972) has pointed out that in years of severe drought in Uganda troops of olive baboons having old females survive somewhat better than those without old females. "Old" in this context means of the order of 30 to 35 years. There are also data (Slobodkin, 1962) indicating that groups of elk containing old animals survive better during periods of severe snow than others do. There are two possible mutually nonexclusive explanations of this. One is that only in a very well-fed troop in excellent condition can an old animal survive, and therefore the animals are in better shape to deal with hard times. Another explanation may be one suggested by Rowell. Old females remember the procedures followed during the previous severe drought or previous heavy snow, and by following these old animals the entire troop survives better.

Consider the baboons. What is required in a bad drought year is going to water holes which still contain water but may not contain all the amenities normally surrounding a baboon water hole. Good water holes have food sources nearby and resting trees near them that are relatively safe from predators while the animals sleep. In a drought year the primary requisite of a water hole is that it contain water, and under the stress of drought what might have been thoroughly unattractive water holes become extremely valuable. The frequency of drought in East Africa is such that a 35-year-old animal will probably have survived at least one and perhaps two severe droughts in the past. If this animal remembers the procedure for finding water in bad years, this memory would be of enormous social value to her progeny, and to the degree that the capacity for this type of memory is related to hereditary material, there would be a selective pressure for an enhancement of the capacity for memory. It may also be the case that the capacity to learn is incompatible with rigidity of response.

But there is a problem related to the learning process which will bring one back to the problem of identity. Assume that the baby baboon follows her mother or grandmother to a bad water hole in a bad drought year; the following year the rains come, the situation returns to normal, the bad water hole is abandoned, and there is no further reinforcement of the information about that water hole. Thirty or more years later the drought comes, and the normal water holes are no longer useful. The baby baboon, now a matriarch, is capable of returning to the bad water hole. Such an animal will have learned something at a very early age and then used that information decades later.

How is learned information retained? There is a possibility that the reinforcement of this information is internal, at least in the baboon. This is not to suggest that, of necessity, long-term memory storage must involve internal reinforcement. It is possible, for example, to consider imprinting phenomena in some organisms, but the primates' strong selective pressure for learning capacity may make an imprinting-based memory more difficult and require another kind of reinforcement. One mechanism for such reinforcement may well be the development of some kind of internal self-image, which puts the organism in the position of reinforcing its own previous learning by an internal dialogue (Slobodkin, 1978).

The evidence, at any rate, for the internal self-image is present in the higher primates before narrative language. There seems to be no evidence of straight storytelling among higher primates other than man, whatever the other types of communication may be. I would like to suggest that the capacity to tell a story came with the development of language, but long before that the stories were present internally. This would have had immediate selective value in terms of reinforcing information over a long time interval, so that it could be used when appropriate conditions for its use reoccurred.

If this is at all valid, it would raise the problem that the organism would have to fit such things as mating into its own internal narrative, thereby raising the possibility of misconstruction of species identity as reported in the papers on dogs, apes, and humans. To misconstrue is a highly complex intellectual act. To be in danger from one's own misconstructions is to be almost human.

Gallup has suggested in this volume (Chapter 11) and elsewhere that providing sign-language capacity to chimpanzees and mirrors to the great apes does not seem to give them a vehicle for learning so much as a method of expressing what they already knew. The internal narrative which must be involved in the situation of objective self-awareness as demonstrated in the great apes probably has a precursor in other mammals which cannot be so objectively demonstrated. The advent of language may have provided a medium for expressing the stories that animals were already telling themselves.

Quite clearly this is the sketch of an outline only. It is substantive in the sense that I am asserting self-awareness did not arise because of a necessity for a species' identity awareness; rather, the problem of abnormalities of species identification results from the development of self-awareness, which occurred for other reasons.

Implicit in this and in the other chapters in this volume is the assertion that as self-awareness develops, the degree of predictability of behavior from genotyic considerations is correspondingly reduced. When a puppy can develop a sexual relationship with a vacuum-cleaner bag, as reported in one of the cases cited by Scott (Chapter 6), it becomes very difficult indeed to tie explicit behaviors to genetics. This assertion is contrary to the current tendency to use insect models as explanations of mammalian and even of human behavior. See, for example, the assertions by Trivers (1974), Wilson (1975), and more particularly by Weinreich (1977) who propose general rules of sociology based on optimization criteria and population-genetics machinery extrapolated from the insects. The material in this volume, in fact, leads to the impression that the development of self-awareness is an enormous liberating force breaking down the degree to which the historical future of man and of the higher apes is in any way predictable on the basis of evolutionary analogies with other species. The distinction between history and evolution has been discussed elsewhere (Slobodkin, 1978).

Species identity as a behavioral event turns out to be much simpler than species identity as an internal or psychological event. In general, sexual partners match as to species. For some this is essentially automatic. For others, a social and ecological context enters into species identity in fascinating and half-known ways.

References

Detheir, V. G. 1954. Evolution of feeding preference in phytophagous insects. *Evolution 8*, 33–54.

Fisher, R. A. 1930. *The genetical theory of natural selection.* Oxford: Clarendon Press.

Lerner, J. W. 1976. *Children with learning disabilities: Theories, diagnosis, teaching strategies.* Boston: Houghton, Mifflin.

Meglitsch, P. A. 1972. *Invertebrate zoology* (2nd ed.). Fairlawn, N.J.: Oxford University Press, 711–746.

Rowell, T. 1972. *Social behavior of monkeys.* London: Penguin.

Slobodkin, L. B. 1962. *Growth and regulation of animal populations.* New York: Holt, Rinehart and Winston.

Slobodkin, L. B. and A. Rapoport. 1974. An optimal strategy of evolution. *Qu. Rev. Biol. 49*, 181–200.

Slobodkin, L. B. 1978. Is history a consequence of evolution? In P. G. Bateson and P. Klopfer (Eds.), *Perspectives in ethology.* London: Plenum.

Slobodkin, L. B. 1978. The peculiar evoluationary strategy of man. *Proc. Boston Colloq. Philos. Sci. 31.*

Stearns, S. C. 1976. Life-history tactics: A review of the ideas. *Q. Rev. Biol. 51*, 3–47.

Thompson, D'A. W. 1948. *Growth and form.* New York: Macmillan.

Trivers, R. 1974. Parent–offspring conflict. *Am. Zool. 14*, 249–264.

Weinreich, J. D. 1977. Human sociobiology: Pair-bonding and resource predicability (effects of social class and race). *Behav. Ecol. Sociobiol. 2*, 91–118.

Wells, M. J. and J. Z. Young. 1975. The subfrontal lobe and touch learning in the octopus. *Brain Res. 92*, 103–121.

Wilson, E. O. 1975. *Sociobiology.* Cambridge, Mass.: Harvard University Press.

Chapter 20

Some Issues in Prezygotic Species-typic Identification and Isolation: A Comparative Evaluation

Slobodan B. Petrovich
Eckhard H. Hess

In the 1850s Charles Darwin and Alfred Russel Wallace independently proposed a "new" process of creation—evolution by natural selection. Their contribution planted the seeds of the powerful scientific and intellectual conceptualization that is still unfolding; evolution is becoming comprehensible as a process, inescapable as a fact, and all-embracing as a concept.

The influence of evolutionary processes on behavior and behavior on evolution is a major concern in ethology. The orientation to date has been bimodal, emphasizing phylogenetic relatedness and comparative studies of homologous behavior patterns, as contrasted to investigations of the ecological adaptation of behavior and the study of teleonomy. Moreover, the ethological approach to behavioral analysis also allows us to see the issues of proximate and ultimate behavioral causation in perspective. On this planet, for example, all animals have the same code of life. In addition the lower the level of intraorganismic structure and function, the greater the phylogenetic similarity (e.g., comparative analyses of human and infrahuman tissues, cells, and hormones such as vasopressin). The similarity of structure and function also increases across all levels of organization as the proximity to a common ancestor increases (e.g., comparative analyses of human and other primate intraorganismic and organismic processes). Similar behavioral functions, furthermore, among phylogenetically unrelated or distantly related forms may result from similar selection pressures caused by similar ecological contingencies (e.g., comparative analyses of territorial aggression). Thus similar behavioral functions in different species often have identical survival value and provide us with information as to how similar selection pressures generate, through parallel or convergent evolution, similar behavioral outcomes.

Ethologists have also demonstrated that each species often reveals unique patterns of behavior. What is obscure, rudimentary, or nonexistent in one species may be extremely exaggerated in another. By dispassionately cataloguing the behavioral phenomenology of different species, it is possible to filter out the significance of specific behavioral phenomena. Any behavior pattern, moreover, however deceptively expressed—ranging from obscure to exaggerated—thus stands a better chance of being objectively described and understood from both phylogenetic and ontogenetic standpoints. Classificatory models derived from data produced by the ethological approach to behavioral analysis can establish interspecies comparisons of homologous behaviors (behavioral similarities due to common descent) and of functionally analogous behaviors (behavioral similarities based on commonality of function rather than on the genotype or on the structure) (e.g., Brown, 1975; Eibl-Eibesfeldt, 1975).

In light of these introductory considerations, can we have a meaningful discussion and comparative evaluation of species-typic identification and isolation processes as well as their behavioral consequences?

Isolating Mechanisms: Lessons from Cricket Bioacoustics

The scientific investigation of processes maintaining species identity and the integrity of the species-specific gene pool has preoccupied naturalists and evolutionists of this century. These processes have historically been packaged under the label of *isolating mechanisms*. Isolating mechanisms are numerous and varied. The principal ones form two basic groups, *prezygotic* (those preventing fertilization and zygote formation) and *postzygotic* (those allowing fertilization but producing inviable, weak, or sterile hybrids). Among prezygotic isolating mechanisms we find (1) geography and habitat, where populations occupy different habitats and, even though reproductively compatible, do not come in contact with each other; (2) seasonal or temporal factors, where populations mature reproductively at different times; (3) ethology, where there is difference and incompatibility in the behavior of these isolated populations; and (4) mechanics, where anatomical differences prevent reproduction.

As an example, we will briefly review the literature on the songs of crickets. The research is representative of ethological methodology. It includes the type of questions that were asked, the experimental subjects employed, the nature of the behavioral response under investigation and its measurement, and comparative analysis within and across species as well as the consideration of ecological and evolutionary factors. Such an example also illustrates the process characteristic of ontogeny, causation, function, and evolution of species-typic isolation and of identification in simpler invertebrate systems. Furthermore, invertebrates, and insects in particular, tell an interesting story (e.g., Wilson, 1971) and their message (even though unheard of in this volume) is important for understanding identification in different classes of vertebrates.

There are approximately 3000 species of crickets, of which field crickets make up a special group of about 400 species. Field crickets are most familiar. They are relatively large, yellowish brown insects known for their rather loud, musical chirping. Male crickets produce sounds by rubbing together stridulating areas located on the forewings and utilize a rapid fluttering motion to produce a typical vibrato chirp. The receiving auditory organs are tympana located within slits on the forelegs. Most cricket species chirp at night, some during the day, and others both day and night. In general, understanding of neurophysiological mechanisms involved in cricket bioacoustics has few parallels, if any, in the animal literature (Alexander, 1968; Bentley and Hoy, 1974; Ewing and Hoyle, 1964; Huber, 1962).

In a southeastern region of the United States during summer there are as many as 20 different species of tree crickets producing discrete sounds, mostly the male's calling song, the function of which is to attract the female for mating. How does a female distinguish the sounds of a conspecific? Studies have demonstrated that males of each species have a particular pulse rate in their song, and it is this pulse rate that provides a female with discriminative cues. It is also interesting to note that the metabolic and physiological processes in a cricket are functionally affected by outside temperatures, so that a pulse rate in the song changes with temperature, earning some species the appropriate label of "thermometer crickets." The refinement of the evolved system is remarkable when one considers that physiological mechanisms which determine females' responsiveness to a signal change at the same time in a fashion that parallels the males' pulse rate.

The sound-producing repertoire of the male cricket serves a number of functions: (1) facilitating and establishing sexual contact (the calling song); (2) mediating sexual attraction at a relatively short distance (the courtship song); (3) signaling departure of a courted female (the courtship-interruption song); (4) repelling or dominating other males (the aggressive sound); (5) maintaining contact between a mated pair (the postcopulatory song); and (6) a wide range of what appear to be recognition sounds (Alexander, 1966, 1968).

How does this brief commentary on cricket bioacoustics illustrate the importance of acoustic communication in cricket speciation and evolution? What are some of the factors that maintain the species-specific integrity of a gene pool of some 20 different species of tree crickets that are not geographically isolated? The species-specific characteristics of the male calling song and the recognition of that song by a conspecific female were identified as an important isolating mechanism (Alexander, 1966; Walker, 1957).

Viewed in the context of our understanding of some of the evolutionary processes, crickets tell an interesting overt and covert evolutionary story. Among the 3000 species, many are isolated by their geography and habitat. When a number of species occupy the same habitat, then temporal, ethological, or mechanical isolating mechanisms maintain species integrity. Thus one species will chirp at night and another during the day (temporal isolation). If more than one species occupy the same habitat and "sing" at the same time, then the differences in the pulse rate (ethological isolation) maintain species identity. Acoustic signals and communication serve in the prezygotic isolation of closely related species.

The literature on the ontogeny of acoustic communication in crickets also deserves more attention from behavioral scientists than it has received to date. It should be kept in mind that many insects mature without hearing the signals of their own species, and they sense many sounds that have absolutely no resemblance to any signals that they as mature adults must eventually produce. As Alexander (1968) has pointed out, there must have been intensive selection pressure for resistance to irrelevant acoustic influences and toward fixed relationship between acoustic genotype and acoustic phenotype. Experiments investigating the genetic correlates of communication signals in several species of crickets offer further support to this thesis (e.g., Alexander, 1966, 1968; Bentley and Hoy, 1974; Fulton, 1933). Fulton, for example, as early as 1933 hybridized *Nemobius allardi* and *Nemobius tinnulus.* These are two sibling species of ground crickets that mature at the same time, overlap geographically and ecologically, but sing different songs. Fulton was able to develop F_1 and F_2 hybrids, carry out F_1 backcrosses with parental species, and analyze the songs of various crosses. Fulton's results were generally clearcut and straightforward. Pulses in the song of F_1 hybrids were delivered at a rate intermediate between those in the songs of the two parental generations. The songs of backcross progeny were more like the parent utilized in the backcross. Subsequent literature on other species has further elucidated the genetic determination of the song pattern of each cricket species. The songs are phenotypic expressions of different genotypes thereby offering evidence that links together genetic information, developmental processes, structural and functional organization of the neuroendocrine system, and behavior (e.g., Bentley and Hoy, 1974).

In summary, it should be kept in mind that crickets are sensitive to information in all five sensory channels: acoustic, chemical, visual, tactile, and thermal. This review selects one example to demonstrate how discrete acoustic signals function in species-typic isolation and identification, while it also offers overt and covert evidence for the proximate and the ultimate causation of such behaviors.

A Discussion of Some Conceptual Issues Relevant to Understanding Vertebrate Identification

Given an example of an ethological isolating mechanism in cricket speciation and evolution, we may be tempted to wonder: How relevant is this cricket–invertebrate example for what one wishes to know about vertebrates, especially primates? The meaningful answer that goes beyond the crude and unsophisticated analogy (e.g., what one sees in crickets is an attenuated or simplified version of what occurs in primates) is by no means easy. Nevertheless, we think that a judicious application of the comparative method points out a way. One of the strong points, for example, of comparative analysis is that it puts the issues of species identification in perspective. In smaller animals, particularly invertebrates with relatively short life spans, selection favored the evolution of inherited, genetically preset behavior patterns adapted to "normal" ecological requirements, one example being the cricket song (e.g., Bentley and Hoy, 1974). This should not be misunderstood to suggest that acoustic experiences in a given context have no effect on cricket signalling. At the same time, however, it is important to emphasize that current evidence clearly supports a relatively fixed relationship between acoustic genotype and acoustic phenotype.

Now let us turn to vertebrates. McCann (Chapter 2, this volume) in his review of species identification in fish concludes, "Nearly all indications support the assumption that species identification in fish is primarily an innate mechanism, a conclusion which applies to all their social behaviors." In a similar vein Froese (Chapter 3) summarizes his review on reptiles by stating, "The role of experience in the identification process in reptiles is virtually unknown." As both authors indicate, a much broader and detailed data base is required before one can make definitive claims. Nevertheless, we wish to suggest that, given the current knowledge of evolution and ecology in these two classes of vertebrates, future research findings will most likely follow closely in the footsteps of the cricket model.

In many birds and mammals, the ontogeny of species-typic identification often provides a classic example of the genotype-dependent process involving the coaction of inherited predispositions and learning. Moreover, examples given by Cooke (Chapter 5), Shapiro (Chapter 4), Guyot, Cross, and Bennett (Chapter 7), and Scott (Chapter 6) support the thesis of species-typic, phylogenetic, and ecological constraints on such learning experiences, and at the same time underscore the fact that these animals learn, on the basis of their "biological preparedness," to form certain associations that appear to be characteristic of the natural history of these species.

Toward the other end of the continuum are those vertebrates (particularly primates and some birds) whose life spans are relatively long and start with a period of infancy that often involves extensive social interactions among parents, offspring, and siblings. In such circumstances opportunities for modification of behavioral responses are abundant, and the advantages of preset, genetically fixed response patterns involved in species identification are reduced. Natural selection appears to have favored specialization toward a plasticity represented by the evolution of complex hierarchies of learning and cognition, one consequence of which appears to be the emergence of *animal tradition* (Lorenz, 1969). It seems that these inferences are also supported by the data presented in this volume.

Scott's and Gallup's chapters particularly reflect our views and provoke an additional commentary. In many vertebrate species, the evolution of learning (e.g., imprinting, social learning, and cognition) appears to have played an important role in the unfolding of species-typic identification. The emergence of complex forms of social interactions requiring not only the recognition of various social strata (e.g., groups and families), but also the recognition of an individual, has contributed to selection pressures in the direction of environmental-stimulus control and away from some of the preset, genetically programmed modes of exchange. The recognition of an individual appears to be one aspect of the complexity of social evolution observed in these vertebrates. Some species have consequently evolved and employ an array of acoustic, chemical, visual, tactile, and thermal signals pertaining to individual recognition. Once established biologically and culturally, the increased flexibility of these processes appears to have paved the way for the more remarkable forms of identification, including those of the existence and the transient nature of *the self*. In contrast other vertebrates and invertebrates, having neither the complexity of the neuroendocrine system nor the social and ecological requirements for this degree of interindividual recognition, apparently rely mostly on relatively simple, genetically programmed exchange of signals.

Imprinting, Attachment, and Species-typic Identification and Isolation: Another Viewpoint

The scientific study of the imprinting phenomenon has been plagued by conflicting experimental results that have nourished diverse and divisive interpretations. Investigators with different perspectives and orientations presently cannot even agree on a consensually valid definition of imprinting. The inclusion and the critical assessment of many variables that have contributed to the current climate are not only beyond the scope of this discussion, but would only repeat the content of other chapters in this volume, Shapiro's in particular. Nevertheless, a few comments airing some of the concerns are in order.

Different laboratories often obtain their subjects from a variety of settings and sources. In most cases animals are maintained under different conditions and exposed to different stimuli and experiences prenatally, postnatally, during imprinting as well as between imprinting, and during the test period involved in the evaluation of the consequences of the imprinting experience. Procedures and instrumentation utilized to imprint subjects vary greatly from one laboratory to the next. There is no agreement about measuring the features of the imprinting process itself and, of course, there is much disagreement as to what constitutes the most useful measurement of the consequences of imprinting. The proliferation of different theoretical positions has not been followed by judicious application of methodological techniques designed to control for experimental bias. At times the problem appears to have been compunded even further by misreading or at best selective reading of previous literature (Petrovich and Hess, 1977).

Understanding these issues is helpful, particularly if one disagrees with some statements and commentaries in this volume. A brief and limited overview from a historical perspective of important features of imprinting and attachment is more constructive than quibbling over the points of contention.

The term *imprinting* ("stamping in") is the English translation of the German word *Prägung,* which Lorenz used in 1935 to describe a process by which newly hatched young of many species of birds (e.g., greylag goslings, jackdaws, mallards, pheasants, and partridges) form a relatively rapid social attachment to a biologically appropriate conspecific or, in its absence, a surrogate object. Thus imprinting is a label, a heuristic tool, an abstraction conceived to vary along a dimension of theoretical and analytic abstractions that are comprised of various levels of behavioral analysis, ranging from those on a microlevel (e.g., biochemical and physiological analyses) to those on a macrolevel (e.g., social and ecological analyses).

Lorenz's original treatise on imprinting was confined to avian species. Lorenz, moreover, pointed out that a demonstration of the imprinting process in one species of birds in no way justifies unwarranted generalizations to other species regardless of their phylogenetic relatedness. Various concepts of imprinting have subsequently served as heuristically useful models for the investigation of early mammalian socialization (e.g., Bowlby, 1969; Hess, 1973).

Hess defined imprinting "as a type of process in which there is an extremely rapid attachment, during a specific critical period, of an innate behavior pattern to specific objects which thereafter become important elicitors of that behavior" (Hess, 1973, p 65). Moreover, Hess sees imprinting as

> a particular type of learning process . . . which may be used by a species for the formation of a filial–maternal bond, pair formation, environment attachment, food preferences, and perhaps other cases involving some sort of object–response relationship. It is, furthermore, a genetically programmed learning, with some species-specific constraints upon the kind of object that may be learned and upon the time of learning. In other words, imprinting is a genotype-dependent ontogenic process (Hess, 1973, p. 351).

Definitions help bring to light the relationship between theoretical assumptions and the outcome of research operations which test the hypothesis generated by those assumptions (Petrovich and Hess, 1977). Hess's early findings supported Lorenz, who in 1935 indicated that imprinting took place during a highly limited, critical period early in the life of the organism, it had lasting effects, the imprinting process selected species-typic, biologically appropriate cues, only specific response patterns of the young animal were imprinted by the specific object of attachment, and imprinting could affect behaviors not yet performed by the young animal (such as sexual behavior) when such behaviors were later expressed.

The investigation of imprinting in both precocial and altricial species (avian and mammalian) has necessitated the reconsideration and updating of some original claims (e.g., Hess, 1973; Hess and Petrovich, 1977; Hoffman and Ratner, 1973; Hoffman and DePaulo, 1977; Immelmann, 1972). Further research, for instance, revealed that different species exploit different sensory modalities for developing the attachment bond: these modalities are chemical, auditory, visual, tactile, and thermal. When some of these ecological, contextual, or species-specific differences were, however, filtered out, the emerging picture supported the thesis that many animals imprint and that the development of such a bond promotes the survival of the individual and the species. The strength of this *functional analogy,* coupled with independent evidence of resemblances of the characteristics of the process, allowed Bowlby to make a case for human imprinting: "So far as is at present known, the way in which attachment behavior develops in the human infant and becomes focused on a discriminated figure is sufficiently like the way in which it develops in other mammals, and in birds, for it to be included legitimately, under the heading of imprinting" (1969, p 223).

Following Bowlby, many investigators have considered the imprinting phenomenon a model for the understanding of normal and abnormal attachment behaviors as well as of the role of attachment in the development of early socialization in canids, monkeys, and man (e.g., Hinde, 1974; Hoffman and Ratner, 1973; Scott and Senay, 1973).

Bowlby's (1969, 1973) masterful synthesis of the literature has had a widespread influence on views of human and animal attachment (e.g., Ainsworth, 1969; Hinde, 1974; Scott, 1971). Given this justly deserved recognition, a few cautionary remarks about imprinting and attachment are in order. From an ethological perspective, it is important to keep in mind that many avian and mammalian species develop social attachments through behavioral mechanisms other than imprinting (see also Chapters 6, 8, 10, 11, and 13). *Imprinting,* in fact, appears to be a particular kind of response system characteristic of the social evolution of precocial species. The best empirical evidence in support of the imprinting phenomenon has come from studies of animals such as ducks, geese, quail, goats, and sheep. The human example, however, has a much better ecological and behavioral analogue in the development of attachment in the altricial species. The young are helpless at birth, and attachment and socialization require extensive periods of interaction between parents and offspring, thereby manifesting parental attachment to the young. Detailed observations, for example, of many avian and mammalian species demonstrate that in each case there is a unique constellation of specific sequences of maternal behaviors around hatching and birth and for some days or even months subsequently. Human data also suggest that affectional bonds begin to emerge prior to delivery. More importantly they are quite delicate and in some cases can be disrupted shortly after parturition, thereby producing some very deleterious consequences for perinatal care (Klaus and Kennell, 1976). Given this dyadic nature of human attachment and socialization, as well as the species-typic plasticity of *Homo sapiens,* prudence requires that the ethological–psychoanalytic theoretical formulations and propositions be complemented by the empirical evidence stemming from classical learning, social learning, and cognitive theories (e.g., Cairns, 1966, 1977; Gewirtz, 1972; Hoffman and DePaulo, 1977; Piaget, 1971; see also Roy's wide-ranging discussion in Chapter 14).

In light of these considerations, let us return to the important question: What are the causal relationships among imprinting, attachment, and species-typic isolation and identification? Regretably, we are left with speculations rather than definitive answers.

Consider the following: In some species the striking features of "distorted" imprinting and attachment processes provide provocative demonstrations of sexuality channeled in the direction of the biologically inappropriate object or of animal and human simulation of object fixation. The experimental demonstration of a critical period for imprinting and attachment may be viewed as functionally compatible with theoretical propositions advanced by the sequential-stage theorists such as Freud, Erikson, and Piaget. Forced separation from objects of attachment has deleterious effects on avian and mammalian (including human and infrahuman-primate) infant behavior. If separation is prolonged or leads to privation and deprivation, the consequences may be even more severe (see chapters by Scott, Roy, and Meier and Dudley-Meer). Prolonged separation and deprivation may progressively lead to depression plus related withdrawal from social interactions (e.g., Bowlby, 1973). Human, avian, and monkey data, on the other hand, indicate that the young indulge in many more adaptive behaviors in the presence of the object of attachment. More importantly these findings also suggest that the "psychological state" of infants plays an overriding role in fostering behaviors such as the exploration of one's *umwelt,* including one's relationship to others. A series of isolation studies by Harlow, Mason, and their associates has shown that in monkeys and chimpanzees, the appropriate early social experiences are essential for the normal development of sexual behavior. Human data on the development of sexuality and gender identity (Money, 1965; Money and Ehrhardt, 1972) also show that sexual phenotypes programmed by the chromosomes and genes at the moment of conception may be drastically modified and altered by various environmental agents of the genes, such as chemical byproducts of other genes, hormones, or exposure to biologically and psychologically inappropriate stimuli during critical developmental periods.

Interesting extrapolations emerge from the consideration of these diverse findings. In a given ecological context biologically appropriate stimuli are required for the induction of biologically appropriate behavior. The ultimate adaptations express themselves through individuals, and the significance of the individual in the context of biological evolution is measured by the extent to which one perpetuates the genes responsible for adaptive phenotypes (Wilson, 1975). Imprinting and attachment behaviors offer covert evidence for one of the very important crossroads in species-typic isolation and identification. Under ecologically appropriate conditions, in general, the processes of imprinting, attachment, and subsequent socialization produce a highly predictable and reliable species-typic behavioral outcome that ultimately may culminate in the production of zygotes and offspring. In contrast, the exposure to biologically inappropriate stimuli during the developmental period of species-typic identification distorts subsequent socialization and thereby may functionally isolate such an individual from a reproductive population.

References

Ainsworth, M. D. S. 1969. Object relations, dependency and attachment: A theoretical review of the infant–mother relationship. *Child Dev. 40,* 969–1025.

Alexander, R. D. 1966. The evolution of cricket chirp. *Nat. Hist. 75,* 26–31.

Alexander, R. D. 1968. Arthropods. In T. A. Sebeok (Ed.), *Animal communication.* Bloomington: Indiana University Press.

Bentley, D., and R. R. Hoy. 1974. The neurobiology of cricket song. *Sci. Am. 231,* 34–44.

Bowlby, J. 1969. *Attachment and loss.* Vol. 1: *Attachment.* New York: Basic Books.

Bowlby, J. 1973. *Attachment and loss.* Vol. 2: *Separation anxiety and anger.* New York: Basic Books.

Brown, J. L. 1975. *The evolution of behavior.* New York: Norton.

Cairns, R. B. 1966. Attachment behavior of mammals. *Psychol. Rev. 73,* 409–426.

Cairns, R. B. 1977. Beyond social attachment: The dynamics of interactional development. In T. Alloway, P. Pliner, and L. Krames (Eds.), *Attachment behavior.* New York: Plenum Press.

Eibl-Eibesfeldt, I. 1975. *Ethology: The biology of behavior.* New York: Holt, Rinehart & Winston.

Ewing, A., and Hoyle, G. 1965. Neuronal mechanism underlying control of sound production in a cricket: *Acheta domesticus. J. Exp. Biol. 43,* 139–153.

Fulton, B. B. 1933. Inheritance of song in hybrids of two subspecies of *Nemobius fasciatus* (Orthoptera). *Ann. Entomol. Soc. Am. 26,* 368–376.

Gewirtz, J. L. (Ed.). 1972. *Attachment and dependency.* Washington, D.C.: Winston.

Hess, E. H. 1973. *Imprinting: Early experience and the developmental psychobiology of attachment.* New York: Van Nostrand.

Hess, E. H. and S. B. Petrovich (Eds.). 1977. *Imprinting: Benchmark papers in animal behavior.* Stroudsburg, Pa.: Dowden, Hutchinson and Ross.

Hinde, R. A. 1974. *Biological bases of human social behavior.* New York: McGraw-Hill.

Hoffman, H. S. and P. DePaulo. 1977. Behavioral control by an imprinting stimulus. *Am. Sci. 65,* 58–66.

Hoffman, H. S. and A. M. Ratner. 1973. A reinforcement model of imprinting: Implications for socialization in monkeys and man. *Psychol. Rev. 80,* 527–544.

Huber, F. 1962. Central nervous control of sound production in crickets and some speculations on its evolution. *Evolution 16,* 429–442.

Immelmann, K. 1972. Sexual and other long-term aspects of imprinting in birds and other species. *Adv. Study Behav. 4,* 147–174.

Klaus, M. H. and H. J. Kennell. 1976. *Maternal–infant bonding.* St. Louis: Mosby.

Lorenz, K. 1969. Innate bases of learning. In K. Pribram (Ed.), *On the biology of learning.* New York: Harcourt Brace Jovanovich.

Money, J. 1965. *Sex research: New developments.* New York: Holt, Rinehart & Winston.

Money, J. and A. A. Ehrhardt. 1972. *Man and woman, boy and girl.* Baltimore: Johns Hopkins University Press.

Petrovich, S. B. and E. H. Hess. 1977. Introduction: A perspective on some of the issues of imprinting. In E. H. Hess and S. B. Petrovich (Eds.), *Imprinting.* Stroudsburg, Pa.: Dowden, Hutchinson & Ross.

Piaget, J. 1971. *Biology and knowledge.* Chicago: University of Chicago Press.

Scott, J. P. 1971. Attachment and separation in dog and man: Theoretical propositions. In H. R. Schaffer (Ed.), *The origins of human social relations.* New York: Academic Press.

Scott, J. P. and C. C. Senay (Eds.). 1973. *Separation and depression: Clinical and research aspects.* Washington, D. C.: American Association for the Advancement of Science.

Walker, T. J. 1957. Specificity in the response of female tree crickets to calling songs of the males. *Ann. Entomol. Soc. Am. 50,* 626–636.

Wilson, E. D. 1971. *The insect societies.* Cambridge, Mass.: Harvard University Press.

Wilson, E. D. 1975. *Sociobiology: The new synthesis.* Cambridge, Mass.: Harvard University (Belknap Press).

Author Index

Species Index

Subject Index